Classical Theory of Crystal Dislocations

From Iron to Gallium Nitride

Classical Theory of Crystal Dislocations

From Iron to Gallium Nitride

Hiroyasu Saka

Nagoya University, Japan

World Scientific

NEW JERSEY • LONDON • SINGAPORE • BEIJING • SHANGHAI • HONG KONG • TAIPEI • CHENNAI • TOKYO

Published by

World Scientific Publishing Co. Pte. Ltd.
5 Toh Tuck Link, Singapore 596224
USA office: 27 Warren Street, Suite 401-402, Hackensack, NJ 07601
UK office: 57 Shelton Street, Covent Garden, London WC2H 9HE

British Library Cataloguing-in-Publication Data
A catalogue record for this book is available from the British Library.

CLASSICAL THEORY OF CRYSTAL DISLOCATIONS
From Iron to Gallium Nitride

ISBN 978-981-4749-16-9

Typeset by Stallion Press
Email: enquiries@stallionpress.com

Printed in Singapore

Foreword

A crystal consists of atoms or molecules that are arranged regularly in three dimensions. However, in real crystals, this regularity is disturbed, giving rise to disorders. These disorders in crystals are referred to as lattice imperfections (lattice defects). Among the lattice defects, linear defects are referred to as dislocations. Most of the crystals we treat practically contain dislocations. In metallic crystals, dislocations provide plasticity and strength. Dislocations are *friendly* in metallic crystals. In contrast, dislocations in semiconductors are nothing else but *adversary*, deteriorating the optical and electrical properties.

Therefore, the first barrier to overcome in developing semiconductors is to grow crystals as perfect as possible. In the case of silicon, which is by now a commodity in the industry, a vast amount of energy was to be spent before obtaining dislocation-free perfect crystals in its early stage of development. In the case of gallium nitride, which was more recently developed successfully, it is no exaggeration to say that growth of high-quality crystals with a low dislocation density was the key to the success in developing high-performance light-emitting devices and electronic devices.

Science of dislocations, i.e., theory of dislocations, has been well established to explain plasticity of metallic materials. A detailed knowledge on dislocations is essential in the field of semiconductor engineering as well. However, theory of dislocations is quite a difficult science, mastering of which is time-consuming and painstaking. Therefore, those researchers and engineers who are actually being engaged in research and development of novel semiconductor materials have awaited a systematic and yet easy textbook of theory of dislocations.

In view of this, Dr. Hiroyasu Saka published a textbook of dislocation theory that covers fundamentals to cutting-edge applications. The author has studied dislocations by means of transmission electron microscopy over half a century. Thus, in the textbook some of the transmission electron microscopy observations are used, together with generous original diagrams. This is the first textbook that deals with dislocations in cutting-edge semiconductors such as GaAs, SiC, and GaN. In conclusion, it provides the classical theory of dislocation covering from fundamentals to cutting-edge applications in such a style that even beginners can read easily.

I would like to recommend this book as a *bible* to win a battle against dislocations, which will last forever in developing novel materials.

February 2016 Isamu Akasaki *Nagoya* Japan

Preface

Theory of dislocations is one of the most difficult sciences for students in materials science and engineering (MSE) for the following three reasons:

(1) Geometrical consideration of dislocation, typical of which is Burgers vector: The definition of Burgers vector appears quite simple, however, it is rather difficult to master its significance completely. Many beginners stumble here.
(2) Elasticity theory: The concept of dislocations stems from applied mathematics. Thus, it has been taken as granted that advanced mathematics is indispensable to understand the theory of dislocations. This is not necessarily so. It is true that solid state physicists have contributed much towards developing theory of dislocations, but too much mathematics has been required as a prerequisite condition for the beginners to master theory of dislocations. Many, if not all, students in MSE are not good enough in advanced mathematics. This is another reason why students of MSE stay away from theory of dislocations.
(3) Crystallography: Dislocations are crystal defects. Therefore, it is needless to say that a detailed knowledge of crystallography is indispensable. However, knowledge of traditional crystallography is of little use. It is the crystallography for dislocations, such as Thompson tetrahedron, that is necessary.

Theory of dislocations has contributed much toward understanding plastic behaviour of metallic materials which are important as structural materials. However, there is still a gap between theory and practice not only in a quantitative sense but also in a qualitative sense. For instance, with regard to controversy on the nature of dislocations in Si, shuffle-glide controversy, the gap is hopelessly wide. Supercomputer is expected to bridge this gap. However, needless to say, in order to make full use of the computer's ability, comprehensive knowledge on classical theory of dislocations is crucial.

On the other hand, a new development is emerging. Importance of dislocations in functional materials is now being recognized. It is well known that reducing the dislocation density in a compound semiconductor GaN was the key to the invention of an LED. Reducing the dislocation density in SiC, a promising candidate for a power device, is also an urgent issue. These materials are quite brittle at room temperature, however, they are surprisingly ductile at high temperatures. In other words, in a high temperature region where these functional materials are processed, behaviour of dislocations is quite spectacular. 'Dislocations are living' literally! It is this understanding that leads to realization of novel functional materials.

In view of these backgrounds, this book is intended:

(1) to summarize elements of classical theory of crystal dislocations with mathematics kept to a minimum
(2) to provide a detailed knowledge on behaviour of dislocations in cutting-edge materials, especially semiconductors.

The first seven chapters are devoted to fundamental properties of dislocations in general. Chapters 8–11 are devoted to detailed description of dislocations in different crystal structures of importance. Chapter 12 is devoted to macroscopic strength of materials, in particular, structural materials. A variety of strengthening mechanisms are presented. Chapter 13 is devoted to behaviour of dislocations in thin crystals such as epilayers and transmission electron microscopy (TEM) foil specimens.

The expected audience is twofold. One is students of MSE. The other is researchers and/or engineers who are struggling in reducing the dislocation density in the products they are developing and/or manufacturing. Fortunately, the author has studied dislocations using TEM, some of which are used in this book. He would like to express his hearty thanks to his graduate students and colleagues who took these TEM pictures. Finally, he owes much to Dr. Takashi Saka for discussions in Chapters 3 and 4 and also Dr. Ichiro Yonenaga for discussion on polarity in Chapter 11.

February 2016 Hiroyasu Saka *Nagoya* Japan

Contents

Foreword v

Preface vii

1. Elements of Crystallography 1

1.1 Unit Cell . . . 1
1.2 Crystallographic Directions and Planes . . . 1
1.3 Hexagonal Indices . . . 3
1.4 The Stereographic Projection and the Standard Projection . . . 6

2. Geometry of Dislocations 9

2.1 Mechanical Properties of Crystals (Stress–Strain Curve) . . . 9
2.2 Ideal Strength of Crystals (Frenkel's Model) . . . 12
2.3 Definition of Dislocations . . . 15
2.4 Burgers Circuit (FS/RH(Perfect)) . . . 16
2.5 Edge Dislocation, Screw Dislocation and Mixed Dislocation . . . 20
2.5.1 Definition . . . 20
2.5.2 Slip plane . . . 23
2.6 Kirchhoff's Law . . . 24
2.7 Reaction Between Two Dislocations . . . 25

2.7.1 Reaction between two edge dislocations with Burgers vectors of $+b$ and $-b$ 25
2.7.2 Screw dislocations of $+b$ and $-b$ 27
2.7.3 Dislocation reaction with different Burgers vectors (1) 28
2.7.4 Dislocation reaction with different Burgers vectors (2) Crossing of dislocations 28
2.8 Prismatic Dislocation 32
2.8.1 Formation of prismatic dislocations by precipitation of point defects 32
2.8.2 Formation of a prismatic loop by pencil glide 35
2.9 Climb Motion of Dislocations 35
2.9.1 Climb of edge dislocations 35
2.9.2 Climb of screw dislocations 37

3. Fundamentals of Elasticity Theory 41

3.1 Displacement, Strain and Stress 41
3.2 Matrix Notation of Hooke's Law 44
3.3 Transformation of Stress and Strain 46
3.4 Hooke's Law in Isotropic Solids 51
3.4.1 Elastic constants 51
3.4.2 Plane stress 57
3.4.3 Plane strain 57
3.5 Cylindrical and Spherical Coordinates 58
3.5.1 Cylindrical coordinate 58
3.5.2 Spherical coordinate 60

4. Elasticity Theory of Dislocations 63

4.1 Screw Dislocation 63
4.1.1 Displacement, strain and stress 63
4.1.2 Strain energy 64
4.1.3 Force on the screw dislocation 66
4.1.4 Image force and Eshelby twist 67
4.2 Edge Dislocation 69
4.2.1 Displacement and stress 69
4.2.2 Strain energy 71
4.2.3 Force on an edge dislocation 73

4.2.4 Image force 75
4.3 Mixed Dislocation 75
4.4 Application of Peach–Koehler Formula 78
4.4.1 Parallel screw dislocations 78
4.4.2 Parallel edge dislocations 80
4.4.3 Perpendicular screw dislocations 83
4.4.4 Edge and screw perpendicular with each other (Fig. 4.21) 85
4.5 Dislocations in Anisotropic Crystals 88

5. Elastic Interaction between Dislocations and Solute Atoms 91

5.1 Isotropic strain (Cottrell effect) 91
5.1.1 General theory 91
5.1.2 Edge dislocation 93
5.1.3 Screw dislocation 95
5.2 Anisotropic Strain 95

6. Motion (Peierls Force) and Multiplication (Frank-Read Source, Bardeen-Herring Source) of Dislocations 103

6.1 Peierls Force 103
6.1.1 General theory 103
6.1.2 The Peierls force in BCC metals 105
6.1.3 The Peierls force in Si, Ge, GaAs 105
6.1.4 Overcoming of P–N potential 105
6.2 Dislocation Sources 106
6.2.1 Frank-Read source 106
6.2.2 Modifications of Frank-Read source 111
6.2.3 Bardeen-Herring source 111

7. Dislocation Groups 115

7.1 Pileup of Dislocations 115
7.2 Polygonization and Small Angle Tilt Boundary 117
7.3 Twist Boundary 120

8. Dissociated Dislocations in FCC Structure 123

8.1 Dissociated Dislocations 123
8.1.1 Thompson tetrahedron 123

8.1.2 Dissociated dislocations and partial dislocations 126
8.1.3 Stacking fault 127
8.1.4 Burgers vectors of dissociated dislocations: Thompson vectors 132
8.2 Dissociation Distance 136
8.3 Jogs on Dissociated Dislocations and Stair-rod Dislocations 138
8.3.1 Stair-rod dislocations 138
8.3.2 Jogs on a screw dislocation 141
8.3.3 Stacking fault tetrahedron 146
8.4 Reactions between Dissociated Dislocations 148
8.4.1 Dislocation dissociated on the same slip plane 148
8.4.2 Reactions between dissociated dislocations on different slip planes: Stair-rod dislocations 150
8.5 Climb Motion of Dissociated Dislocations 154
8.6 Composition and Temperature Dependence of the Dissociation Distance 160
8.6.1 Composition dependence of SFE 161
8.6.2 Temperature dependence of SFE 165
8.7 Interaction between the Stacking Fault and Solute Atoms 170
8.7.1 Suzuki effect 170
8.7.2 Radiation Induced Segregation 174

9. Dissociated Dislocations in HCP 177

9.1 Stacking Faults in Hexagonal Close Packed Structure 177
9.1.1 Frank type stacking fault in HCP 179
9.1.2 Burgers vectors of Shockley partials 181
9.2 Composition Dependence of SFE in HCP Alloys . . . 185

10. Dislocations in Ordered Alloys and Intermetallic Compounds and the Inverse Temperature Dependence of Strength 189

10.1 General Theory 189

10.1.1 Superdislocations and superpartial dislocations 189
10.1.2 Reaction of superdislocations with APB 193
10.2 Dislocations in Ordered Alloys 194
10.2.1 B2 structure 194
10.2.2 DO_3 structure 195
10.2.3 $L1_2$ structure 195
10.3 Strengthening by Ordering 197
10.3.1 Reactions with grown-in APB 197
10.3.2 Strengthening by APB tube formed as a result of cutting superdislocations 199
10.4 Inverse Temperature Dependence of the Strength . . . 200
10.4.1 $L1_2$ structure 200
10.4.2 Inverse temperature dependence of strength in β'-brass 205
10.4.3 Summary 211

11. Dislocations in Diamond, Zincblende, Wurtzite Structures and SiC 215

11.1 Dislocations in Diamond Structure (Shuffle-Set and Glide-Set Dislocations) 215
11.1.1 Theoretical consideration 215
11.1.2 Experimental results 220
11.2 Dislocations in the Zincblende Structure 228
11.2.1 Polarity 228
11.2.2 Shockley partial dislocations 238
11.3 Shockley Partial Dislocations in the Wurtzite Structure 239
11.4 Dislocations in SiC 241
11.4.1 Polytypes in SiC 241
11.4.2 Dislocations in 4H–SiC 243

12. Dislocations and Macroscopic Strength 253

12.1 Geometry of Yielding of a Single Crystal 253
12.1.1 Critical resolved shear stress — Schmid law 253
12.1.2 Derivation of shear stress and shear strain in the uniaxial deformation 255

12.2 Phenomenology of Yielding of a Single Crystal 260
12.2.1 Mobile dislocation density is constant (Johnston-Gilman theory) 264
12.2.2 Modification to Johnston-Gilman theory . . . 267
12.3 Thermally Activated Process of Dislocation Motion — Internal Stress and Effective Stress 270
12.3.1 General theory 270
12.3.2 Peierls–Nabarro force 282
12.4 Yielding of BCC Metals 282
12.4.1 Peierls force in BCC metals 282
12.4.2 $\psi - \chi$ curve 287
12.4.3 Failure of Schmid law 291
12.4.4 $(0\bar{1}1)$ anomalous slip 296
12.5 Initial Mobile Dislocation Density is Zero 299
12.5.1 Whisker 299
12.5.2 Yielding of a mild steel 300
12.6 Work Hardening of a Single Crystal 301
12.6.1 General description of work hardening of a single crystal 301
12.6.2 Mechanism of work hardening 303
12.7 Solid-solution Hardening, Precipitation Hardening and Dispersion Hardening 306
12.7.1 Solid-solution hardening 306
12.7.2 Precipitation hardening 309
12.7.3 Radiation hardening 313

13. Dislocations in Thin Foils 317

13.1 Misfit Dislocations in Epitaxial Layers 317
13.1.1 Critical thickness 317
13.1.2 Estimate of the critical thickness 318
13.1.3 Energy consideration 322
13.1.4 Onset of slip 324
13.1.5 Summary 327
13.2 Threading Dislocations in Epilayers 328
13.3 Dissociated Dislocations in Thin Crystals 331
13.3.1 Bending of dislocations in the vicinity of crystal surface 331
13.3.2 Effect of surface on the dissociation distance of a dissociated dislocation 334

A. Appendices 341

A.1 Reducing Dislocation Density 341
A.2 Crystal Model of HCP 343
A.3 Model of a Screw Dislocation 343
A.4 Stereoprojection 344
A.5 Determining the Slip System 350
A.5.1 Slip direction 350
A.5.2 Slip plane 350
A.6 Thompson Tetrahedra 353
A.7 Transformation of Stress (Eq. 3.18) 354
A.8 Inverse Transformation of Stress (Eq. 3.21) 354

Index 355

Chapter 1

Elements of Crystallography

1.1 Unit Cell

A crystal consists of atoms or molecules arranged in a pattern that repeats periodically. The unit of the repetition is referred to as a unit cell. Figure 1.1(a) shows the lattice constants defining the dimension and shape of the unit cell, that is, a, b, c, α, β and γ. In this book, the face-centered cubic (FCC) structure, the body-centered cubic (BCC) structure, the diamond structure, the zincblende structure and the hexagonal close-packed (HCP) structures will be treated mostly, the unit cell of which is shown in Figs. 1.1(b)–(f). Except for the HCP structure, others belong to the cubic lattices. That is, the lengths of a, b and c are equal and the axes angles α, β and γ are 90° and can be described most easily with rectangular Cartesian coordinates. Needless to say, there are more complex crystal structures, but only a few of them will be treated in this book.

Exercise 1.1 Iron (Fe) crystallizes in two structures. i.e., the FCC (γ-phase) and the BCC (α-phase) structures. The respective lattice constants are 0.3571 nm[1] and 0.28664 nm (1Å = 0.1 nm). Calculate their densities. Here, the atomic weight of Fe is 55.845.

1.2 Crystallographic Directions and Planes

A crystallographic direction is indicated by a vector such as shown in Fig. 1.2(a). A set of crystallographically equivalent directions is indicated

[1] γ-phase is stable in a temperature range between 911 and 1394°C. This value is extrapolated from the high-temperature values to room temperature.

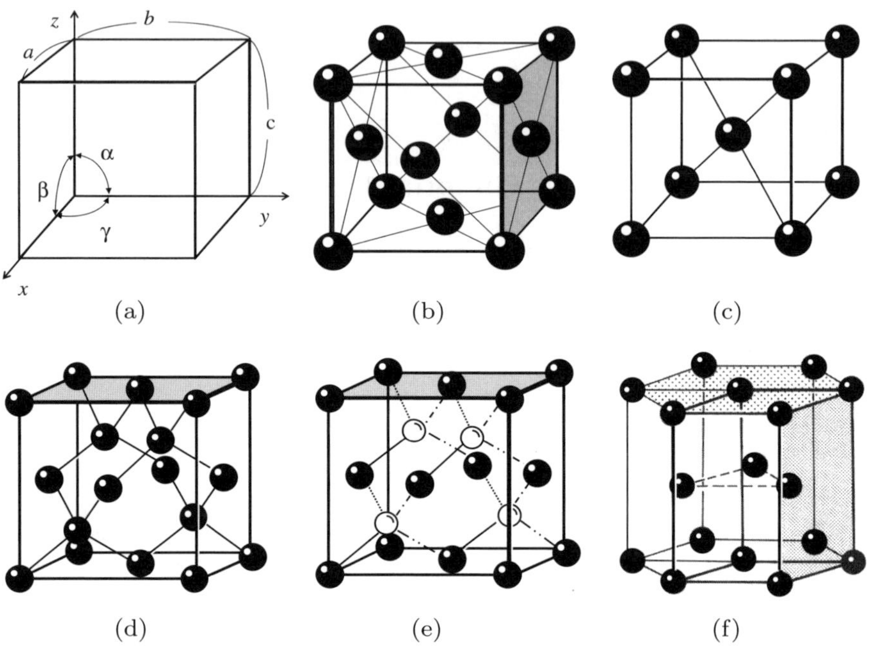

Fig. 1.1 Typical crystal structures. (a) Cubic lattice, (b) FCC structure, (c) BCC structure, (d) Diamond structure, (e) Zincblende structure, (f) HCP structure.

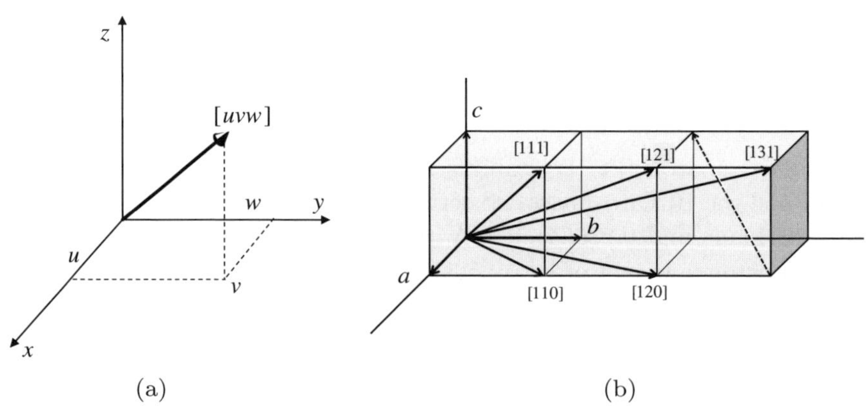

Fig. 1.2 (a) Indication of crystallographic directions, (b) Some examples.

by $\langle u, v, w \rangle$. A particular direction is indicated by $[u, v, w]$. To indicate the sense of the direction, a notation like $[uvw\rangle$ or $\langle uvw]$ is used [1]. Figure 1.2(b) shows some directions.

Miller indices are used to indicate crystallographic planes. A particular plane is indicated by (hkl). A set of crystallographically equivalent planes is indicated by $\{hkl\}$. Figure 1.3(a) shows the definition of (hkl) plane. The intercepts of the (hkl) plane on the x, y and z-axes are $l/h, 1/k$ and $1/l$, respectively. The sense of a plane is such that the normal drawn from the origin is positive and vice versa. Figure 1.3(b)–(d) show some examples. If a plane has Miller index such as $(h0l)$, the plane is parallel to the y-axis. Figure 1.3(e) shows some examples of $(hk0)$ planes. If the z-axis is perpendicular to the plane of the paper, all of these planes are parallel to the z-axis, that is, perpendicular to the plane of the paper.

The negative index is indicated by placing a minus sign above the index: that is, $-h = \bar{h}$. For instance, $\bar{1}$ indicates -1.

$$(\bar{h}kl) = (-hkl).$$

Exercise 1.2 Indicate the direction of the broken line shown in Fig. 1.2(b).

Exercise 1.3 Name all the planes indicated by $\{110\}$.

Exercise 1.4 Draw (100), (110), $(1\bar{1}0)$, (111), $(1\bar{1}1)$, (112) and, $(11\bar{2})$ planes.

Exercise 1.5 In cubic crystals, the normal to (hkl) plane is coincident to the direction $[hkl]$. Confirm this (Fig. 1.4). This does not necessarily hold for crystals with low symmetry, for instance, a tetragonal crystal where $a = b \neq c, \alpha = \beta = \gamma = 90°$. In this case the normal to (101) plane is not coincident to $[101]$. Prove this.

1.3 Hexagonal Indices

Indices based on the three axes, $\boldsymbol{a}_1$ and $\boldsymbol{a}_2$ and $\boldsymbol{c}$ are generally used for hexagonal crystals but are open to the objection that equivalent planes do not have similar indices. For instance, the planes (100) and $(\bar{1}10)$ are both prismatic planes and are equivalent. The same objection applies to the indices of directions. For these reasons, the hexagonal indices such as

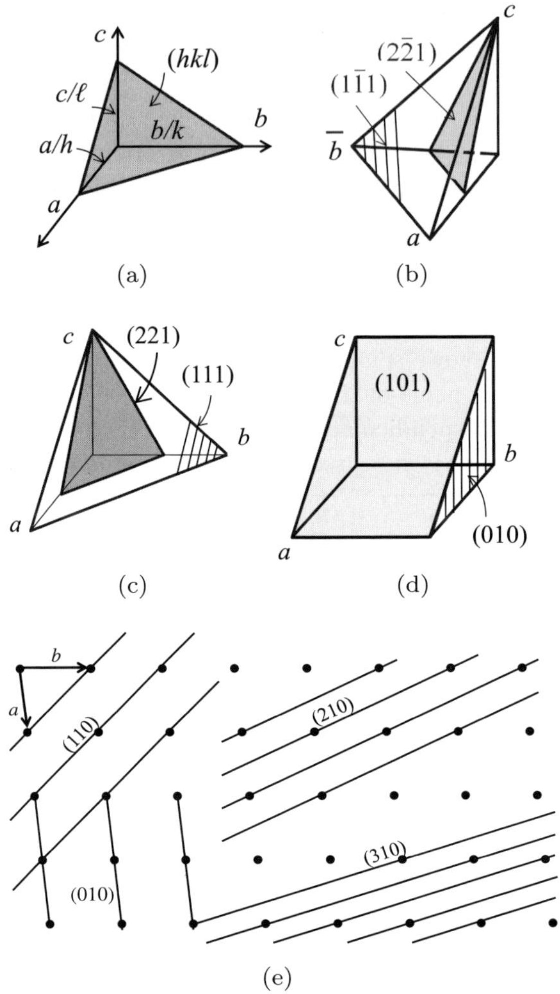

Fig. 1.3 (a)–(e) Miller indices of some important planes.

shown in Fig. 1.5 are used. In addition to the usual $\boldsymbol{a}_1$ and $\boldsymbol{a}_2$ axes, a third axis $\boldsymbol{a}_3$ is introduced and planes and directions are indicated by $(hkil)$ and $[uvtw]$, respectively. Here, h, k, i or u, v, t are indices corresponding to the $\boldsymbol{a}_1$, $\boldsymbol{a}_2$, and $\boldsymbol{a}_3$-axes. l and w are indices corresponding to the $\boldsymbol{c}$-axis. h, k, i and u, v, t are related by the equations,

$$h + k = -i, \quad u + v = -t. \tag{1.1}$$

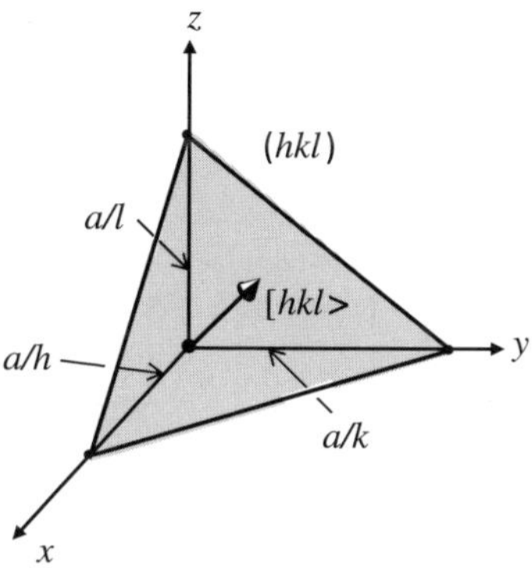

Fig. 1.4 In cubic crystals, [hkl] is perpendicular to (hkl) plane. For crystals other than cubic structure, this does not necessarily hold.

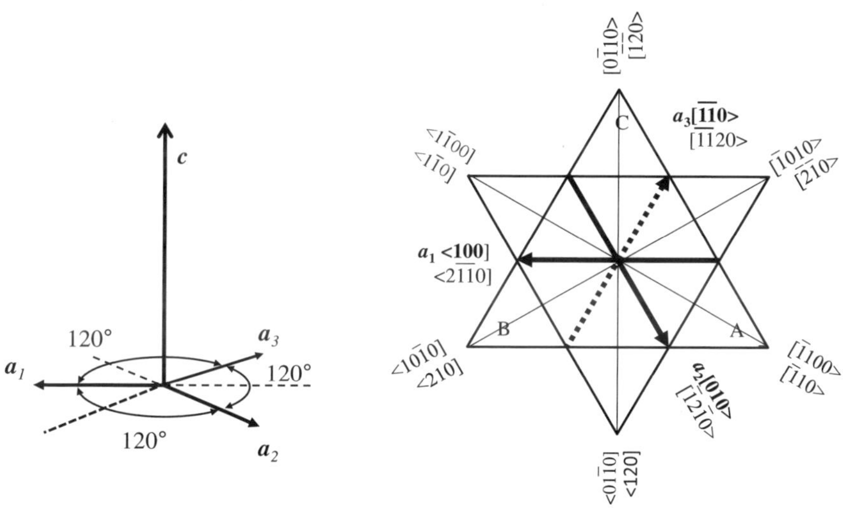

Fig. 1.5 Indices for a hexagonal system.

Transformations between the three and four indices, i.e., between (HKL) and $(hkil)$ and between $[UVW]$ and $[uvtw]$ are given by

$$H = h, \quad K = k, \quad L = l, \quad i = -(h+k) = -(H+K), \tag{1.2}$$
$$U = 2u + v, \quad V = u + 2v, \quad W = w,$$

$$u = \frac{1}{3}(2U - V), \quad v = \frac{1}{3}(2V - U), \quad w = W, \quad t = -(u+v). \tag{1.3}$$

Figure 1.6 shows some typical directions and planes in a hexagonal crystal.

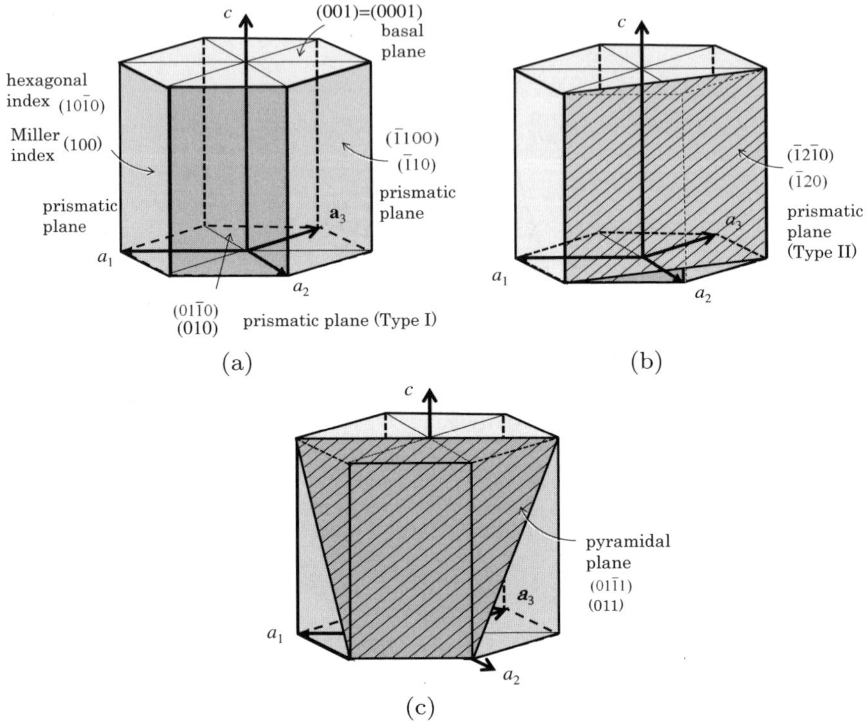

Fig. 1.6 Indices of some planes in a hexagonal system.

1.4 The Stereographic Projection and the Standard Projection

Crystallographic axes, planes and angles are described conveniently by a sphere, known as the reference sphere or polar sphere. This sphere, the crystal globe, can be compared to the terrestrial globe of the earth. Let the radius of the globe be unity. In the following, only the cubic system is considered. A line erected vertically from the origin of the globe is the Z-axis, that is, [001]. This corresponds to the north arctic pole on the earth. The south anti-arctic pole is [00$\bar{1}$]. Draw a direction from the origin of the globe to a point on the equator and let this be the X-axis, i.e., [100]. The circle on the surface of the globe passing through the Z-axis and the X-axis is the great circle of the latitude 0°. On the earth, this great circle passes through Royal Greenwich Observatory in London. The X-axis passes

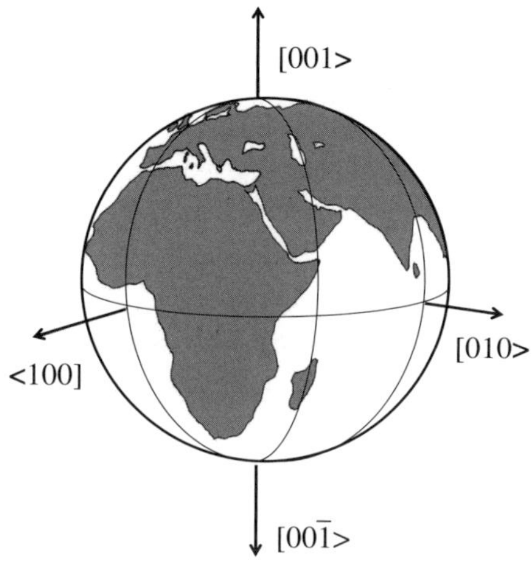

Fig. 1.7 The Earth is compared to the crystal globe.

through Gulf of Guinea on African West Coast. In the right hand system the Y-axis is at the latitude 0° and at the east longitude 90° (on the earth near Bay of Bengal between Sumatra Island and Sri Lanka).

Crystallographic direction is defined as a vector drawn from the origin to a point on the surface of crystal globe. Bordeaux in France is located at east latitude (EL) $=\sim$ 0° and north altitude (NA) $=\sim$ 45°. Thus, the *crystallographic direction* of Bordeaux is [101].

Exercise 1.6 Galápagos Islands are located at west latitude (WL) = 89° 16' and south latitude (SA) = 1° 36'. This corresponds to nearly [0$\bar{1}$0]. Confirm this.

A two-dimensional (2D) map of the north semi-sphere projected along the north arctic pole is the standard [001] projection, and shown in Fig. 1.8. In the cubic system, $\langle 001\rangle, \langle 110\rangle$ and $\langle 111\rangle$ have high symmetry and because of their importance they are indicated by ■, ⬬ and ▲, respectively.

Reference

1. Hirth, J.P. and Lothe, J. *Theory of Dislocations*, 1st ed., McGraw-Hill, (1967); 2nd ed., Wiley, (1982).

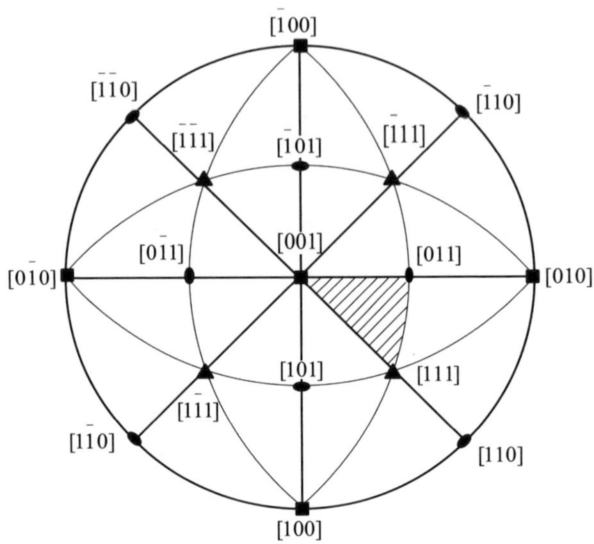

Fig. 1.8 Standard 001 projection. Unit stereographic triangle is hatched.

Further readings

1. Barrett, C.S. and Massalski, T.B. *Structure of Metals, Crystallographic Properties, Principles, and Data*, 3rd ed., Pergamon Press (1980).
2. Saka, H. *Electron Microscopy of Crystals* (in Japanese), Uchidarokakuho, (1997).

Chapter 2

Geometry of Dislocations

2.1 Mechanical Properties of Crystals (Stress–Strain Curve)

The most convenient methods to measure quantitatively mechanical properties of materials are tensile and compression tests.[1] An external force is applied uniaxially to a specimen with a uniform cross-section, and is plotted as a function of the amount of deformation (Fig. 2.1).

Stress is defined as the external force divided by the cross sectional area and usually indicated by σ*[2]: The dimension is [force F/area A] and Pa (= N/m^2; 1 kgf/mm^2 =9.8 MPa) is used as the unit. The amount of deformation is defined by a strain. A strain (usually indicated by ε) is a change in longitude length of the sample divided by the original length, and therefore dimensionless. A 10 % elongation or compression used in everyday life means $\varepsilon = 0.1$.

$$\sigma = F/A, \quad \varepsilon = \frac{\Delta L}{L}. \tag{2.1}$$

In all the materials, irrespective of whether crystal or not, in the initial stage of deformation a stress σ is proportional to a strain ε (Hooke's law), and the specimen returns to the original length when the external stress is removed. This is referred to as elastic deformation (Fig. 2.1(a)). Elastic deformation is a linear reversible process.

---

[1] Besides these, there are bending, torsion, shear deformations, however, for the time being, we restrict ourselves to tensile and compression tests.

[2] For strict definition of stress and strain refer to Chapter 3. Often, an axial stress in tensile and compression tests is expressed by σ and a shear stress by τ.

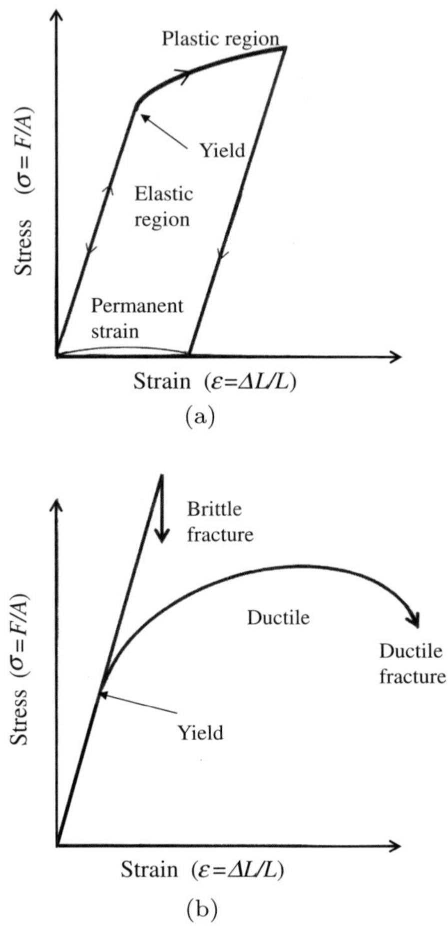

Fig. 2.1 Stress–strain curve.

On further increase in the external force, a metal shows a large amount of deformation and does not return to the original shape, leaving a permanent strain after the force is removed completely. This is a nonlinear irreversible process and referred to as plastic deformation. The transition from elastic to plastic deformation is defined as yielding. A crystal yields to an external force and deforms to a large extent.

By contrast, brittle materials such as ceramics fracture without plastic deformation. This is referred to as brittle fracture. Needless to say, even metals fracture eventually after plastic deformation. This is referred to as ductile fracture (Fig. 2.1(b)).

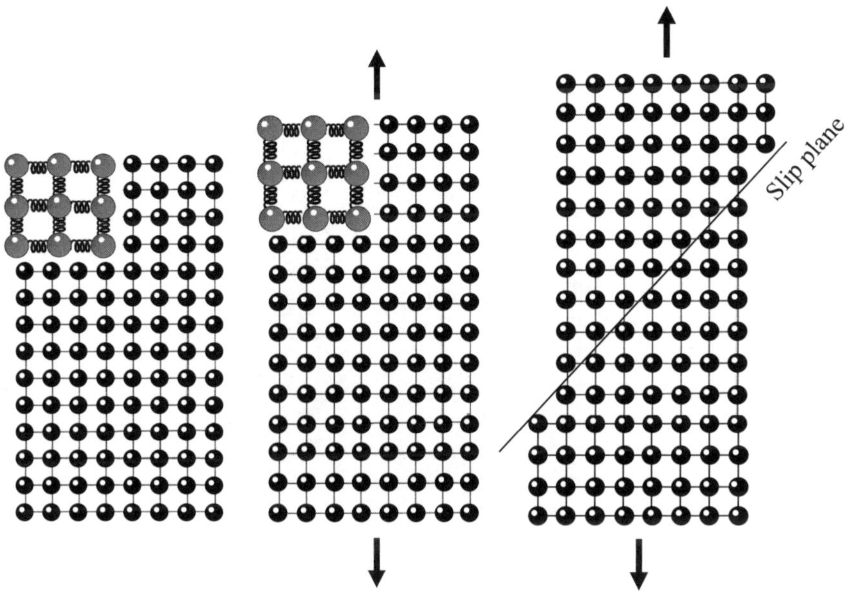

(a) (b) (c)

Fig. 2.2 Atomic pictures of deformation of a crystal. (a) Before deformation, (b) Elastic deformation, (c) Plastic deformation. Insets in (a) and (b) illustrate schematically that each of the atoms are connected with springs. In (b) note elongation and shrinkage of the springs.

Elemental process on atomic scale of plastic deformation of a crystal is shown in Fig. 2.2. For the sake of simplicity, let us consider a single crystal being stretched. Atoms composing the crystal are supposed to connect with each other by springs. When the external force is small, these springs will be elongated along the direction of the external force. This is elastic deformation. The crystal returns to the original shape when the external force is removed (Fig. 2.2(b)). On further increase in the external force, atoms displace across a plane which is inclined approximately by 45° with respect to the direction of the external force (Fig. 2.2(c)). This is referred to as a slip, glide or slide. This process results in formation of a step with one atomic distance at the external surface of the crystal. Inside the crystal, however, the arrangement of atoms across the slip plane before and after slip remains unchanged, so that on removing the external force the crystal does not return to the original shape. This is the elemental process of plastic deformation. The intersection of a slip plane with the external crystal

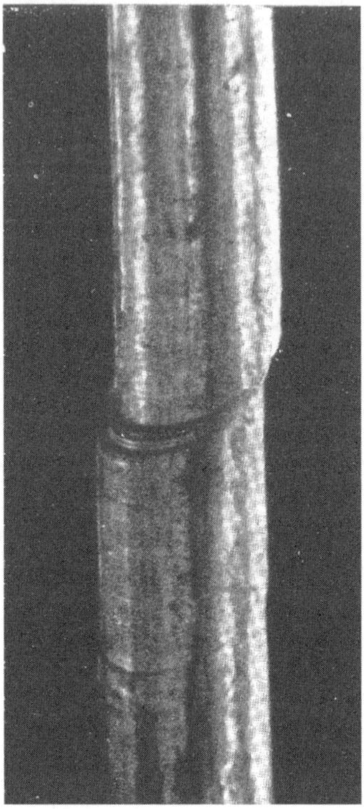

Fig. 2.3 Slip bands in a Cd single crystal (After Bilby, reproduced with permission from [1].)

surface is referred to as a slip line. Figure 2.3 is a photograph of the slip lines on a Cd single crystal (crystal structure; HCP).

2.2 Ideal Strength of Crystals (Frenkel's Model)

Let us consider the process of a slip in more detail. The upper half crystal in Fig. 2.4(a) (indicated by square in broken line) is displaced by $\boldsymbol{b}$, a simple lattice-translation vector (from position A to position B), with respect to the lower half crystal, as shown in Fig. 2.4(b)–(f)). The potential energy of the system is plotted as a function of the displacement x (broken line in Fig. 2.4(g)). It must show minimums both at the start and at the finish of the displacement. The corresponding force against the motion of atoms (solid line in Fig. 2.4(g)) is also a periodic function with the atomic distance b. For the sake of simplicity, let the periodic function be expressed by a sinusoidal function of x with the period of b (Fig. 2.4(g)).

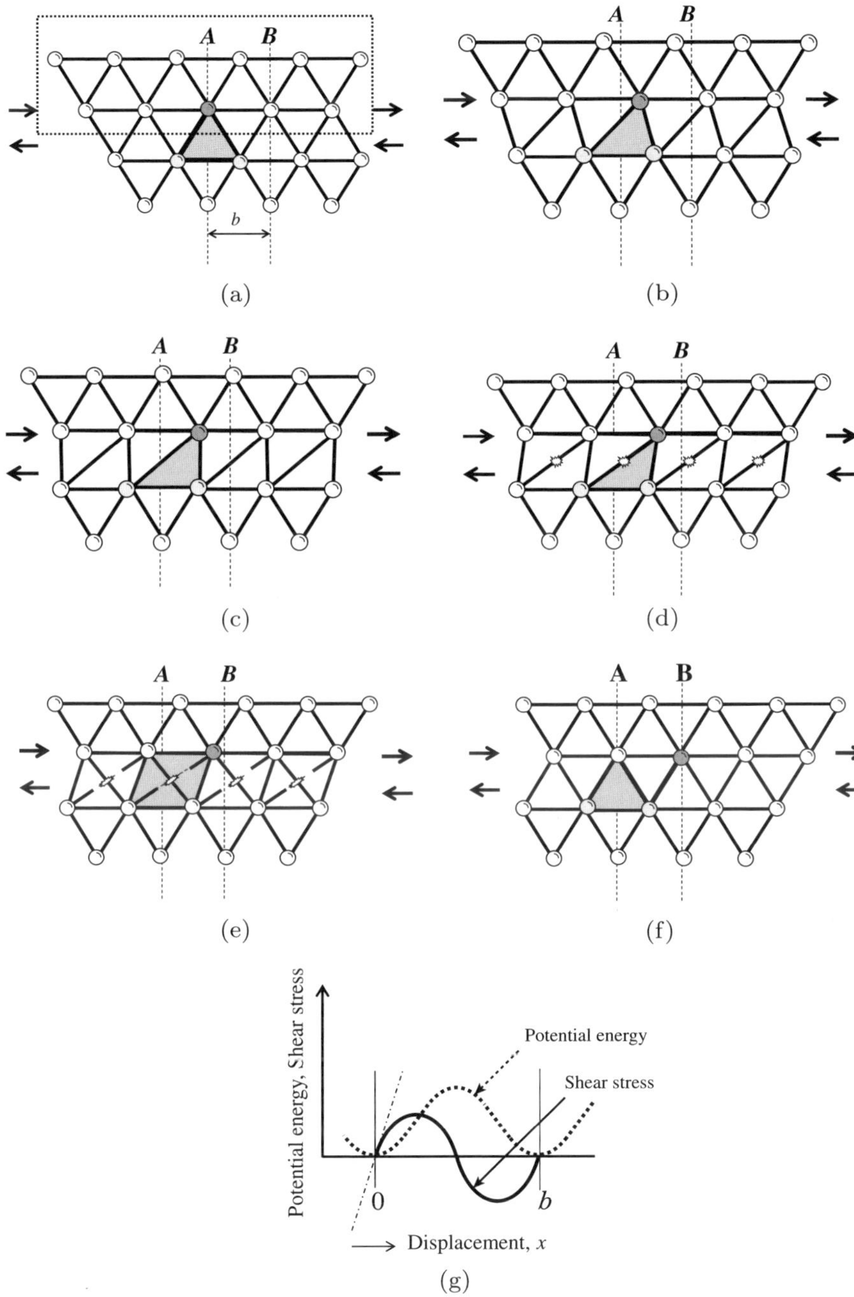

Fig. 2.4 (a)–(g) Frenkel's model of shear deformation. Hatched atom glides from A to B. Note deformation of a hatched triangular lattice. (g) Solid and broken lines indicate shear stress and potential energy, respectively.

That is,

$$\sigma = A \sin\left(\frac{2\pi x}{b}\right). \tag{2.2a}$$

At the very beginning of the shear ($x \ll 1$), the Hooke's law (a dash-chained line) in Fig. 2.4(g) must hold, so that the stress (shear) σ, which is required to give rise to the displacement, is proportional to the shear strain (x/a). That is,

$$\sigma = \mu\left(\frac{x}{a}\right). \tag{2.2b}$$

Here, μ is the proportional coefficient in Hooke's law and referred to as an elastic constant.[3]

Equating Eqs. (2.2a) and (2.2b) gives a relation

$$\left.\frac{d\sigma}{dx}\right|_{x=0} = A\frac{2\pi}{b} = \frac{\mu}{a}.$$

Putting $a \simeq b$, the maximum force in Eq. (2.2a), i.e., the yield stress (the ideal strength) is expressed as,

$$\sigma_{\max} = A \sin\left.\left(\frac{2\pi x}{b}\right)\right|_{\max} = A = \frac{\mu}{2\pi}. \tag{2.2c}$$

Exercise 2.1 The shear modulus of aluminium and iron are 2.65×10^{11} dyne/cm^2 and 8×10^{11} dyne/cm^2, respectively. Calculate the ideal strength using Eq. (2.2c) and compare the measured values. (Hint): Ordinary steels, i.e., mild steels have strength of ~40 kgf/mm^2. (1 kgf/mm^2 = 9.8 MPa = 9.8×10^7 dyne/cm^2). Even specially strengthened steels have strength of 100–150 kgf/mm^2 at most.

Thus, the theoretical strength (ideal strength) calculated from Eq. (2.2c) is much far larger than the real value. The sinusoidal curve may be oversimplified, but even modifying the force-displacement curve as shown in Fig. 2.5, the resultant stress amounts to:

$$\sigma_{\max} = \frac{\mu}{30}. \tag{2.3}$$

[3]Deformation modes and elastic constants. Each of the elastic constants has its own traditional name depending on the deformation mode, and a particular letter is assigned (see Chapter 3). In shear deformation, the elastic constant is referred to as the shear modulus and indicated customarily by μ or G.

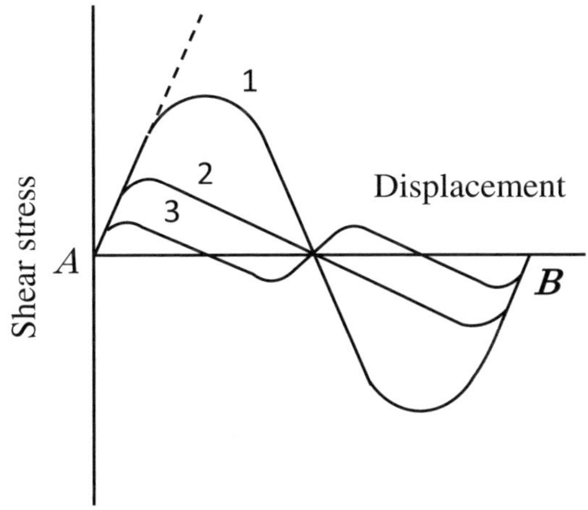

Fig. 2.5 Some modifications (2),(3) to stress-displacement curve (1) (Reproduced with permission from [2]).

It is evident that the gap between the ideal strength and the real strength in metals cannot be explained as an extension of elastic deformation.

2.3 Definition of Dislocations

In the foregoing discussion, it is postulated that all the atoms above a slip plane move with respect to the atoms beneath the slip plane simultaneously. Here, let us assume that, for some reason, a slip initiates at some part of a slip plane and spreads over its total area (Fig. 2.6). The boundary between the slipped and the unslipped regions is defined as a dislocation.

In Frenkel's model, all of the bonds between atoms across a slip plane must be broken simultaneously as shown in Fig. 2.4(e). On the other hand, in the slip by the motion of a dislocation, only atoms adjacent to the dislocation move and the other atoms remain unchanged. Therefore, the net force required to move the dislocation is much lower than the ideal strength given by Eq. (2.3).

Consecutive motion of a dislocation from site-to-site along the slip plane is analogous to the crawling of a caterpillar (Fig. 2.7). In a sort of reverse tug-of-war, the atoms ahead of the dislocation oppose the motion while those behind promote it (Cottrell).[4] In the tug-of-war, a slight difference in the pulling force between the two teams controls the game. A dislocation

[4]The situation can be compared to the scrum in a rugby football game.

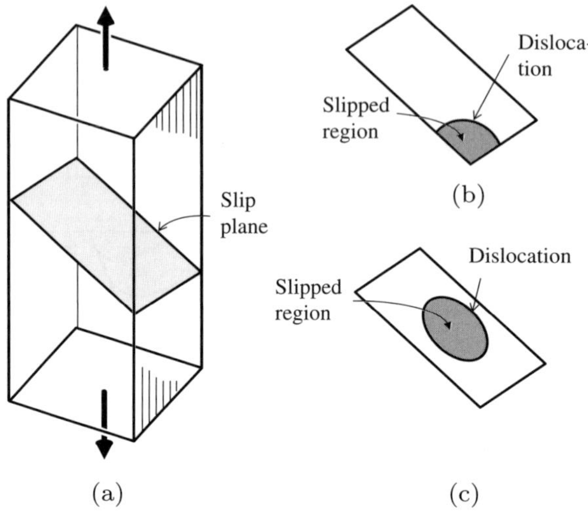

Fig. 2.6 Definition of a dislocation.

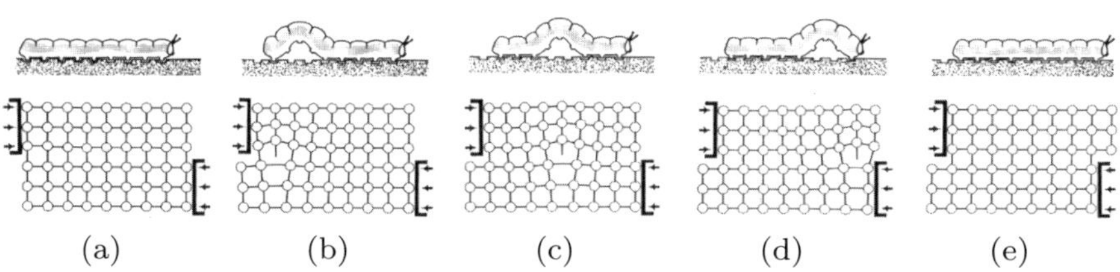

Fig. 2.7 Caterpillar model of a dislocation motion (Reproduced with permission from [3]).

moves by a small force. In other words, this mechanism can explain why the real strength of a crystal is much smaller than the ideal one.

2.4 Burgers Circuit (FS/RH(Perfect))

In the plastic deformation caused by a slip, a vector defining the direction and the amount of slip is referred to as Burgers vector and denoted by $\boldsymbol{b}$. This vector defines both the direction and the amount of slip, so that this is a vector. But strictly speaking, Burgers vector is not a vector but a pseudo-vector because it does not define the sense of direction.

Figure 2.8(a) shows a crystal divided into four pieces. Suppose that the upper right crystal (B) is displaced rightward by $\boldsymbol{b}$ (atomic distance), then an edge dislocation is created in the middle. The sense of displacement can

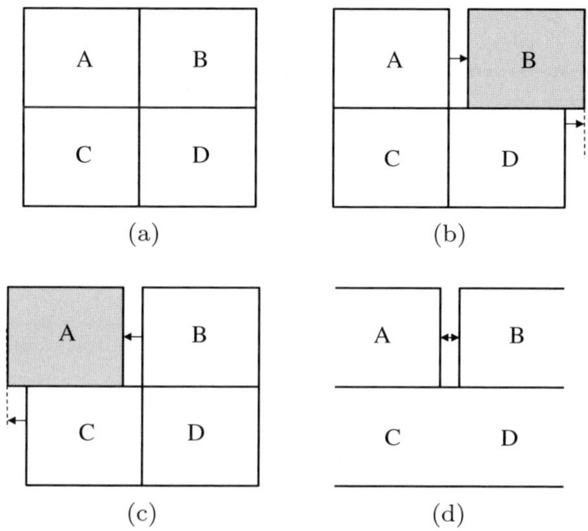

Fig. 2.8 Divide a crystal into four portions A, B, C and D. Displace A or B, while fixing C and D. As shown in (b) and (c), when the change in the outer shape is observed, one can tell which of A and B displaced to which direction. However, when the change of the outer shape is not observed as shown in (d), one cannot tell which of A or B displaced.

be determined from the step formed at the crystal surface (Fig. 2.8(b)). The same dislocation can be created by displacing the left upper crystal (A) leftward by $\boldsymbol{b}$ (Fig. 2.8(c)). Excepting whether a step is formed on the right side of the crystal or left side, the atomic arrangements around these two edge dislocations are exactly same.

In other words, if geometry of a step formed on the crystal surface by a dislocation under consideration is not known, it is impossible to distinguish between Figs. 2.8(a) and 2.8(b). In order to avoid such an ambiguity the Burgers circuit is used to define Burgers vector. The definition of the Burgers circuit is as follows: (Fig. 2.9).

Definition 1

(1) First of all, stand on a dislocation line and define its sense. It is not important whether the sense is positive or negative. Important is that the sense is defined.
(2) Draw in the real crystal containing the dislocation clockwise a Burgers circuit which lies entirely in good material and encloses the dislocation (Fig. 2.9(a)).

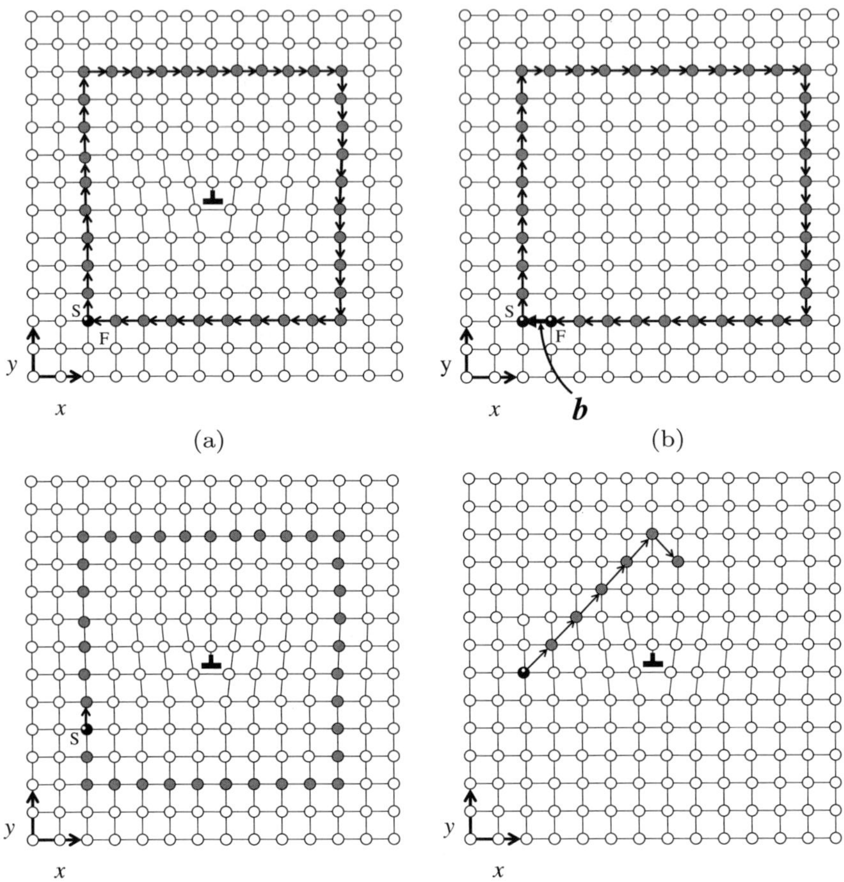

Fig. 2.9 (a)–(d) Definition of Burgers vector, viewing into the paper.

(3) Then draw the same circuit in the perfect reference lattice, as shown in (Fig. 2.9(b)).
(4) In the perfect reference lattice a gap is present.
(5) Draw the vector connecting from F to S in the perfect reference lattice. This vector FS is the Burgers vector ($\boldsymbol{b}$) (Fig. 2.9(b)).

This definition of $\boldsymbol{b}$ is referred to as FS/RH(perfect) convention. FS means that the closure vector is drawn from F to S, RH (right hand) means that the sense of the circuit is that of a right handed screw, and (perfect) means that $\boldsymbol{b}$ is defined in the perfect reference lattice.

Exercise 2.2 When the Burgers vector $\boldsymbol{b}$ is defined using *FS*/LH(perfect) for the dislocation shown in Fig. 2.9(a) and 2.9(b), then the resultant $\boldsymbol{b}$ is reversed. Confirm this. LH (left hand) means to draw the Burgers circuit counterclockwise.

Exercise 2.3 In Fig. 2.9, show that reversing the sense of the dislocation line ξ causes $\boldsymbol{b}$ to reverse its sense of direction.

Exercise 2.4 If one changes the starting and finishing points of a Burgers circuit, the resultant $\boldsymbol{b}$ remains unchanged (Fig. 2.9(c)).

Exercise 2.5 Draw a Burgers circuit along <110> instead of <100> and confirm that the resultant $\boldsymbol{b}$ is the same (Fig. 2.9(d)).

Definition 2

In *FS*/RH (perfect) convention, if one views the slipped area at right, the upper crystal is displaced by $\boldsymbol{b}$ with respect to the lower crystal, which is fixed.

Exercise 2.6 Confirm this.

It is to be emphasized that in defining $\boldsymbol{b}$ it is of vital importance to determine the direction along which one views the dislocation, i.e., to determine the sense of the direction of a dislocation line. One cannot define $\boldsymbol{b}$ without defining the sense of the direction of a dislocation line ξ.

Both $\boldsymbol{b}$ and ξ are indicated by arrows. In order to avoid confusion, throughout this book $\boldsymbol{b}$ will be indicated by a single-headed arrow ($\rightarrow$) and ξ by a double-headed arrow ($\twoheadrightarrow$).

Note 1: Definition of the Burgers vector is very important. It is no exaggeration to say that about one-third of dislocation theory can be understood if one understands the Burgers circuit. Inversely if one doesn't realize the importance of the Burgers circuit, one is far from understanding dislocation theory.

Note 2: *FS*/RH(perfect) convention is employed by [4], [5]. Thompson's simplified vector notation, which is used extensively for face-centered cubic (FCC) crystals, also employs this convention (see Chapter 8). However, the opposite convention has been used by [6], [7]. Thus, one must be careful about defining the convention for $\boldsymbol{b}$ in using results from the literature [4].

2.5 Edge Dislocation, Screw Dislocation and Mixed Dislocation

2.5.1 *Definition*

By definition a dislocation is the boundary between a slipped area and an unslipped area. Therefore, a dislocation must close within a crystal forming a loop, or both ends must emerge at the surface(s) of crystal. A dislocation line never ends within a crystal (Fig. 2.10).

Now let us consider a dislocation loop (Fig. 2.11). Take the sense of direction of the dislocation line ξ clockwise when viewing the dislocation

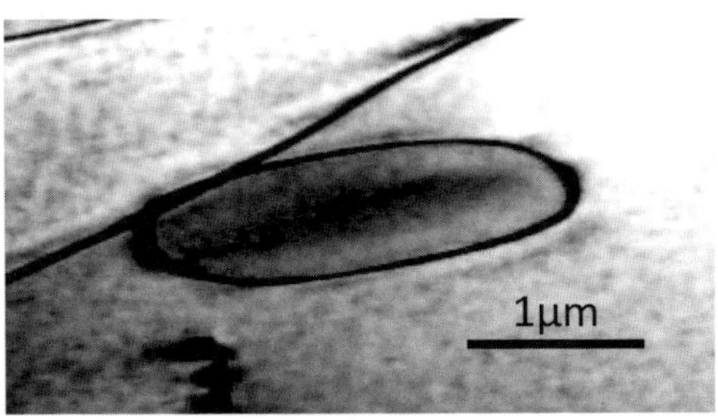

Fig. 2.10 Transmission electron micrograph of a dislocation loop.

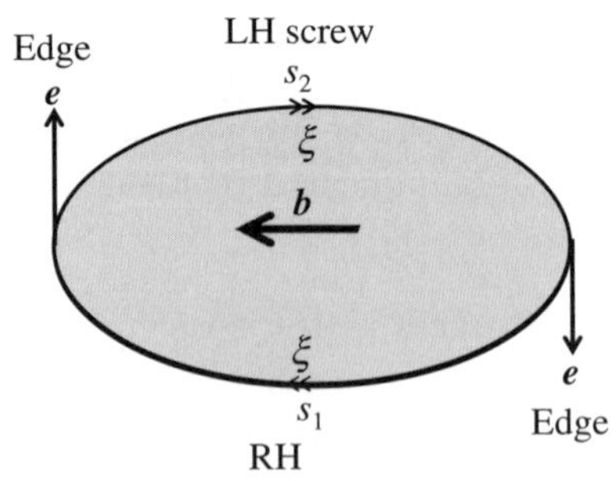

Fig. 2.11 Definition of an edge dislocation and a screw dislocation.

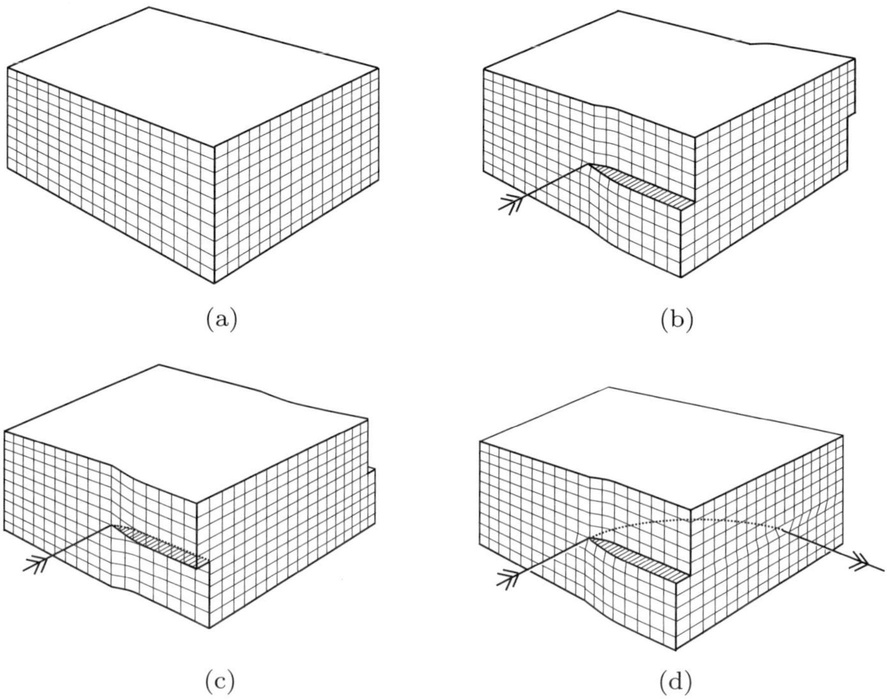

Fig. 2.12 (a) is a perfect crystal free from dislocation, (b) contains a RH screw dislocation, (c) contains a LH screw dislocation, (d) contains a mixed dislocation.

loop from above. Part of the dislocation line where ξ and $\boldsymbol{b}$ are perpendicular is referred to as an edge dislocation. Part where ξ and $\boldsymbol{b}$ are parallel is referred to as a screw dislocation.

The screw dislocation is illustrated schematically in Figs. 2.12(b) and 2.12(c). Atomic planes within the crystal are analogous to the screw staircase. This is the reason why this type of dislocation is called a screw dislocation.

Figures 2.12(b) and 2.12(c) show a RH and a LH screw dislocation, respectively.

Exercise 2.7 Draw Burgers circuits in Figs. 2.12(a) and 2.12(b) and determine $\boldsymbol{b}$.

Exercise 2.8 Demonstrate that s_1 is a RH and s_2 is a LH screw dislocation in Fig. 2.11.

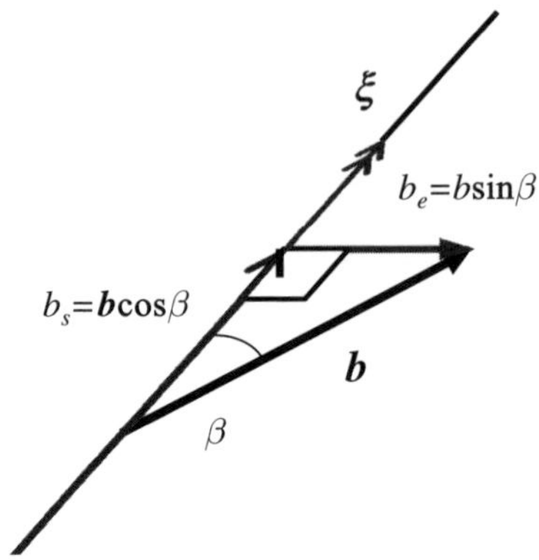

Fig. 2.13 Burgers vector of a mixed dislocation.

When the senses of $\boldsymbol{b}$ and ξ are same (parallel), the screw dislocation is RH. When the senses of $\boldsymbol{b}$ and ξ are opposite (anti-parallel), the screw dislocation is LH.

Exercise 2.9 Confirm this using a model in Appendix A-3.

A mixed dislocation is between a pure edge dislocation and a pure screw dislocation (Fig. 2.12(d)). For a mixed dislocation line with an angle of β between $\boldsymbol{b}$ and ξ, the following relations hold (Fig. 2.13).

$$\begin{aligned} \boldsymbol{b} &= \boldsymbol{b}_e + \boldsymbol{b}_s \\ b_e &= b\sin\beta \\ b_s &= b\cos\beta, \end{aligned} \tag{2.4}$$

where b, b_e and b_s are the absolute values of $\boldsymbol{b}$ and $\boldsymbol{b}_e$ and $\boldsymbol{b}_s$, respectively.

The displacement, stress and strain of a single mixed dislocation can be expressed by overlapping those of a single pure edge dislocation with $b\sin\theta$ and those of a single pure screw dislocation with $b\cos\theta$. The detail will be described in Sec. 4.3.

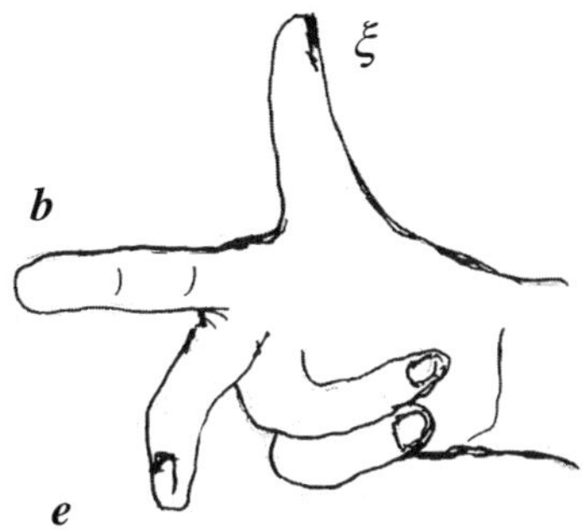

Fig. 2.14 Definition of $\boldsymbol{e}$.

2.5.2 *Slip plane*

The slip plane contains both Burgers vector $\boldsymbol{b}$ and the direction of the dislocation line ξ. The unit vector normal to the slip plane $\boldsymbol{e}$ is given by Fig. 2.14.

$$\boldsymbol{e} = \frac{\xi \times \boldsymbol{b}}{|\xi \times \boldsymbol{b}|}. \tag{2.5}$$

Exercise 2.10 An edge dislocation has an extra half plane (ehp) as shown by the "leg" of the inverted T in Fig. 2.9(a). An ehp at an edge dislocation is inserted from either above or below the crystal surface, depending on the sign of the edge dislocations, as shown in Fig. 2.11. Show that $\boldsymbol{e}$ in Eq. (2.5) points in the direction of ehp from the edge dislocation to the surface of a crystal.

Taking ξ into the paper, an edge dislocation with its ehp above the slip plane is referred to as a positive edge dislocation and that with its ehp below the slip plane as a negative edge dislocation.

Burgers vector of a screw dislocation is parallel to the dislocation direction ξ, that is, $\boldsymbol{b}//\zeta$, so that the slip plane $\boldsymbol{e}$ cannot be defined. Any plane containing $\boldsymbol{b}$ (planes belonging to the zone axis of $\boldsymbol{b}$) can be a slip plane. Therefore, a screw dislocation can change its slip plane easily, that is, can cross-slip easily (Fig. 2.15).

Exercise 2.11 Suppose that a screw dislocation is present in a single crystal shown in Fig. 2.16. Suppose that when this crystal is compressed, the screw dislocation moves rightward. Determine its $\boldsymbol{b}$. What about when it moves leftward?

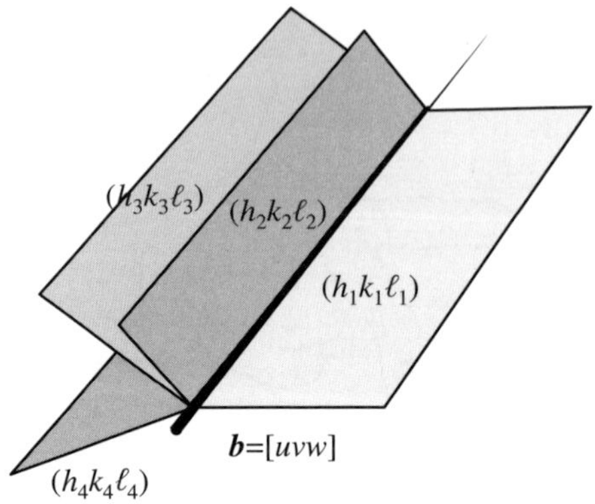

Fig. 2.15 Cross-slip of a screw dislocation. Supposing that $\boldsymbol{b} = [uvw]$, all the planes (hkl) containing $\boldsymbol{b}$ satisfy the relation $uh + vk + wl = 0$. The group of (hkl) planes belongs to the $[uvw]$ zone axis.

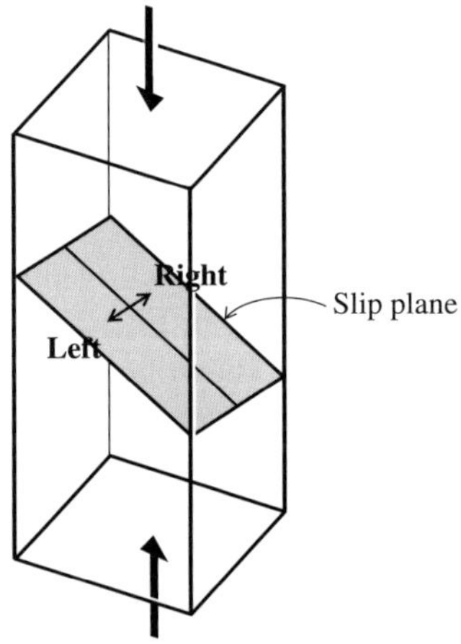

Fig. 2.16

2.6 Kirchhoff's Law

When a dislocation ① reacts with a second dislocation ② to form a third dislocation ③, the following relation holds.

$$\boldsymbol{b}_1 = \boldsymbol{b}_2 + \boldsymbol{b}_3$$

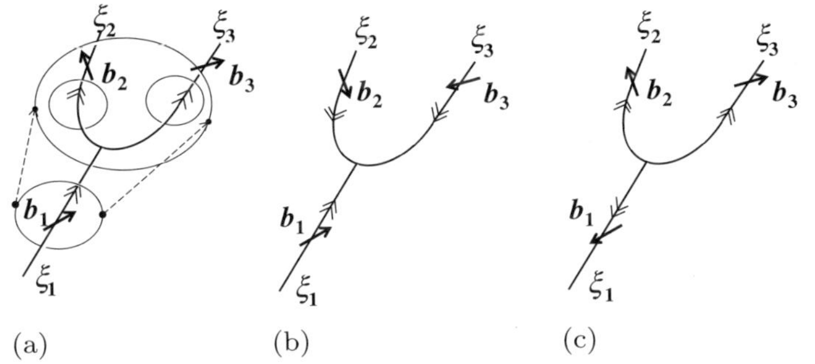

Fig. 2.17 Kirchhoff's law.

where $\boldsymbol{b}_1$, $\boldsymbol{b}_2$ and $\boldsymbol{b}_3$ are Burgers vectors of dislocations ①, ② and ③, respectively (with the directions of ξ_1, ξ_2 and ξ_3 as shown in Fig. 2.17(a)). This can be easily understood by drawing the respective Burgers circuits.

If one takes ξ in such a way that the three dislocations flow into the meeting point, which is referred to as a dislocation node, (Fig. 2.17(b)), or flow away from the node (Fig. 2.17(c)),

$$\boldsymbol{b}_1 + \boldsymbol{b}_2 + \boldsymbol{b}_3 = 0. \tag{2.6}$$

This is referred to as Kirchhoff's law because of the analogy to the Kirchhoff's law of electrical current flow.

Note 3: In transmission electron microscope (TEM) observation it is often required to determine the Burgers vectors of dislocation segments composing a dislocation network such as shown in Fig. 2.18. It is recommended to confirm that the obtained results satisfy Kirchhoff's law. If Kirchhoff's law is not satisfied, it is likely that there are some errors in analyzing TEM observations. Figure 2.18 shows an example of dislocation reaction in β'-CuZn. The reaction, $[111] = [010] + [101]$, satisfies Kirchhoff's law (see also Chapter 10).

2.7 Reaction Between Two Dislocations

2.7.1 *Reaction between two edge dislocations with Burgers vectors of $+b$ and $-b$*

On the same slip plane

As shown in Fig. 2.19, when the respective ehp's of positive and negative edge dislocations meet, they annihilate, so do the two edge dislocations.

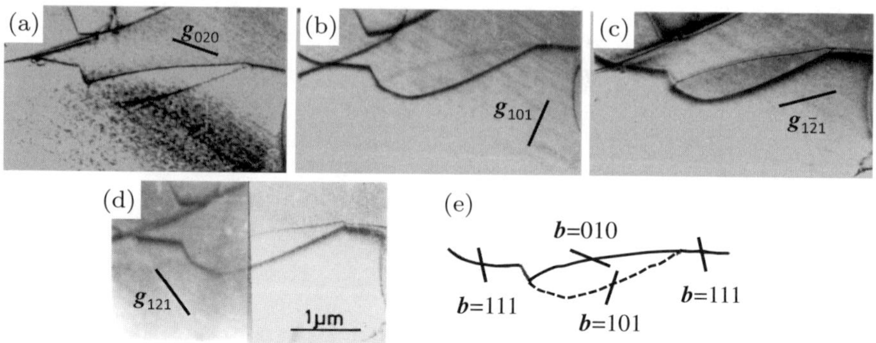

Fig. 2.18 Dislocation reaction in β'–CuZn (Reproduced with permission from [8]). $\boldsymbol{g}$ indicates the reflection vector under which the dislocations are imaged. When $\boldsymbol{g} \cdot \boldsymbol{b} = 0$, the contrast of dislocation is weak. By making use of this criterion, $\boldsymbol{b}$ can be determined. (Ref. [9]).

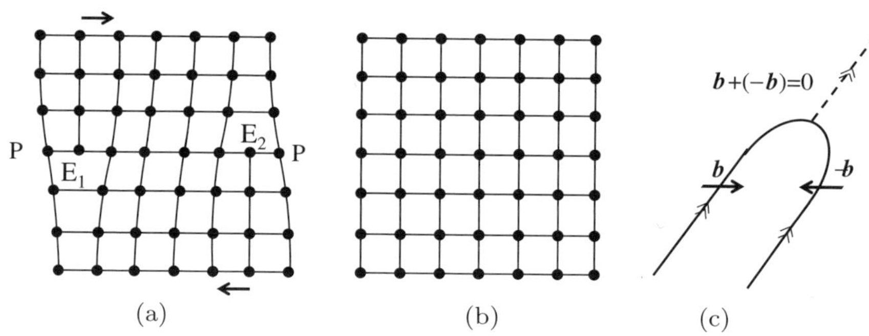

Fig. 2.19 Annihilation of a positive edge dislocation (E_1) and a negative dislocation (E_2).

Applying the Kirchhoff's law, one obtains

$$(+\boldsymbol{b}) + (-\boldsymbol{b}) = 0.$$

Slip planes with one atomic distance apart

When an edge dislocation of $+\boldsymbol{b}$ (E_1) slips on P_1 and another edge dislocation of $-\boldsymbol{b}(E_2)$ slips on P_2, a gap with one atomic distance is formed between the respective ehp's as shown in Fig. 2.20(b), resulting in formation of a column of vacancies.

Exercise 2.12 When, inversely, a dislocation of $+\boldsymbol{b}$ (E_1) is below and a dislocation of $-\boldsymbol{b}(E_2)$ is above, a column of interstitials is formed. Confirm this.

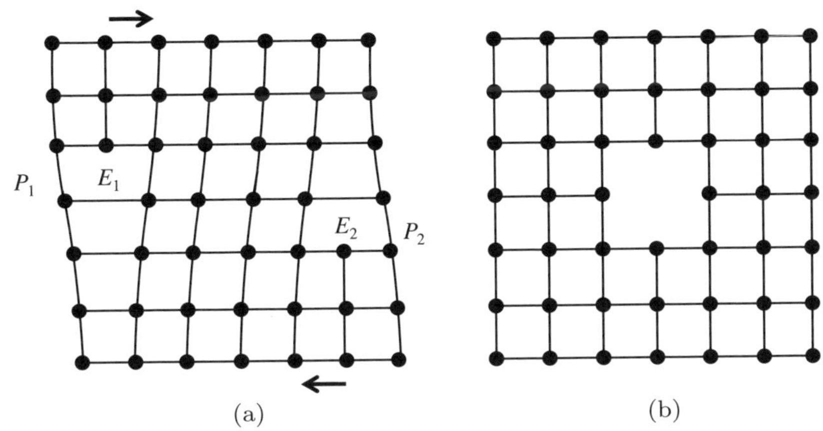

Fig. 2.20

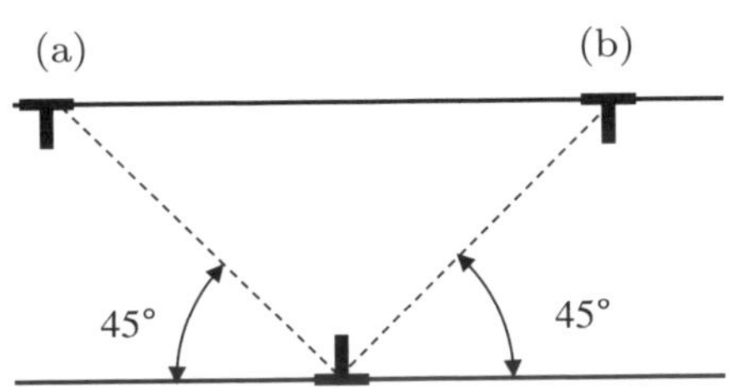

Fig. 2.21 Dislocation dispole.

Slip planes separated by more than one atomic distance

Elastic interaction between edge dislocations of $\pm\boldsymbol{b}$ hinders the two dislocations from moving (Sec. 4.4.2 for details). Dislocations with the opposite Burgers vectors (positive and negative) standing nearby are analogous to the electric dipole with positive and negative charges and are referred to as a dislocation dipole (Fig. 2.21). Edge dislocations composing a dipole make an angle of 45°, as shown in Figs. 2.21(a) and 2.21(b).

2.7.2 *Screw dislocations of $+b$ and $-b$*

Screw dislocations can slip on planes belonging to the zone axis of $\boldsymbol{b}$, so that they attract each other (Sec. 4.4.1 for details) and disappear according to Kirchhoff's law:

$$(+\boldsymbol{b}) + (-\boldsymbol{b}) = 0.$$

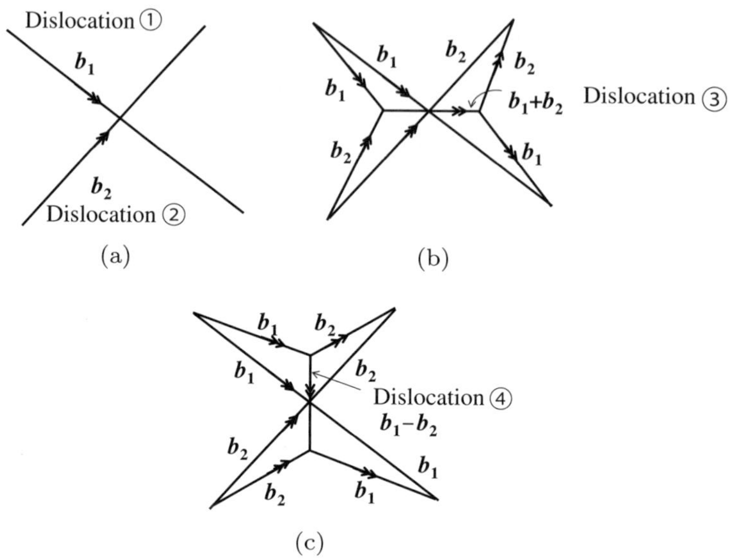

Fig. 2.22

2.7.3 *Dislocation reaction with different Burgers vectors (1)*

Generally speaking when two dislocations ① and ② react (Fig. 2.22(a)), two types of reactions can take place (Figs. 2.22(b) and 2.22(c)). The type of reactions that take place depends on the energy of the resultant configuration.

2.7.4 *Dislocation reaction with different Burgers vectors (2) Crossing of dislocations*

When two dislocations cross each other, jog(s) can be formed. The jogs hinder the motion of the dislocations. Details of jogs will be described in Chapter 8. Here, the geometry of crossing is considered, applying definition 2, that is, viewing the slipped area at right, the upper crystal is displaced by $\boldsymbol{b}$ with respect to the fixed lower crystal.

Suppose that dislocation ① is fixed and dislocation ② approaches from right behind (Fig. 2.23(a)), the direction of the dislocation ② is indicated by a double-headed arrow. That is, dislocation ② leaves behind the slipped area. The direction of $\boldsymbol{b}$ of dislocation ① is not important.

Supposing that dislocation ② is a screw, there are two possibilities, i.e., RH and LH screw dislocations, $\boldsymbol{b}$ of which is $\boldsymbol{b}_R$ or $\boldsymbol{b}_L$, respectively.

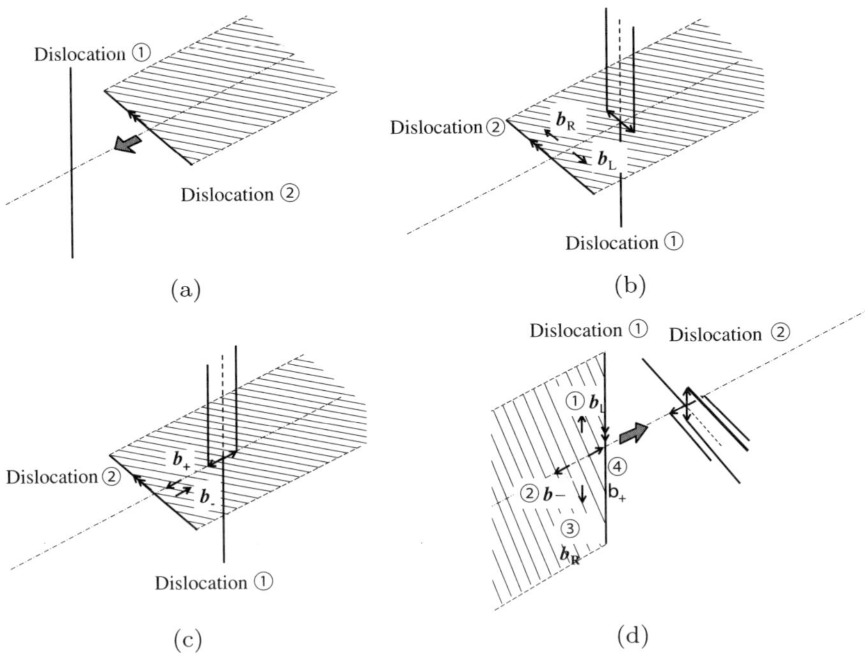

Fig. 2.23 Crossing of dislocations.

When dislocation ② crosses dislocation ①, a jog is formed on dislocation ① depending on the sense of $\boldsymbol{b}$ ($\boldsymbol{b}_R$ or $\boldsymbol{b}_L$) of dislocation ②, as shown in Fig. 2.23(b).

When dislocation ② is an edge dislocation, jogs such as shown in Fig. 2.23(c) are formed. Inversely, consider the case where dislocation ① cuts a fixed dislocation ② (Fig. 2.23(d)). Dislocation ① has four possibilities, that is, edge (positive and negative) or screw (RH or LH). Resultant jogs corresponding to these possibilities are shown in Fig. 2.23(d).

Exercise 2.13 When $\boldsymbol{b}$ of dislocation ① in Fig. 2.23(d) is ①, ②, ③ and ④, determine geometry of bends formed on dislocation ②.

Exercise 2.14 Figure 2.24 illustrates Fig. 79 (p.173) in Cottrell's textbook "Dislocatioms and plastic flow in crystals" [2]. Use the aforementioned operation to confirm this.

Figure 2.25 is an example of the crossing of screw dislocations in Fe as observed by an *in-situ* deformation experiment in a high-voltage electron

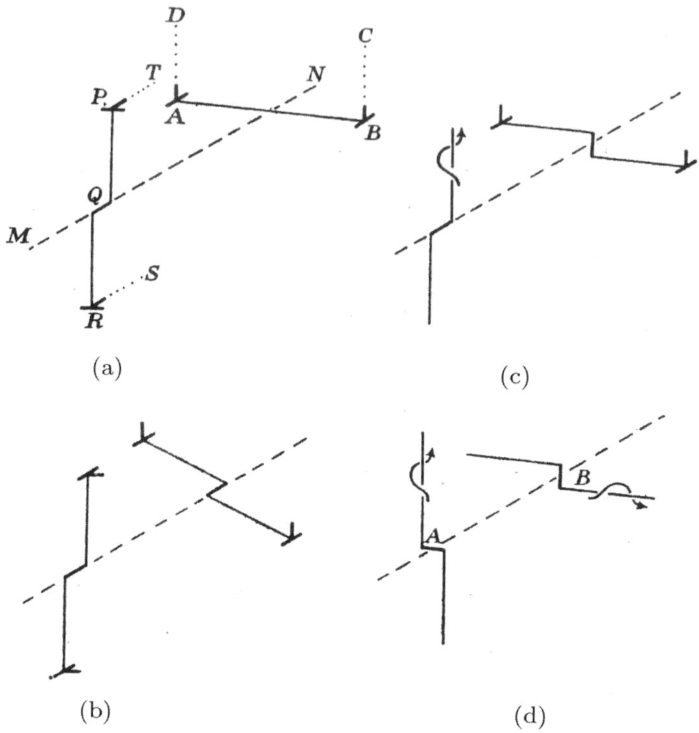

Fig. 2.24 The crossing of dislocations (Reproduced with permission from [2]).

microscope (HVEM). Screw dislocation A is moving from left to right.[5] Dislocation A meets dislocation B which is moving on a different slip system ((b),(c),(d)). Immediately after the crossing, cusps are formed both on A and B, presumably caused by jogs formed by the crossing of A and B.

Jog J_B on dislocation B immediately moves along B and disappears outside of crystal. Jog J_A stays on dislocation A for a while, then moves to the direction opposite to that in which J_B moved and disappears outside of crystal. During these processes, screw dislocation A stays unmoved, but once relieved from jogs (J_A, J_B) it moves leftward (f),(g) until it meets another screw dislocation $C(h-j)$. By the crossing of A and C, jogs J'_A and J_C are formed (k). Thus, jogs formed by the crossing of screw dislocations hinder motions of the screw dislocations involved.

[5]Fe crystallizes in the BCC structure at room temperature and dislocations are aligned along screw dislocations (See Chapter 12).

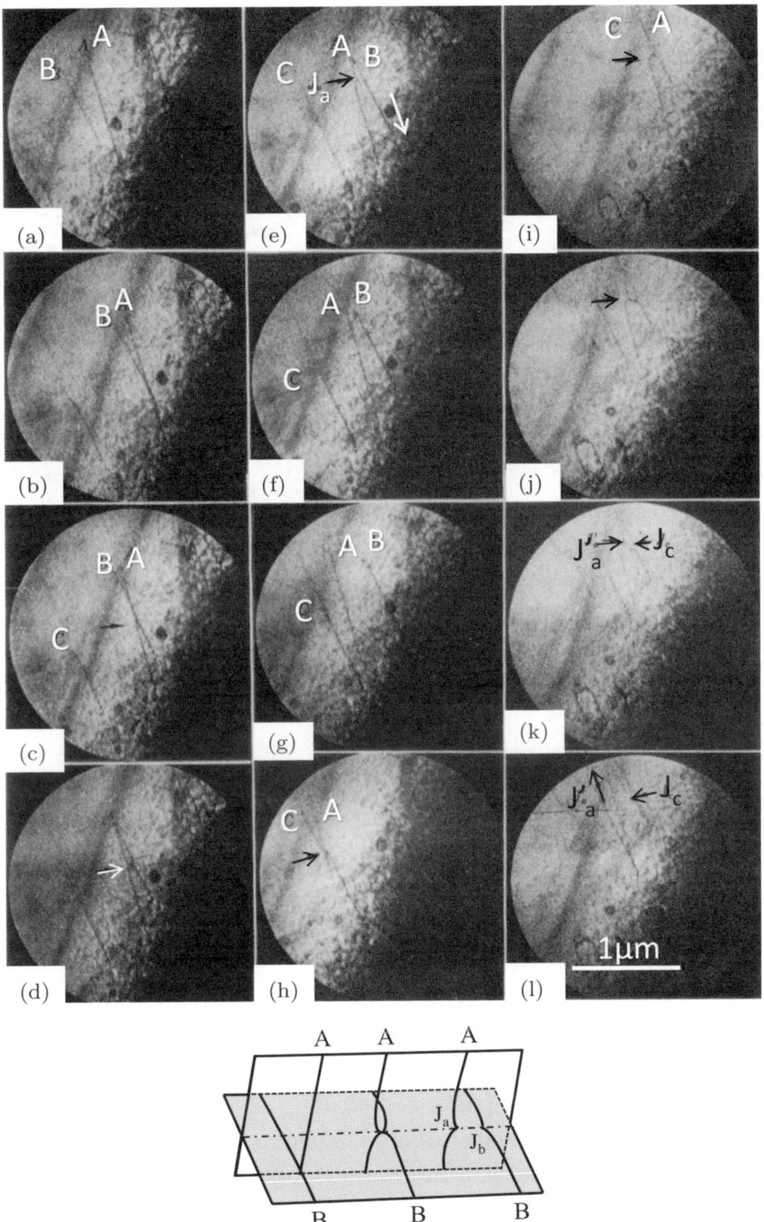

Fig. 2.25 Formation of jogs by the crossing of screw dislocations in Fe (Reproduced with permission from [10]). (m) Schematic illustration.

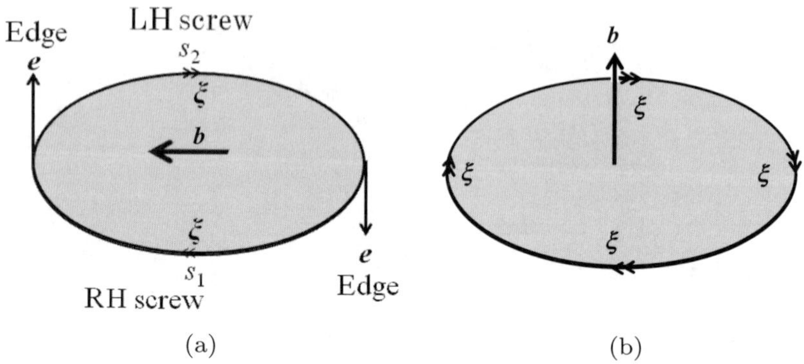

Fig. 2.26 (a) Glide dislocation and (b) Prismatic dislocation.

2.8 Prismatic Dislocation

So far a dislocation is defined as a boundary between slipped and unslipped areas, so that Burgers vector $\boldsymbol{b}$ lies on that plane where the dislocation loop lies (i.e., slip plane) (Fig. 2.26(a)). This type of dislocation is referred to as a glide dislocation. On the other hand, when $\boldsymbol{b}$ is out of the slip plane, the dislocation is referred to as a prismatic dislocation (Fig. 2.26(b)).

2.8.1 *Formation of prismatic dislocations by precipitation of point defects*

A dislocation is one type of a lattice defect. Because of its linear dimension it is referred to as a linear defect. On the other hand, there are point defects.

Typical point defects are a vacancy where an atom is removed from a lattice site and an interstitial which is inserted into an interstitial position (Fig. 2.27). Vacancies can exist at thermal equilibrium. The equilibrium concentration of vacancies increases with increasing temperature exponentially. Therefore, if a crystal is cooled from high temperatures, the equilibrium concentration of vacancies decreases drastically. If cooled very rapidly (quenched), vacancies get supersaturated and lose their sites on crystal lattices. Some of these supersaturated vacancies wander in the crystal and eventually reach the crystal surface and disappear.

Vacancies can exist at thermal equilibrium but interstitials cannot. However, both vacancies and interstitials can be formed by plastic deformation or radiation damage. These point defects precipitate from the matrix and coagulate to form disc-like dislocation loops. Figure 2.28 shows atomic

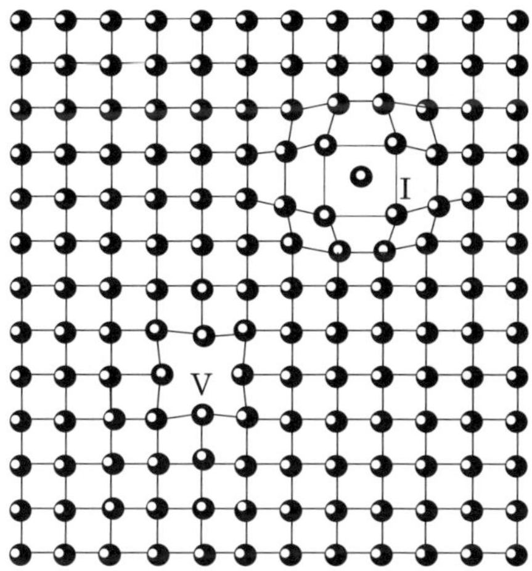

Fig. 2.27 Point defects. Vacancy (V) and interstitial (I).

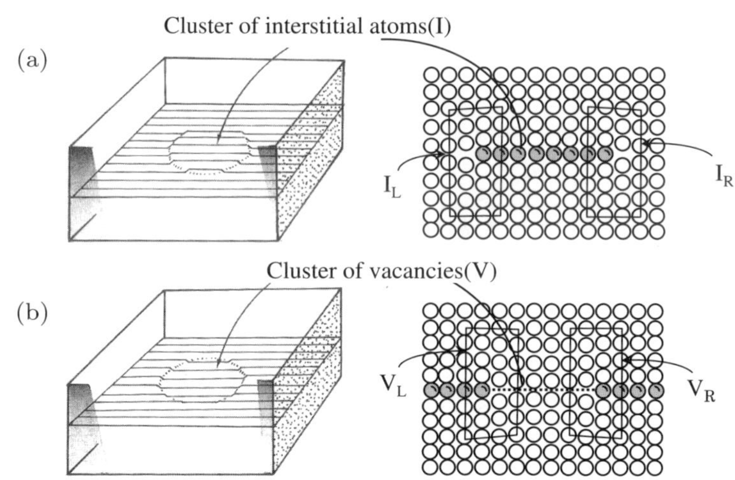

Fig. 2.28 (a),(b) Formation of dislocation loops by precipitation of point defects (Reproduced with permission from [11]).

arrangements around such dislocation loops. Figure 2.28(a) shows an interstitial type and Fig. 2.28(b) shows a vacancy type. For both of them, $\boldsymbol{b}$ is nearly perpendicular to the plane on which the dislocation loops lie, that is, prismatic loops.

Figure 2.29(a) shows an example of prismatic loops. The geometry is shown in Fig. 2.29(b). $\boldsymbol{b}$ is perpendicular to the paper and the reflection

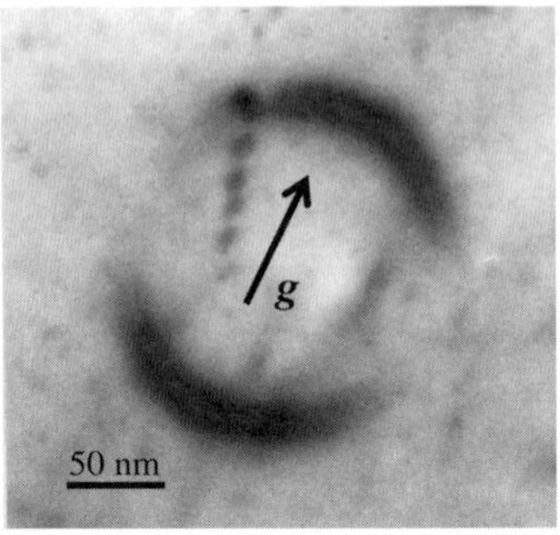

(a)

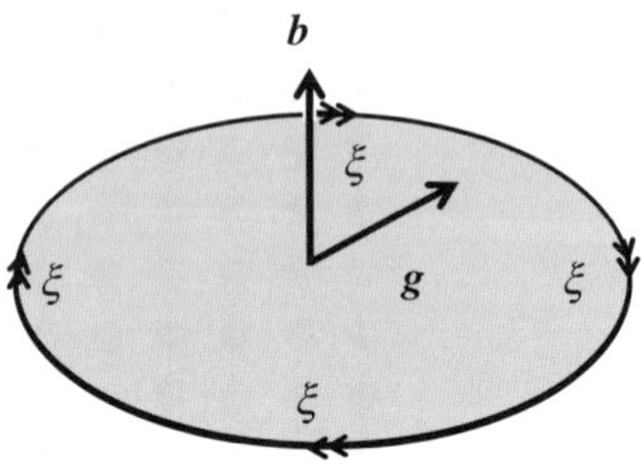

(b)

Fig. 2.29 (a),(b) An example of a prismatic loop in $SrTiO_3$ (Reproduced with permission from [12]).

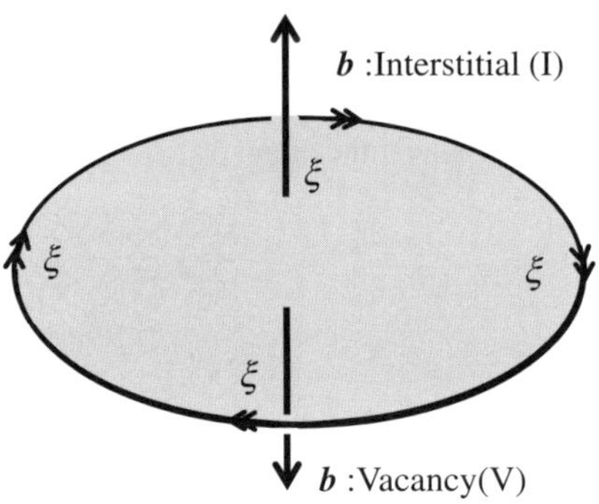

Fig. 2.30 Burgers vector of a prismatic loop.

vector $\boldsymbol{g}$ is parallel to it. Thus, $\boldsymbol{g} \cdot \boldsymbol{b} = 0$, so that the contrast of the dislocation loop is very weak. But a strain (displacement) field of an edge dislocation is so complicated that even in the case of $\boldsymbol{g} \cdot \boldsymbol{b} = 0$ a weak contrast (the residual contrast) appears when $\xi \cdot [\boldsymbol{g} \times \boldsymbol{b}] \neq 0$.

Exercise 2.15 Draw Burgers circuits (I_R, I_L, V_R, V_L) at the right and left sides in Fig. 2.28(a). Take the direction of the dislocation (ξ) in such a way as is shown in Fig. 2.30(b).

When the relation between $\boldsymbol{b}$ and ξ of a prismatic loop is a RH screw, the prismatic loop is of vacancy type. When it is a LH screw, the loop is of interstitial type.

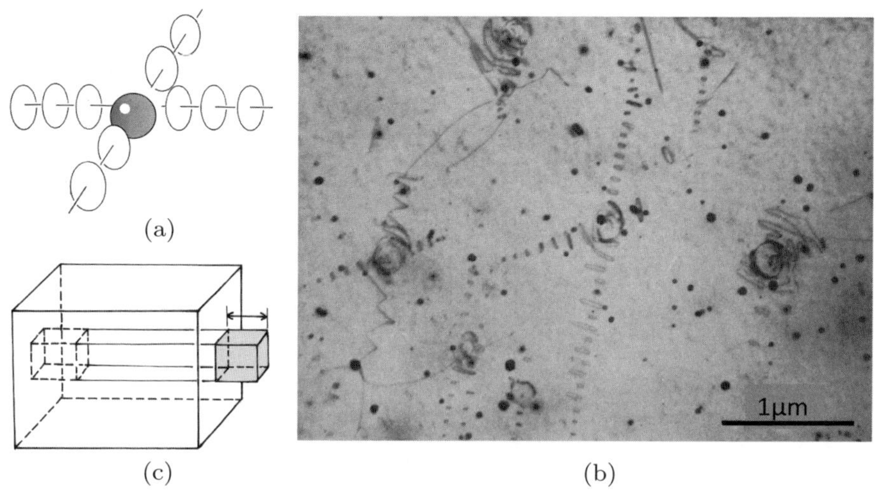

Fig. 2.31 Rosette patterns of prismatic loops. (a) Schematic illustration, (b) Example in Mg-0.7% Mn alloy (Reproduced with permission from [13]) (c) Pencil glide.

2.8.2 *Formation of a prismatic loop by pencil glide*

Metallic materials have inclusions. These inclusions are non-metallic particles such as oxides and carbides formed during refinement. These inclusions have thermal expansion coefficients different from that of the matrix metal. Thus, on cooling from high temperatures a strain field is generated around an inclusion, leading to punching-out of dislocation loops (Fig. 2.31(a)). These dislocation loops are punched out along crystallographically equivalent directions and these patterns are referred to as rosette patterns. Figure 2.31(b) shows rosette patterns composed of arrays of punched dislocation loops in a Mg alloy. The motion of edge dislocations in the punching-out is referred to as a pencil glide because slip takes place like that in a bundle of pencils sheared in the direction of their lengths (Fig. 2.31(c)).

2.9 Climb Motion of Dislocations

2.9.1 *Climb of edge dislocations*

As far as an edge dislocation moves on a slip plane defined by Eq. (2.5), total numbers of atoms remains unchanged. That is, the number of atoms is conserved (Fig. 2.32). Such a motion is referred to as a conservative motion. Let us suppose that vacancies meet an edge dislocation. An edge dislocation accompanies an ehp. If vacancies gather just beneath an ehp,

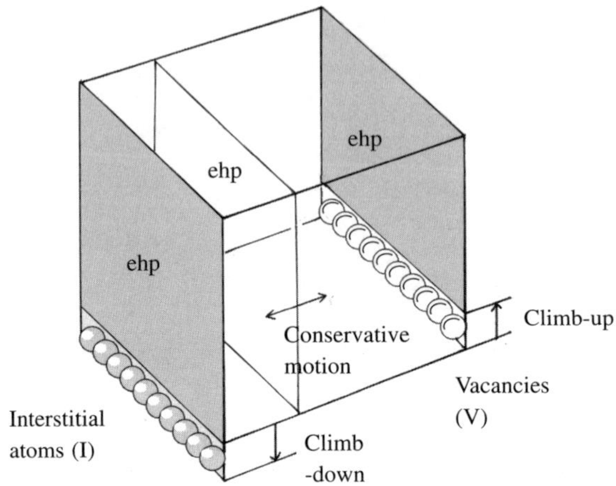

Fig. 2.32 Conservative motion.

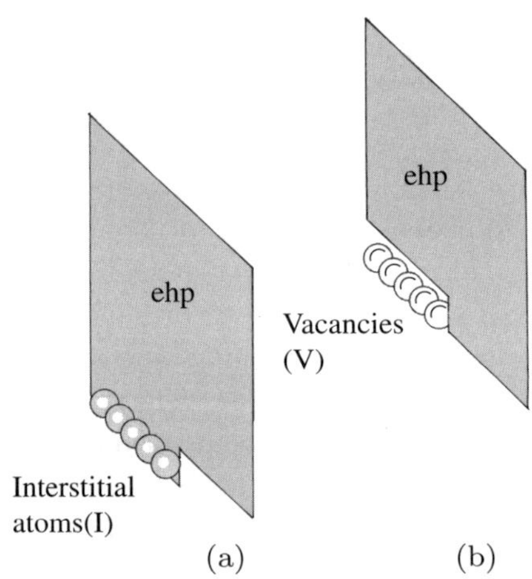

Fig. 2.33 Climb motion. (a) climb-down (b) climb-up.

the ehp moves upwards by one atomic distance. This is referred to as a climb motion (Fig. 2.33(a)).

When interstitials gather around an edge dislocation, the ehp moves downward (Fig. 2.33(b)). This is referred to as a climb-down to distinguish from a climb-up motion. In both cases an edge dislocation is a sink for point defects.

Exercise 2.16 Consider a process in which vacancies are emitted from an ehp. Determine whether the ehp climbs up or climbs down. (Hint): Emitting vacancies is equivalent to absorbing interstitials.

2.9.2 *Climb of screw dislocations*

Screw dislocations do not accompany ehp, so that screw dislocations are not supposed to climb in the sense edge dislocations do. However, in reality, screw dislocations react with point defects, transforming themselves into helical dislocations (or helices).

Do not confuse screw dislocations with helical dislocations.

Now, consider the process by which a screw dislocation transforms itself into a helix. Let us assume that the screw dislocation is bent occasionally, so that short segments with edge components are present along its length. Such a bend is referred to as a jog. This jog has an edge component which absorbs point defects and climbs (Fig. 2.34). Now, let ehp's

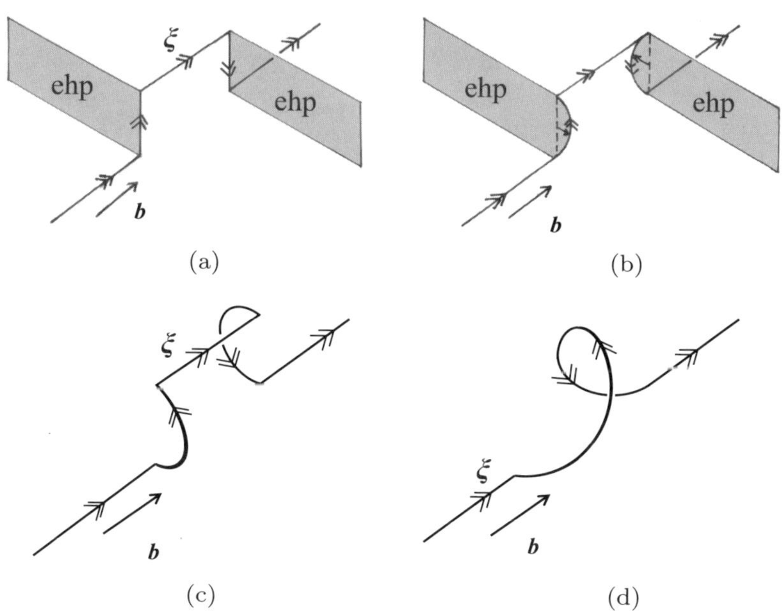

Fig. 2.34 Transformation of a screw dislocation into a LH helical dislocation by absorbing interstitials.

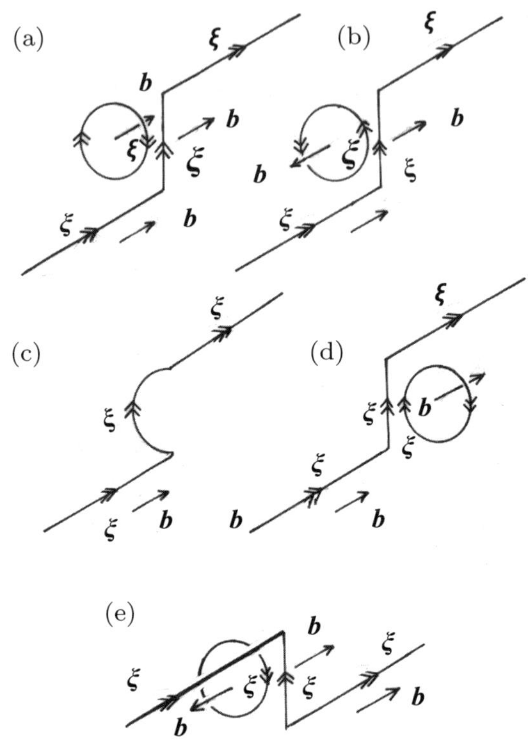

Fig. 2.35 Formation of a helical dislocation through reaction with a prismatic dislocation loop.

associated with the short edge components be as shown in Fig. 2.34(a). If interstitials are absorbed onto this ehp, then the jog will climb-down as shown in Fig. 2.34(b). Eventually a LH helical dislocation will be formed (Fig. 2.34(c)).

Exercise 2.17 Show that when vacancies (V) are absorbed, a screw dislocation transforms into a RH helix.

The aforementioned process assumes that individual point defects are absorbed onto a jog. Let us consider the case where a prismatic loop is already present, which reacts with ehp of a jog on a nearby screw dislocation. Figure 2.35 shows the reaction between a jog with an edge component on a RH screw dislocation and a dislocation loop of vacancy type.

The dislocation directions ξ and Burgers vectors $\boldsymbol{b}$ of the prismatic loop and the screw dislocation are shown in Fig. 2.35. Now, let the prismatic loop be located at the left side of the jog, which has an edge component (Fig. 2.35(a)). The ξ of edge-component-containing jog and that of the nearby prismatic loop (running almost parallel to the jog) are anti-parallel. In order to facilitate the application of Kirchhoff's law let the direction ξ of the prismatic loop be reversed. Needless to say, the sign of $\boldsymbol{b}$ of the loop is reversed (Fig. 2.35(b)). Applying the Kirchhoff's law to the two parallel dislocations (the prismatic loop on the left, plus screw dislocation on the right), the product has the Burgers vector,

$$\boldsymbol{b} + (-\boldsymbol{b}) = 0$$

that is, the two segments are annihilated (Fig. 2.35(c)) and a RH helix is formed.

Exercise 2.18 Show that this result is in keeping with Fig. 2.34.

Exercise 2.19 If the prismatic loop is located at the right side of the screw as shown in Fig. 2.35(d), the aforementioned annihilation does not take place. Discuss the reason for this difference.

Exercise 2.20 Consider a jog shown in Fig. 2.35(e) and show that a LH helix is formed.

We assume that the screw dislocation is not perfectly straight and contains jogs with the edge component. Reactions with a perfectly straight pure screw dislocation with a prismatic loop will be discussed in Sec. 4.4.4.

The following rules hold:

$$\begin{aligned}
\text{RH screw} + \text{vacancy} &= \text{RH helix}\\
\text{RH screw} + \text{interstitial} &= \text{LH helix}\\
\text{LH screw} + \text{vacancy} &= \text{LH helix}\\
\text{LH screw} + \text{interstitial} &= \text{RH helix.}
\end{aligned}$$

Figure 2.36 shows a pair of stereo-electron micrographs of helical dislocations in a Ag–Al alloy [14].

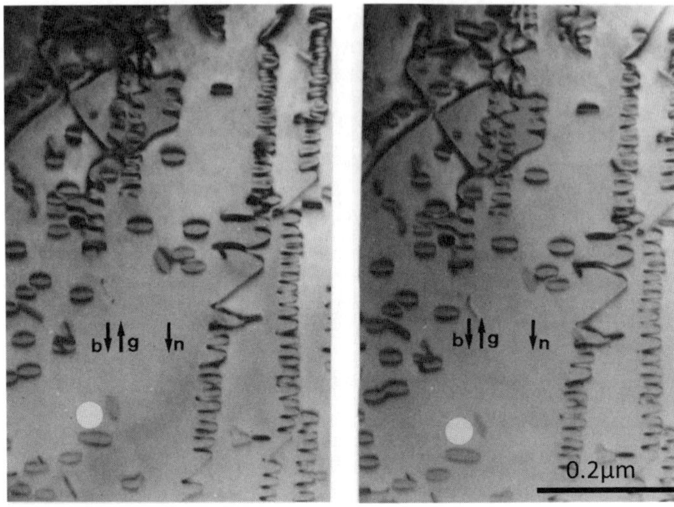

Fig. 2.36 A pair of stereo-electron micrographs of helical dislocations. Place a partition between the two micrographs and see the right-sided picture with right eye and the left-sided picture with left eye. The images appear in 3D [14].

References

1. Bilby, B.A. cited in [2, Fig. 3].
2. Cottrell, A.H. *Dislocations and Plastic Flow in Crystals*, Oxford University Press, 1954.
3. Cotterill, R. *The Cambridge Guide to the Material World*, Cambridge University Press, 1985.
4. Hirth, J.P. and Lothe, J. *Theory of Dislocations*, 1st ed., McGraw-Hill, (1967); 2nd ed., Wiley, (1982).
5. Friedel, J. *Dislocations*, Pergamon Press, Oxford, 1964.
6. Read, Jr.W.T. *Dislocations in Crystals*, McGraw-Hill, New York,1953
7. Weetman, J. and Weertman, J.R. *Elementary Dislocation Theory*, MaMillan, New York, 1964.
8. Saka, H. and Zhu, Y.M. *Phil. Mag. A* **51** (1985), 629.
9. Saka, H. *Phil. Mag. A* **49** (1984), 327.
10. Saka, H., Noda, K., and Imura, T. *Trans ISIJ.* **13** (1973), 318.
11. Saka, H. *Electron Microscopy of Crystals* (in Japanese), Uchidarokakuho, 1997.
12. Nishigaki, J., Kuroda, K., and Saka, H. *Phys. Stat. Sol(A)* **128** (1991) 319.
13. Challis, M.G. in *The World through the Electron Microscope, Metallurgy IV*, compiled by S. Koda, (Japan JEOL,1968), p. 62.
14. Nakada, Y. and Saka, H. unpublished work.

Chapter 3

Fundamentals of Elasticity Theory

A dislocation strains the nearby crystal lattices, giving rise to an elastic stress. The most important properties of a dislocation are its elastic properties. An understanding of elasticity theory is a necessary prerequisite for understanding elastic properties of dislocations. This chapter is devoted to understanding fundamentals of elasticity theory. The point of elasticity theory is to master calculation of matrices. Especially, the treatment of subscripts is tedious. In order to cope with this the reader is recommended to highlight subscripts in different colour such as red, blue etc. The account of this chapter on elasticity theory is brief and only fundamental. For more comprehensive considerations the reader is referred to the textbooks of elasticity theory. Much of this chapter is based on Hirth–Lothe [1]. It is worth reminding the reader of their message, "*No amount of exposition can take the place of hard study in mastering the formulation of elasticity theory.* "

3.1 Displacement, Strain and Stress

Now let us consider that a point $\boldsymbol{r}$ moves under an external force. The amount of movement $\boldsymbol{u}$ is referred to as a displacement. The dimension of the displacement is [L].

Let $\boldsymbol{u}$ be expressed as $(\boldsymbol{u}_1, \boldsymbol{u}_2, \boldsymbol{u}_3) = (\boldsymbol{u}, \boldsymbol{v}, \boldsymbol{w})$ in orthogonal Cartesian coordinates $(\boldsymbol{x}_1, \boldsymbol{x}_2, \boldsymbol{x}_3)$ (or equivalently in $(\boldsymbol{i}, \boldsymbol{j}, \boldsymbol{k})$ or $(\boldsymbol{x}, \boldsymbol{y}, \boldsymbol{z})$) (Fig. 3.1).

$$\begin{aligned}\boldsymbol{u} &= u_1\boldsymbol{x} + u_2\boldsymbol{y} + u_3\boldsymbol{z} \\ &= u_1\boldsymbol{i} + u_2\boldsymbol{j} + u_3\boldsymbol{k} \\ &= u_1\boldsymbol{x}_1 + u_2\boldsymbol{x}_2 + u_3\boldsymbol{x}_3 \end{aligned} \tag{3.1}$$

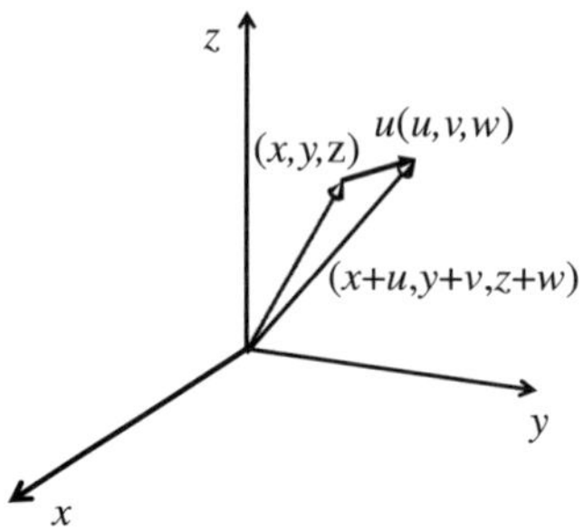

Fig. 3.1 Definition of a displacement $\boldsymbol{u}$.

$\boldsymbol{u}$ can be different from position to position, that is, $\boldsymbol{u}$ is a function of the position (x_1, x_2, x_3). If $\boldsymbol{u}$ is not a function of the position with a constant value R, the crystal is displaced by R as a whole: This is of no interest in elasticity theory. That $\boldsymbol{u}$ is different from position to position gives rise to a strain. Strain is defined as.[1]

$$\varepsilon_{ij} = \frac{1}{2}\left(\frac{\partial u_i}{\partial x_j} + \frac{\partial u_j}{\partial x_i}\right), \tag{3.2}$$

$\varepsilon_{ij} (i \neq j)$ is referred to as shearing as shown in Fig. 3.2. In engineering a shear strain is defined as

$$\gamma_{ij} = 2\varepsilon_{ij}. \tag{3.3}$$

As can be seen from Fig. 3.2, γ_{ij} is related to the angles of shearing which can be measured.

$$\gamma_{ij} = \phi_1 + \phi_2 \tag{3.4}$$

[1] Takeuchi [2], Landau–Lifshits[3], Nye [4] employ this definition.
Timoshenko–Goodier [5] express shearing strain by γ.

$$\gamma_{ij} = \frac{\partial u_i}{\partial x_j} + \frac{\partial u_j}{\partial x_i} \quad (3.2') = (3.3)$$

However, this is consistent with Eq. 3.3 in this book which appears later.
On the other hand, Suzuki [6] and Love [7] employ the following expression;

$$\varepsilon''_{ij} = \frac{\partial u_i}{\partial x_j} + \frac{\partial u_j}{\partial x_i}. \quad (3.2'')$$

Here, in order to distinguish from the definition by Eq. (3.2), double-prime ($''$) is attached.

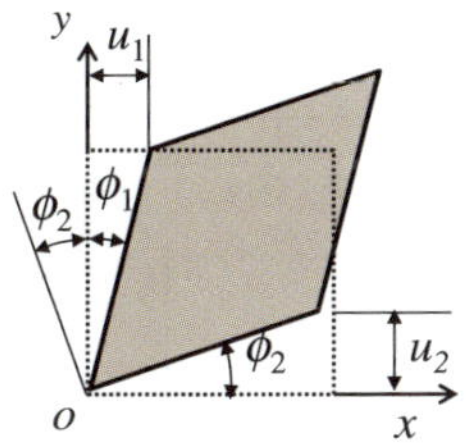

Fig. 3.2 Definition of a shear strain.

Once a strain is defined, a stress is defined by Hooke's law as follows.

$$\sigma_{ij} = \sum_{kl} c_{ijkl}\varepsilon_{kl}. \tag{3.5}$$

Here, c_{ijkl} are the proportionality coefficients referred to as elastic constants. Subscripts ij indicate the coordinates of the stress and kl the coordinates of the strain.

It follows from Eq. (3.2) that

$$\varepsilon_{kl} = \varepsilon_{lk}.$$

Therefore,

$$c_{ijkl} = c_{jikl} = c_{ijlk} = c_{jilk}. \tag{3.6}$$

Substituting Eq. (3.5) to Eq. (3.2),

$$\sigma_{ij} = \sum_{kl} c_{ijkl}\frac{\partial u_k}{\partial x_l}, \tag{3.7}$$

σ_{ij} is shown in Fig. 3.3, where σ_{ij} is ith component of the force per unit area on a plane whose outward-drawn normal is parallel to the positive x_j direction.

When the element of volume is at rest and no internal torque is present,

$$\sigma_{ij} = \sigma_{ji} \tag{3.8}$$

holds.

Exercise 3.1 Draw other stresses in Fig. 3.3.

Stress consists of a pair of force. Therefore, the sign of a stress is defined by the sign of the force along the outward-drawn normals to (100), (010) (shadowed in Fig. 3.3) and (001) planes. Same is true for a strain.

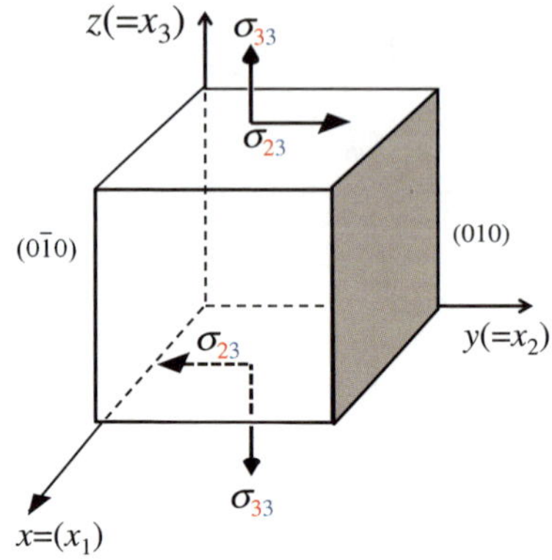

Fig. 3.3 Definition of a stress.

3.2 Matrix Notation of Hooke's Law

In Hooke's law,

$$\sigma_{ij} = \sum_{kl} c_{ijkl}\varepsilon_{kl} \tag{3.5 again}$$

i, j, k, l are one of $1, 2, 3$. Therefore, each of $\sigma_{ij}, \varepsilon_{kl}$ has nine elements, so that c_{ijkl} has $9 \times 9 = 81$ elements.

Equation (3.5) is expressed explicitly as follows,

$$\begin{aligned}
\boldsymbol{\sigma}_{11} &= \boldsymbol{c}_{1111}\boldsymbol{\varepsilon}_{11} + \boldsymbol{c}_{1122}\boldsymbol{\varepsilon}_{22} + \boldsymbol{c}_{1133}\boldsymbol{\varepsilon}_{33} + \boldsymbol{c}_{1123}\boldsymbol{\varepsilon}_{23} + \boldsymbol{c}_{1131}\boldsymbol{\varepsilon}_{31} + \boldsymbol{c}_{1112}\boldsymbol{\varepsilon}_{12}\\
&\quad + \boldsymbol{c}_{1132}\boldsymbol{\varepsilon}_{32} + \boldsymbol{c}_{1113}\boldsymbol{\varepsilon}_{13} + \boldsymbol{c}_{1121}\boldsymbol{\varepsilon}_{21}\\
\boldsymbol{\sigma}_{22} &= \boldsymbol{c}_{2211}\boldsymbol{\varepsilon}_{11} + \boldsymbol{c}_{2222}\boldsymbol{\varepsilon}_{22} + \boldsymbol{c}_{2233}\boldsymbol{\varepsilon}_{33} + \boldsymbol{c}_{2223}\boldsymbol{\varepsilon}_{23} + \boldsymbol{c}_{2231}\boldsymbol{\varepsilon}_{31} + \boldsymbol{c}_{2212}\boldsymbol{\varepsilon}_{12}\\
&\quad + \boldsymbol{c}_{2232}\boldsymbol{\varepsilon}_{32} + \boldsymbol{c}_{2213}\boldsymbol{\varepsilon}_{13} + \boldsymbol{c}_{2221}\boldsymbol{\varepsilon}_{21}\\
\boldsymbol{\sigma}_{33} &= \boldsymbol{c}_{3311}\boldsymbol{\varepsilon}_{11} + \boldsymbol{c}_{3322}\boldsymbol{\varepsilon}_{22} + \boldsymbol{c}_{3333}\boldsymbol{\varepsilon}_{33} + \boldsymbol{c}_{3323}\boldsymbol{\varepsilon}_{23} + \boldsymbol{c}_{3331}\boldsymbol{\varepsilon}_{31} + \boldsymbol{c}_{3312}\boldsymbol{\varepsilon}_{12}\\
&\quad + \boldsymbol{c}_{3332}\boldsymbol{\varepsilon}_{32} + \boldsymbol{c}_{3313}\boldsymbol{\varepsilon}_{13} + \boldsymbol{c}_{3321}\boldsymbol{\varepsilon}_{21}\\
\boldsymbol{\sigma}_{23} &= c_{2311}\boldsymbol{\varepsilon}_{11} + \boldsymbol{c}_{2322}\boldsymbol{\varepsilon}_{22} + \boldsymbol{c}_{2333}\boldsymbol{\varepsilon}_{33} + \boldsymbol{c}_{2323}\boldsymbol{\varepsilon}_{23} + \boldsymbol{c}_{2331}\boldsymbol{\varepsilon}_{31} + \boldsymbol{c}_{2312}\boldsymbol{\varepsilon}_{12}\\
&\quad + \boldsymbol{c}_{2332}\boldsymbol{\varepsilon}_{32} + \boldsymbol{c}_{2313}\boldsymbol{\varepsilon}_{13} + \boldsymbol{c}_{2321}\boldsymbol{\varepsilon}_{21}\\
\boldsymbol{\sigma}_{31} &= \boldsymbol{c}_{3111}\boldsymbol{\varepsilon}_{11} + \boldsymbol{c}_{3122}\boldsymbol{\varepsilon}_{22} + \boldsymbol{c}_{3133}\boldsymbol{\varepsilon}_{33} + \boldsymbol{c}_{3123}\boldsymbol{\varepsilon}_{23} + \boldsymbol{c}_{3131}\boldsymbol{\varepsilon}_{31} + \boldsymbol{c}_{3112}\boldsymbol{\varepsilon}_{12}\\
&\quad + \boldsymbol{c}_{3132}\boldsymbol{\varepsilon}_{32} + \boldsymbol{c}_{3113}\boldsymbol{\varepsilon}_{13} + \boldsymbol{c}_{3121}\boldsymbol{\varepsilon}_{21}\\
\boldsymbol{\sigma}_{12} &= c_{1211}\boldsymbol{\varepsilon}_{11} + \boldsymbol{c}_{1222}\boldsymbol{\varepsilon}_{22} + \boldsymbol{c}_{1233}\boldsymbol{\varepsilon}_{33} + \boldsymbol{c}_{1223}\boldsymbol{\varepsilon}_{23} + \boldsymbol{c}_{1231}\boldsymbol{\varepsilon}_{31} + \boldsymbol{c}_{1212}\boldsymbol{\varepsilon}_{12}\\
&\quad + \boldsymbol{c}_{1232}\boldsymbol{\varepsilon}_{32} + \boldsymbol{c}_{1213}\boldsymbol{\varepsilon}_{13} + \boldsymbol{c}_{1221}\boldsymbol{\varepsilon}_{21}
\end{aligned}$$

$$\begin{aligned}
\boldsymbol{\sigma}_{32} &= \boldsymbol{c}_{3211}\boldsymbol{\varepsilon}_{11} + \boldsymbol{c}_{3222}\boldsymbol{\varepsilon}_{22} + \boldsymbol{c}_{3233}\boldsymbol{\varepsilon}_{33} + \boldsymbol{c}_{3223}\boldsymbol{\varepsilon}_{23} + \boldsymbol{c}_{3231}\boldsymbol{\varepsilon}_{31} + \boldsymbol{c}_{3212}\boldsymbol{\varepsilon}_{12} \\
&\quad + \boldsymbol{c}_{3232}\boldsymbol{\varepsilon}_{32} + \boldsymbol{c}_{3213}\boldsymbol{\varepsilon}_{13} + \boldsymbol{c}_{3221}\boldsymbol{\varepsilon}_{21} \\
\boldsymbol{\sigma}_{13} &= \boldsymbol{c}_{1311}\boldsymbol{\varepsilon}_{11} + \boldsymbol{c}_{1322}\boldsymbol{\varepsilon}_{22} + \boldsymbol{c}_{1333}\boldsymbol{\varepsilon}_{33} + \boldsymbol{c}_{1323}\boldsymbol{\varepsilon}_{23} + \boldsymbol{c}_{1331}\boldsymbol{\varepsilon}_{31} + \boldsymbol{c}_{1312}\boldsymbol{\varepsilon}_{12} \\
&\quad + \boldsymbol{c}_{1332}\boldsymbol{\varepsilon}_{32} + \boldsymbol{c}_{1313}\boldsymbol{\varepsilon}_{13} + \boldsymbol{c}_{1321}\boldsymbol{\varepsilon}_{21} \\
\boldsymbol{\sigma}_{21} &= \boldsymbol{c}_{2111}\boldsymbol{\varepsilon}_{11} + \boldsymbol{c}_{2122}\boldsymbol{\varepsilon}_{22} + \boldsymbol{c}_{2133}\boldsymbol{\varepsilon}_{33} + \boldsymbol{c}_{2123}\boldsymbol{\varepsilon}_{23} + \boldsymbol{c}_{2131}\boldsymbol{\varepsilon}_{31} + \boldsymbol{c}_{2112}\boldsymbol{\varepsilon}_{12} \\
&\quad + \boldsymbol{c}_{2132}\boldsymbol{\varepsilon}_{32} + \boldsymbol{c}_{2113}\boldsymbol{\varepsilon}_{13} + \boldsymbol{c}_{2121}\boldsymbol{\varepsilon}_{21}.
\end{aligned} \tag{3.9}$$

In the matrix notation,

$$\{\sigma_{ij}\} = \begin{matrix} & (k,l) \rightarrow \\ (\boldsymbol{i},\boldsymbol{j}) & \{c_{ijkl}\}\{\varepsilon_{kl}\}, \\ \downarrow & \end{matrix} \tag{3.10}$$

where the matrix $\{c_{ijkl}\}$ is elastic constants having 9×9 elements. The elastic constants are often written in a contracted manner as

$$\begin{matrix} ij & \text{or} & kl & 11 & 22 & 33 & 23 & 31 & 12 & 32 & 13 & 21 \\ m & \text{or} & n & 1 & 2 & 3 & 4 & 5 & 6 & 7 & 8 & 9. \end{matrix} \tag{3.11}$$

Thus, by definition,

$$c_{11} = c_{1111}, \quad c_{12} = c_{1122}, \quad c_{33} = c_{3333}, \quad c_{13} = c_{1133},$$
$$c_{32} = c_{3322}, \quad c_{44} = c_{2323} \cdots\cdots.$$

Using this notation, Eq. (3.11) is written,

$$\begin{pmatrix} \sigma_{11} \\ \sigma_{22} \\ \sigma_{33} \\ \sigma_{23} \\ \sigma_{31} \\ \sigma_{12} \\ \sigma_{32} \\ \sigma_{13} \\ \sigma_{21} \end{pmatrix} = \begin{pmatrix} \sigma_{1} \\ \sigma_{2} \\ \sigma_{3} \\ \sigma_{4} \\ \sigma_{5} \\ \sigma_{6} \\ \sigma_{7} \\ \sigma_{8} \\ \sigma_{9} \end{pmatrix} = \begin{pmatrix} c_{11} & c_{12} & c_{13} & c_{14} & c_{15} & c_{16} & c_{14} & c_{15} & c_{16} \\ c_{21} & c_{22} & c_{23} & c_{24} & c_{25} & c_{26} & c_{24} & c_{25} & c_{26} \\ & & & & & & & & \\ & & & & & & & & \\ & & & & & & & & \\ & & & & & & & & \\ & & & & & & & & \\ & & & & & & & & \\ & & & & & & & & \end{pmatrix} \begin{pmatrix} \varepsilon_{11} \\ \varepsilon_{22} \\ \varepsilon_{33} \\ \varepsilon_{23} \\ \varepsilon_{31} \\ \varepsilon_{12} \\ \varepsilon_{32} \\ \varepsilon_{13} \\ \varepsilon_{21} \end{pmatrix}. \tag{3.12}$$

Because $\sigma_{ij} = \sigma_{ji}, \varepsilon_{ij} = \varepsilon_{ji}$ (for i $\neq$ j), the above 9×9 representation is reduced to 6×6 representation.

$$\begin{pmatrix} \sigma_{11} \\ \sigma_{22} \\ \sigma_{33} \\ \sigma_{23} \\ \sigma_{31} \\ \sigma_{12} \end{pmatrix} = \begin{pmatrix} c_{11} & c_{12} & c_{13} & c_{14} & c_{15} & c_{16} \\ c_{21} & c_{22} & c_{23} & c_{24} & c_{25} & c_{26} \\ & & & & & \\ & & & & & \\ & & & & & \\ & & & & & \end{pmatrix} \begin{pmatrix} \varepsilon_{11} \\ \varepsilon_{22} \\ \varepsilon_{33} \\ 2\varepsilon_{23} \\ 2\varepsilon_{31} \\ 2\varepsilon_{12} \end{pmatrix}. \tag{3.13}$$

Exercise 3.2 Fill c_{ij} in matrices in Eqs. (3.12) and (3.13).

3.3 Transformation of Stress and Strain

Transformations of coordinate axes are useful in the theory of dislocation interactions. Consider two orthogonal coordinates (x_1, x_2, x_3), (x'_1, x'_2, x'_3) (Fig. 3.4). The transformation between these coordinates is given by

$$\begin{bmatrix} x'_1 \\ x'_2 \\ x'_3 \end{bmatrix} = \{T_{ij}\} \begin{bmatrix} x_1 \\ x_2 \\ x_3 \end{bmatrix} = \begin{bmatrix} T_{11} & T_{12} & T_{13} \\ T_{21} & T_{22} & T_{23} \\ T_{31} & T_{32} & T_{33} \end{bmatrix} \begin{bmatrix} x_1 \\ x_2 \\ x_3 \end{bmatrix}, \tag{3.14}$$

where T_{ij} is the direction cosine between x'_i- and x_j-axes.

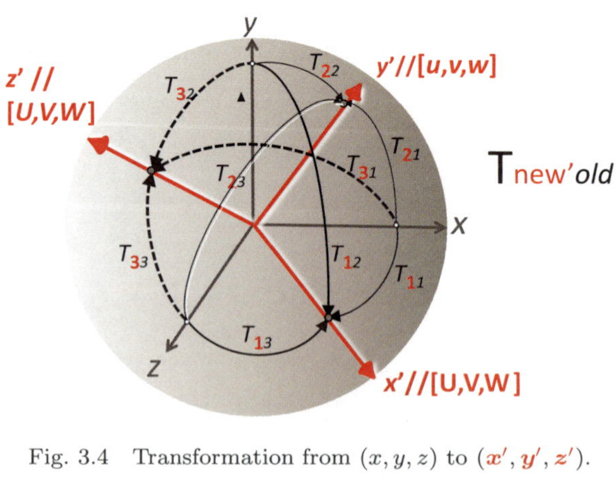

Fig. 3.4 Transformation from (x, y, z) to (x', y', z').

The inversion transformation is given by

$$\begin{bmatrix} x_1 \\ x_2 \\ x_3 \end{bmatrix} = \{T_{ij}^{-1}\} \begin{bmatrix} x'_1 \\ x'_2 \\ x'_3 \end{bmatrix} = \begin{bmatrix} T_{11}^{-1} & T_{12}^{-1} & T_{13}^{-1} \\ T_{21}^{-1} & T_{22}^{-1} & T_{23}^{-1} \\ T_{31}^{-1} & T_{32}^{-1} & T_{33}^{-1} \end{bmatrix} \begin{bmatrix} x'_1 \\ x'_2 \\ x'_3 \end{bmatrix} = \begin{bmatrix} T_{11} & T_{21} & T_{31} \\ T_{12} & T_{22} & T_{32} \\ T_{13} & T_{23} & T_{33} \end{bmatrix} \begin{bmatrix} x'_1 \\ x'_2 \\ x'_3 \end{bmatrix}, \tag{3.15}$$

so that

$$\{\boldsymbol{T}_{ij}^{-1}\} = \{\tilde{\boldsymbol{T}}_{ij}\} = \{\boldsymbol{T}_{ji}\} = \begin{bmatrix} \boldsymbol{T}_{11} & \boldsymbol{T}_{21} & \boldsymbol{T}_{31} \\ \boldsymbol{T}_{12} & \boldsymbol{T}_{22} & \boldsymbol{T}_{32} \\ \boldsymbol{T}_{13} & \boldsymbol{T}_{23} & \boldsymbol{T}_{33} \end{bmatrix}. \tag{3.15'}$$

Usually [100], [010] and [001] axes are taken as (x, y, z) axes, respectively. Let us refer the newly selected coordinates (x', y', z') to the unit vectors in the old (x, y, z) coordinate as follows,[2]

$$\begin{aligned} \mathrm{x}' &= \frac{1}{\sqrt{\mathrm{U}^2 + \mathrm{V}^2 + \mathrm{W}^2}}[\mathrm{U}, \mathrm{V}, \mathrm{W}] \\ \mathrm{y}' &= \frac{1}{\sqrt{u^2 + v^2 + w^2}}[u, v, w], \\ \mathrm{z}' &= \frac{1}{\sqrt{U^2 + V^2 + W^2}}[U, V, W]. \end{aligned} \tag{3.16}$$

Then,

$$\boldsymbol{T}_{ij} = \begin{bmatrix} \boldsymbol{T}_{11} & \boldsymbol{T}_{12} & \boldsymbol{T}_{13} \\ \boldsymbol{T}_{21} & \boldsymbol{T}_{22} & \boldsymbol{T}_{23} \\ \boldsymbol{T}_{31} & \boldsymbol{T}_{32} & \boldsymbol{T}_{33} \end{bmatrix} = \begin{bmatrix} \frac{\mathrm{U}}{\sqrt{\mathrm{U}^2+\mathrm{V}^2+\mathrm{W}^2}} & \frac{\mathrm{V}}{\sqrt{}} & \frac{\mathrm{W}}{\sqrt{}} \\ \frac{\mathrm{u}}{\sqrt{u^2+v^2+w^2}} & \frac{v}{\sqrt{}} & \frac{w}{\sqrt{}} \\ \frac{U}{\sqrt{U^2+V^2+W^2}} & \frac{V}{\sqrt{}} & \frac{W}{\sqrt{}} \end{bmatrix}. \tag{3.16'}$$

In practice, T_{ij} can be written as follows,

$$T_{ij} = \begin{bmatrix} \mathrm{U} & \mathrm{V} & \mathrm{W} \\ u & v & w \\ U & V & W \end{bmatrix} \begin{matrix} \div\sqrt{\mathrm{U}^2 + \mathrm{V}^2 + \mathrm{W}^2} \\ \div\sqrt{u^2 + v^2 + w^2} \\ \div\sqrt{U^2 + V^2 + W^2} \end{matrix}. \tag{3.16''}$$

[2] In other words, $x' // [\mathrm{U}, \mathrm{V}, \mathrm{W}], y' // [u, v, w], z' // [U, V, W]$.

Displacement $\boldsymbol{u}$ is transformed in the same manner.

$$\begin{bmatrix} u_1' \\ u_2' \\ u_3' \end{bmatrix} = \{T_{ij}\} \begin{bmatrix} u_1 \\ u_2 \\ u_3 \end{bmatrix} = \begin{bmatrix} T_{11} & T_{12} & T_{13} \\ T_{21} & T_{22} & T_{23} \\ T_{31} & T_{32} & T_{33} \end{bmatrix} \begin{bmatrix} u_1 \\ u_2 \\ u_3 \end{bmatrix}. \tag{3.17}$$

Strain and stress are transformed as follows:

$$\boldsymbol{\sigma}_{ij}' = \sum_{l,m} \boldsymbol{T}_{il} \boldsymbol{T}_{jm} \boldsymbol{\sigma}_{lm}, \tag{3.18}$$

$$\boldsymbol{\varepsilon}_{ij}' = \sum_{l,m} \boldsymbol{T}_{il} \boldsymbol{T}_{jm} \boldsymbol{\varepsilon}_{lm}. \tag{3.19}$$

Exercise 3.3 Calculate ε_{11}' using Eq. (3.19).

(Solution) ij in the left side are $i = 1, j = 1$. First substitute $i = 1, j = 1$. There are $9\varepsilon_{lm}$ in the right side. Thus, write just ε_{lm} explicitly for all of them for the time being

$$\varepsilon'_{11} = \sum_{lm} T_{1l} T_{1m} \varepsilon_{lm}$$

$$= T_{1l}T_{1m}\varepsilon_{lm} + T_{1l}T_{1m}\varepsilon_{lm} + T_{1l}T_{1m}\varepsilon_{lm}$$
$$+ T_{1l}T_{1m}\varepsilon_{lm} + T_{1l}T_{1m}\varepsilon_{lm} + T_{1l}T_{1m}\varepsilon_{lm}$$
$$+ T_{1l}T_{1m}\varepsilon_{lm} + T_{1l}T_{1m}\varepsilon_{lm} + T_{1l}T_{1m}\varepsilon_{lm}$$

Then, substitute one of

$$\varepsilon_{lm} = (\varepsilon_{11}\ \varepsilon_{22}\ \varepsilon_{33}\ \varepsilon_{23}\ \varepsilon_{31}\ \varepsilon_{12}\ \varepsilon_{32}\ \varepsilon_{13}\ \varepsilon_{21})$$

with concurrent substitution of lm into $T_{1\ell}T_{1_m}$.

$$\rightarrow T_{1l}T_{1l}\varepsilon_{11} + T_{1l}T_{1m}\varepsilon_{lm} + T_{1l}T_{1m}\varepsilon_{lm}$$
$$+ T_{1l}T_{1m}\varepsilon_{lm} + T_{1l}T_{1m}\varepsilon_{lm} + T_{1l}T_{1m}\varepsilon_{lm}$$
$$+ T_{1l}T_{1m}\varepsilon_{lm} + T_{1l}T_{1m}\varepsilon_{lm} + T_{1l}T_{1m}\varepsilon_{lm}$$

$$\rightarrow T_{1l}T_{1l}\varepsilon_{11} + T_{1l}T_{12}\varepsilon_{12} + T_{1l}T_{1m}\varepsilon_{lm}$$
$$+ T_{1l}T_{1m}\varepsilon_{lm} + T_{1l}T_{1m}\varepsilon_{lm} + T_{1l}T_{1m}\varepsilon_{lm}$$
$$+ T_{1l}T_{1m}\varepsilon_{lm} + T_{1l}T_{1m}\varepsilon_{lm} + T_{1l}T_{1m}\varepsilon_{lm}$$

$$
\begin{aligned}
&\rightarrow T_{11}T_{11}\varepsilon_{11}+T_{11}T_{12}\varepsilon_{12}+T_{11}T_{13}\varepsilon_{13}\\
&+T_{1l}T_{1m}\varepsilon_{lm}+T_{1l}T_{1m}\varepsilon_{lm}+T_{1l}T_{1m}\varepsilon_{lm}\\
&+T_{1l}T_{1m}\varepsilon_{lm}+T_{1l}T_{1m}\varepsilon_{lm}+T_{1l}T_{1m}\varepsilon_{lm}
\end{aligned}
$$

$$
\begin{aligned}
&\rightarrow T_{11}T_{11}\varepsilon_{11}+T_{11}T_{12}\varepsilon_{12}+T_{11}T_{13}\varepsilon_{13}\\
&+T_{12}T_{11}\varepsilon_{21}+T_{12}T_{12}\varepsilon_{22}+T_{12}T_{13}\varepsilon_{23}\\
&+T_{1l}T_{1m}\varepsilon_{lm}+T_{1l}T_{1m}\varepsilon_{lm}+T_{1l}T_{1m}\varepsilon_{lm}
\end{aligned}
$$

$$
\begin{aligned}
&\rightarrow T_{11}T_{11}\varepsilon_{11}+T_{11}T_{12}\varepsilon_{12}+T_{11}T_{13}\varepsilon_{13}\\
&+T_{12}T_{11}\varepsilon_{21}+T_{12}T_{12}\varepsilon_{22}+T_{12}T_{13}\varepsilon_{23}\\
&+T_{13}T_{11}\varepsilon_{31}+T_{13}T_{12}\varepsilon_{32}+T_{13}T_{13}\varepsilon_{33}
\end{aligned}
$$

The reader is recommended to use a template in Appendix A.7.

Transformation of a stress is essentially the same.

The inverse transformations of strain ε and stress σ are given by

$$\varepsilon_{ij} = \sum_{l,m} \boldsymbol{T}_{li}\boldsymbol{T}_{mj}\varepsilon'_{lm}, \tag{3.20}$$

$$\sigma_{ij} = \sum_{l,m} \boldsymbol{T}_{li}\boldsymbol{T}_{mj}\sigma'_{lm}. \tag{3.21}$$

Stress has 9 elements and the stress field $\sum$ is written as

$$\sum = \begin{bmatrix} \sigma_{11}\boldsymbol{ii} & \sigma_{12}\boldsymbol{ij} & \sigma_{13}\boldsymbol{ik} \\ \sigma_{21}\boldsymbol{ji} & \sigma_{22}\boldsymbol{jj} & \sigma_{23}\boldsymbol{jk} \\ \sigma_{31}\boldsymbol{ki} & \sigma_{32}\boldsymbol{kj} & \sigma_{33}\boldsymbol{kk} \end{bmatrix}. \tag{3.22}$$

A surface element $\boldsymbol{A}$ in this stress field receives a force $\boldsymbol{F}$

$$\boldsymbol{F} = \sum \cdot \boldsymbol{A} = \boldsymbol{A} \cdot \sum. \tag{3.23}$$

(see Fig. 3.5) Here,

$$\boldsymbol{A} = A\boldsymbol{n} = (A_1\boldsymbol{i} \quad A_2\boldsymbol{j} \quad A_3\boldsymbol{k})$$

A is the area, $\boldsymbol{n}$ is the outward-drawn normal unit vector of the surface element $\boldsymbol{A}$.

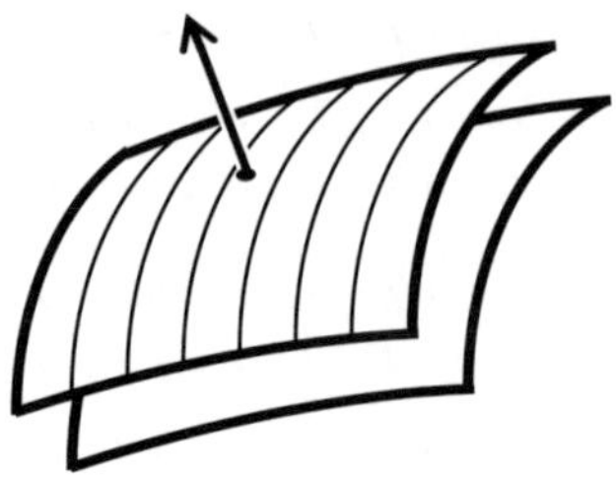

Fig. 3.5 Definition of F_{ij} in curvilinear coordinate.

Thus,

$$\boldsymbol{F} = \begin{bmatrix} \sigma_{11}\boldsymbol{ii} & \sigma_{12}\boldsymbol{ij} & \sigma_{13}\boldsymbol{ik} \\ \sigma_{21}\boldsymbol{ji} & \sigma_{22}\boldsymbol{jj} & \sigma_{23}\boldsymbol{jk} \\ \sigma_{31}\boldsymbol{ki} & \sigma_{32}\boldsymbol{kj} & \sigma_{33}\boldsymbol{kk} \end{bmatrix} \cdot \begin{bmatrix} A_1\boldsymbol{i} \\ A_2\boldsymbol{j} \\ A_3\boldsymbol{k} \end{bmatrix}$$

$$= \begin{bmatrix} A_1\boldsymbol{i} & A_2\boldsymbol{j} & A_3\boldsymbol{k} \end{bmatrix} \cdot \begin{bmatrix} \sigma_{11}\boldsymbol{ii} & \sigma_{12}\boldsymbol{ij} & \sigma_{13}\boldsymbol{ik} \\ \sigma_{21}\boldsymbol{ji} & \sigma_{22}\boldsymbol{jj} & \sigma_{23}\boldsymbol{jk} \\ \sigma_{31}\boldsymbol{ki} & \sigma_{32}\boldsymbol{kj} & \sigma_{33}\boldsymbol{kk} \end{bmatrix}. \tag{3.24}$$

We use Gibbs dyadic notation [1] for $\sum$. In this notation,

$$\sum = \begin{bmatrix} \sigma_{11}\boldsymbol{ii} & \sigma_{12}\boldsymbol{ij} \\ \sigma_{21}\boldsymbol{ji} & \sigma_{22}\boldsymbol{jj} \end{bmatrix},$$

$$\boldsymbol{b} \cdot \sum = \begin{bmatrix} b_1\boldsymbol{i} & b_2\boldsymbol{j} \end{bmatrix} \begin{bmatrix} \sigma_{11}\boldsymbol{ii} & \sigma_{12}\boldsymbol{ij} \\ \sigma_{21}\boldsymbol{ji} & \sigma_{22}\boldsymbol{jj} \end{bmatrix},$$

$$= b_1\sigma_{11}\boldsymbol{iii} + b_2\sigma_{21}\boldsymbol{ijj} + b_1\sigma_{12}\boldsymbol{iij} + b_2\sigma_{22}\boldsymbol{jjj},$$

$$= b_1\sigma_{11}\boldsymbol{i} + b_2\sigma_{21}\boldsymbol{i} + b_1\sigma_{12}\boldsymbol{j} + b_2\sigma_{22}\boldsymbol{j},$$

$$= (b_1\sigma_{11} + b_2\sigma_{21})\boldsymbol{i} + (b_1\sigma_{12} + b_2\sigma_{22})\boldsymbol{j},$$

$$\sum \cdot \boldsymbol{b} = \begin{bmatrix} \sigma_{11}\boldsymbol{ii} & \sigma_{12}\boldsymbol{ij} \\ \sigma_{21}\boldsymbol{ji} & \sigma_{22}\boldsymbol{jj} \end{bmatrix} \begin{bmatrix} b_1\boldsymbol{i} \\ b_2\boldsymbol{j} \end{bmatrix},$$

$$= \sigma_{11}b_1\boldsymbol{iii} + \sigma_{12}b_2\boldsymbol{ijj} + \sigma_{21}b_1\boldsymbol{iji} + \sigma_{22}b_2\boldsymbol{jjj},$$

$$= \sigma_{11}b_1\boldsymbol{i} + \sigma_{12}b_2\boldsymbol{i} + \sigma_{21}b_1\boldsymbol{j} + \sigma_{22}b_2\boldsymbol{j},$$

$$= (\sigma_{11}b_1 + \sigma_{12}b_2)\boldsymbol{i} + (\sigma_{21}b_1 + \sigma_{22}b_2)\boldsymbol{j},$$

The products of $\boldsymbol{ii} = \boldsymbol{jj} = 1$. In general $\Sigma\cdot\boldsymbol{b} \neq \boldsymbol{b}\cdot\Sigma$, but in elasticity theory because $\sigma_{12} = \sigma_{21}$, $\Sigma \cdot \boldsymbol{b} = \boldsymbol{b} \cdot \Sigma$.

3.4 Hooke's Law in Isotropic Solids

3.4.1 *Elastic constants*

Hooke's law is written as

$$\begin{pmatrix} \sigma_{11} \\ \sigma_{22} \\ \sigma_{33} \\ \sigma_{23} \\ \sigma_{31} \\ \sigma_{12} \end{pmatrix} = \begin{pmatrix} c_{11} & c_{12} & c_{13} & c_{14} & c_{15} & c_{16} \\ c_{21} & c_{22} & c_{23} & c_{24} & c_{25} & c_{26} \\ c_{31} & c_{32} & c_{33} & c_{34} & c_{35} & c_{36} \\ c_{41} & c_{42} & c_{43} & c_{44} & c_{45} & c_{46} \\ c_{51} & c_{52} & c_{53} & c_{54} & c_{55} & c_{56} \\ c_{61} & c_{62} & c_{63} & c_{64} & c_{65} & c_{66} \end{pmatrix} \begin{pmatrix} \varepsilon_{11} \\ \varepsilon_{22} \\ \varepsilon_{33} \\ 2\varepsilon_{23} \\ 2\varepsilon_{31} \\ 2\varepsilon_{12} \end{pmatrix}$$

or

$$\begin{pmatrix} \sigma_{11} \\ \sigma_{22} \\ \sigma_{33} \\ \sigma_{23} \\ \sigma_{31} \\ \sigma_{12} \end{pmatrix} = \begin{pmatrix} c_{11} & c_{12} & c_{13} & c_{14} & c_{15} & c_{16} \\ c_{21} & c_{22} & c_{23} & c_{24} & c_{25} & c_{26} \\ & & & & & \\ & & & & & \\ & & & & & \\ & & & & & \end{pmatrix} \begin{pmatrix} \varepsilon_{11} \\ \varepsilon_{22} \\ \varepsilon_{33} \\ \gamma_{23} \\ \gamma_{31} \\ \gamma_{12} \end{pmatrix}.$$

For cubic crystals the independent elastic constants are c_{11}, c_{12}, c_{44}.

Proof. Rotate by 180° around the x axis. Then, the new coordinate systems are written in the old system as $x' = x, y' = -y, z' = -z$. That is, $[\mathrm{UVW}] = [100], [uvw] = [0\bar{1}0], [UVW] = [00\bar{1}]$, so that

Eq. (3.16″) gives

$$T_{ij} = \begin{bmatrix} \mathrm{U} & \mathrm{V} & \mathrm{W} \\ u & v & w \\ U & V & W \end{bmatrix} \begin{matrix} \div\sqrt{\mathrm{U}^2 + \mathrm{V}^2 + \mathrm{W}^2} \\ \div\sqrt{u^2 + v^2 + w^2} \\ \div\sqrt{U^2 + V^2 + W^2} \end{matrix} = \begin{bmatrix} 1 & 0 & 0 \\ 0 & -1 & 0 \\ 0 & 0 & -1 \end{bmatrix}.$$

Substitute into Eq. (3.19)

$$\varepsilon'_{ij} = \sum_{l,m} T_{il} T_{jm} \varepsilon_{lm}$$

$$\varepsilon'_{11} = \varepsilon_{11}, \quad \varepsilon'_{22} = \varepsilon_{22}, \quad \varepsilon'_{33} = \varepsilon_{33}$$

$$\varepsilon'_{23} = \varepsilon_{23}, \quad \varepsilon'_{13} = -\varepsilon_{13}, \quad \varepsilon'_{12} = -\varepsilon_{12}$$

$$\begin{aligned}\sigma'_{12} &= \sum_{l,m} T_{1l} T_{2m} \varepsilon_{lm} = -\sigma_{12}\\ &= -c_{61}\varepsilon_{11} - c_{62}\varepsilon_{22} - c_{63}\varepsilon_{33} - c_{64}\varepsilon_{64} - c_{65}\varepsilon_{65} - c_{66}\varepsilon_{66}.\end{aligned}$$

On the other hand,

$$\begin{aligned}\sigma'_{12} &= c'_{61}\varepsilon'_{11} + c'_{62}\varepsilon'_{22} + c'_{63}\varepsilon'_{33} + c'_{64}\varepsilon'_{64} + c'_{65}\varepsilon'_{65} + c'_{66}\varepsilon'_{66}\\ &= c'_{61}\varepsilon_{11} + c'_{62}\varepsilon_{22} + c'_{63}\varepsilon_{33} + c'_{64}\varepsilon_{64} + c'_{65}\varepsilon_{65} + c'_{66}\varepsilon_{66}\end{aligned}$$

$$\therefore c'_{61} = -c_{61}, \quad c'_{62} = -c_{62}, \quad c'_{63} = -c_{63}, \quad c'_{64} = -c_{64}.$$

whether (') is attached or not, there should be no difference in the orthogonal cubic systems.

$$c'_{61} = c_{61} = 0, \quad c'_{62} = c_{62} = 0, \quad c'_{63} = c_{63} = 0, \quad c'_{64} = c_{64} = 0.$$

Similarly for $\sigma'_{13}, \sigma'_{23}$,

$$c_{51} = c_{52} = c_{53} = c_{54} = c_{45} = c_{46} = 0.$$

At this point $\{c_{ij}\}$ matrix can be written,

$$\begin{bmatrix}\sigma_{11}\\ \sigma_{22}\\ \sigma_{33}\\ \sigma_{23}\\ \sigma_{31}\\ \sigma_{12}\end{bmatrix} = \begin{bmatrix} c_{11} & c_{12} & c_{13} & c_{14} & c_{15} & c_{16}\\ c_{21} & c_{22} & c_{23} & c_{24} & c_{25} & c_{26}\\ c_{31} & c_{32} & c_{33} & c_{34} & c_{35} & c_{36}\\ c_{41} & c_{42} & c_{43} & c_{44} & 0 & 0\\ 0 & 0 & 0 & 0 & c_{55} & c_{56}\\ 0 & 0 & 0 & 0 & c_{65} & c_{66}\end{bmatrix} \begin{bmatrix}\varepsilon_{11}\\ \varepsilon_{22}\\ \varepsilon_{33}\\ \gamma_{23}\\ \gamma_{31}\\ \gamma_{12}\end{bmatrix}$$

Next, rotate by 180° around z-axis,

$$c_{41} = c_{42} = c_{43} = c_{46} = c_{65} = 0.$$

At this point $\{c_{ij}\}$ matrix can be written as

$$\begin{bmatrix}\sigma_{11}\\ \sigma_{22}\\ \sigma_{33}\\ \sigma_{23}\\ \sigma_{31}\\ \sigma_{12}\end{bmatrix} = \begin{bmatrix} c_{11} & c_{12} & c_{13} & c_{14} & c_{15} & c_{16}\\ c_{21} & c_{22} & c_{23} & c_{24} & c_{25} & c_{26}\\ c_{31} & c_{32} & c_{33} & c_{34} & c_{35} & c_{36}\\ 0 & 0 & 0 & c_{44} & 0 & 0\\ 0 & 0 & 0 & 0 & c_{55} & c_{56}\\ 0 & 0 & 0 & 0 & 0 & c_{66}\end{bmatrix} \begin{bmatrix}\varepsilon_{11}\\ \varepsilon_{22}\\ \varepsilon_{33}\\ \gamma_{23}\\ \gamma_{31}\\ \gamma_{12}\end{bmatrix}$$

Next rotate counterclockwise by 90° around y-axis,

$$c_{12} = c_{32}, \quad c_{33} = c_{11}, \quad c_{13} = c_{31}, \quad c_{44} = c_{66}.$$

Then similarly rotate counterclockwise 90° around x-axis,

$$c_{mn} = \begin{bmatrix} c_{11} & c_{12} & c_{12} & 0 & 0 & 0\\ c_{12} & c_{11} & c_{12} & 0 & 0 & 0\\ c_{12} & c_{12} & c_{11} & 0 & 0 & 0\\ 0 & 0 & 0 & c_{44} & 0 & 0\\ 0 & 0 & 0 & 0 & c_{44} & 0\\ 0 & 0 & 0 & 0 & 0 & c_{44}\end{bmatrix}. \tag{3.25}$$

In the elastically isotropic crystal, i.e., if the elastic properties are independent of direction only two independent elastic constants are required.

Now let only $\varepsilon_{11} \neq 0$ among ε_{ij}. Equation (3.25) gives

$$\sigma_{11} = c_{11}\varepsilon_{11}$$

$$\sigma_{22} = \sigma_{33} = c_{12}\varepsilon_{11} = \left(\frac{c_{12}}{c_{11}}\right)\sigma_{11}.$$

Now rotate the coordinate system by 45° about z-axis (Fig. 3.6).

Transformation of the coordinates is

$$[\mathrm{U}, \mathrm{V}, \mathrm{W}] = [110], [u, v, w] = [-1, 1, 0], [U, V, W] = [0, 0, 1],$$

so that Eq. (3.16′) gives $\{T_{ij}\}$ matrix as follows.

$$T_{ij} = \begin{bmatrix} 1/\sqrt{2} & 1/\sqrt{2} & 0\\ -1/\sqrt{2} & 1/\sqrt{2} & 0\\ 0 & 0 & 1\end{bmatrix}.$$

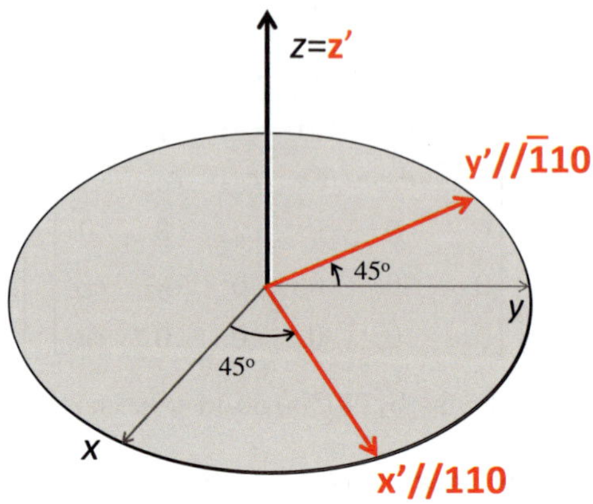

Fig. 3.6 Transformation of coordinates from (x, y, z) to (x', y', z').

Thus, Eq. (3.19) gives,

$$\varepsilon'_{12} = T_{11}T_{21}\varepsilon_{11} = -\frac{1}{2}\varepsilon_{11}.$$

Equation (3.21) gives

$$\begin{aligned}\sigma'_{12} &= \sum_{l,m} T_{1l}T_{2m}\sigma_{lm} \\ &= T_{11}T_{21}\sigma_{11} + T_{12}T_{22}\sigma_{22} + T_{13}T_{23}\sigma_{33} \\ &= -\frac{1}{2}\sigma_{11} + \frac{1}{2}\sigma_{22} \\ &= -\frac{1}{2}c_{11}\varepsilon_{11} + \frac{1}{2}c_{12}\varepsilon_{11} \\ &= (c_{11} - c_{12})\varepsilon'_{12}.\end{aligned}$$

On the other hand, Eq. (3.25) gives

$$\begin{aligned}\sigma'_{12} &= 2c'_{44}\varepsilon'_{12} \\ \therefore c_{11} - c_{12} &= 2c'_{44}\end{aligned}$$

Because of elastic isotropy $c'_{44} = c_{44}$. Thus,

$$c_{11} - c_{12} = 2c_{44} \tag{3.26}$$

Only two of the three elastic constants are independent.

For anisotropic crystals Eq. (3.26) does not hold.

$$A = \frac{2c_{44}}{c_{11} - c_{12}}, \tag{3.27}$$

is referred to as the anisotropy ratio. When $A = 1$, the medium is isotropic. The degree of anisotropy is estimated by deviation of A from 1.

It is usual to use λ (the Lame constant) and μ (the shear modulus) among the independent two elastic constants. Then, Hooke's law is written as follows:

$$\begin{aligned}
\sigma_{11} &= (\lambda + 2\mu)\,\varepsilon_{11} + \lambda\varepsilon_{22} + \lambda\varepsilon_{33}, \\
\sigma_{22} &= \lambda\varepsilon_{11} + (\lambda + 2\mu)\varepsilon_{22} + \lambda\varepsilon_{33}, \\
\sigma_{33} &= \lambda\varepsilon_{11} + \lambda\varepsilon_{22} + (\lambda + 2\mu)\varepsilon_{33}, \\
\sigma_{23} &= 2\mu\varepsilon_{23}, \\
\sigma_{31} &= 2\mu\varepsilon_{31}, \\
\sigma_{12} &= 2\mu\varepsilon_{12}.
\end{aligned} \tag{3.28}$$

Exercise 3.4 Express Eq. (3.28) in matrix.

Comparison with Eq. (3.13) gives

$$\begin{aligned}
\mu &= c_{44} = 1/2(c_{11} - c_{12}), \\
\lambda &= c_{12}, \quad \lambda + 2\mu = c_{11}.
\end{aligned}$$

On the other hand, by solving Eq. (3.28) for ε,

$$\varepsilon_{11} = \frac{\lambda + \mu}{\mu(3\lambda + 2\mu)}\sigma_{11} - \frac{\lambda}{2\mu(3\lambda + 2\mu)}(\sigma_{22} + \sigma_{33}).$$

Young modulus E and Poisson ratio (the ratio of traverse contraction to elongation in simple tension, ideally 1/3) are defined as

$$\begin{aligned}
E &= \frac{\mu(3\lambda + 2\mu)}{\mu + \lambda}, \\
\nu &= \frac{\lambda}{2(\mu + \lambda)},
\end{aligned}$$

$$
\begin{aligned}
\varepsilon_{11} &= \frac{1}{E}[\sigma_{11} - \nu(\sigma_{22} + \sigma_{33})], \\
\varepsilon_{22} &= \frac{1}{E}[\sigma_{22} - \nu(\sigma_{11} + \sigma_{33})], \\
\varepsilon_{33} &= \frac{1}{E}[\sigma_{33} - \nu(\sigma_{11} + \sigma_{22})], \\
\varepsilon_{23} &= \frac{1}{2\mu}\sigma_{23}, \\
\varepsilon_{31} &= \frac{1}{2\mu}\sigma_{31}, \\
\varepsilon_{12} &= \frac{1}{2\mu}\sigma_{12}.
\end{aligned}
\tag{3.29}
$$

Exercise 3.5 Express Eq. (3.29) in matrix.

Exercise 3.6 Prove that the following relations hold.

$$
K = \frac{3}{3\lambda + 2\mu} = -\frac{e}{p} = \frac{1}{B}, \tag{3.30}
$$

where

$$
e = \varepsilon_{11} + \varepsilon_{22} + \varepsilon_{33}
$$

is expansion.

$$
\begin{aligned}
p &= -\frac{1}{3}(\sigma_{11} + \sigma_{22} + \sigma_{33}) = \frac{1}{3}\Delta, \\
\Delta &= -(\sigma_{11} + \sigma_{22} + \sigma_{33}).
\end{aligned}
$$

p is pressure and K is the compressibility and B is the bulk modulus.

$$
\begin{aligned}
E &= \frac{\mu(3\lambda + 2\mu)}{\mu + \lambda} = \frac{9\mu B}{3B + \mu} = 2\mu(1 + \nu), \\
\nu &= \frac{3B - 2\mu}{2(3B + \mu)} = \frac{\lambda}{2(\mu + \lambda)} = \frac{E - 2\mu}{2\mu}, \\
\mu &= \frac{E}{2(1 + \nu)}, \\
\lambda &= \frac{\nu E}{(1 + \nu)(1 - 2\nu)} = \frac{2\nu\mu}{1 - 2\nu}.
\end{aligned}
\tag{3.31}
$$

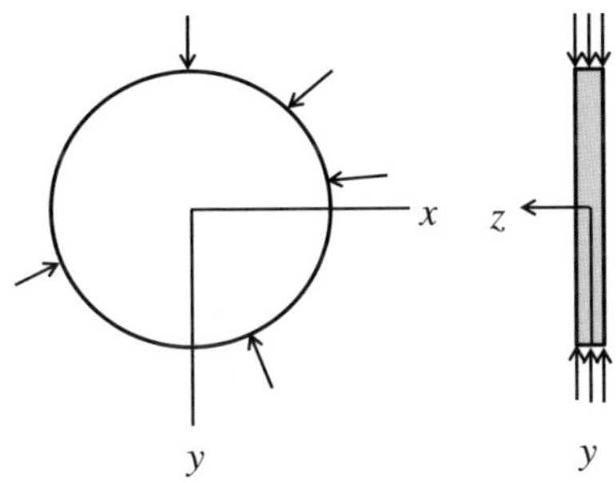

Fig. 3.7 Plane stress.

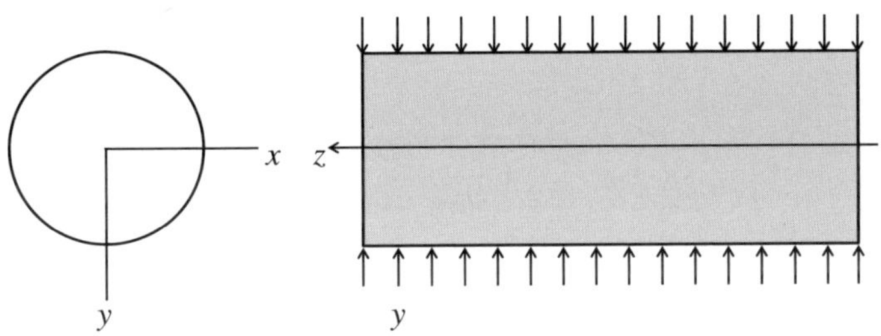

Fig. 3.8 Plane strain.

3.4.2 *Plane stress*

Stresses containing z-component are zero.

$$\sigma_{zz} = \sigma_{xz} = \sigma_{yz} = 0. \tag{3.32}$$

Non-zero stresses σ_{xx}, σ_{yy}, $\sigma_{xy}(\neq 0)$ are not dependent of z.

3.4.3 *Plane strain*

$\boldsymbol{u}$ and $\boldsymbol{v}$ are functions of x, y and independent of z.

$$\begin{aligned} u &= u(x_1, x_2), \\ v &= v(x_1, x_2), \\ w &= 0, \\ \frac{\partial}{\partial x_3} &= 0. \end{aligned}$$

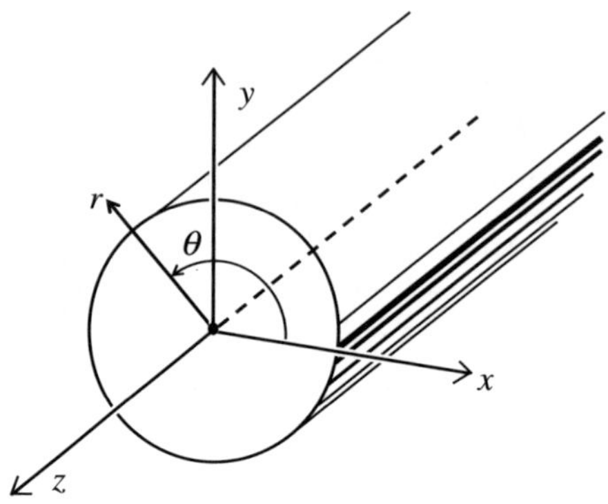

Fig. 3.9 Cylindrical coordinate.

Therefore,

$$\varepsilon_{yz} = \frac{1}{2}\left(\frac{\partial v}{\partial z} + \frac{\partial w}{\partial y}\right) = 0,$$

$$\varepsilon_{xz} = \frac{1}{2}\left(\frac{\partial u}{\partial z} + \frac{\partial w}{\partial x}\right) = 0,$$

$$\varepsilon_{zz} = \frac{\partial w}{\partial z} = 0.$$

Because $\varepsilon_{zz} = 0$,

$$\sigma_{zz} = \nu(\sigma_{xx} + \sigma_{yy}). \tag{3.33}$$

3.5 Cylindrical and Spherical Coordinates

3.5.1 *Cylindrical coordinate*

In cylindrical coordinate (by Landau–Lifshit [3]),

$$\varepsilon_{rr} = \frac{\partial u_r}{\partial r},$$

$$\varepsilon_{\theta\theta} = \frac{1}{r}\frac{\partial u_\theta}{\partial \theta} + \frac{u_r}{r},$$

$$\varepsilon_{zz} = \frac{\partial u_r}{\partial z},$$

$$2\varepsilon_{\theta z} = \frac{1}{r}\frac{\partial u_z}{\partial \theta} + \frac{\partial u_\theta}{\partial z} = \varepsilon''_{\theta z},$$

$$2\varepsilon_{zr} = \frac{\partial u_r}{\partial z} + \frac{\partial u_z}{\partial r} = \varepsilon''_{zr},$$

$$2\varepsilon_{r\theta} = \frac{\partial u_\theta}{\partial r} - \frac{u_\theta}{r} + \frac{1}{r}\frac{\partial u_r}{\partial \theta} = \varepsilon''_{r\theta}, \qquad \text{(3.34 Landau–(Suzuki)}'')$$

$$\sigma_{rr} = (\lambda + 2\mu)\varepsilon_{rr} + \lambda\varepsilon_{\theta\theta} + \lambda\varepsilon_{zz},$$

$$\sigma_{\theta\theta} = \lambda\varepsilon_{rr} + (\lambda + 2\mu)\varepsilon_{\theta\theta} + \lambda\varepsilon_{zz},$$

$$\sigma_{zz} = \lambda\varepsilon_{rr} + \lambda\varepsilon_{\theta\theta} + (\lambda + 2\mu)\varepsilon_{zz},$$

$$\sigma_{\theta z} = \mu\varepsilon''_{\theta z} = 2\mu\varepsilon_{\theta z},$$

$$\sigma_{rz} = \mu\varepsilon''_{rz} = 2\mu\varepsilon_{rz},$$

$$\sigma_{r\theta} = \mu\varepsilon''_{r\theta} = 2\mu\varepsilon_{r\theta}. \qquad \text{(3.35 Landau–(Suzuki)}'')$$

$$\sigma_{rr} = \lambda e + 2\mu\frac{\partial u_r}{\partial r},$$

$$\sigma_{\theta\theta} = \lambda e + 2\mu\left(\frac{1}{r}\frac{\partial u_\theta}{\partial \theta} + \frac{u_r}{r}\right),$$

$$\sigma_{zz} = \lambda e + 2\mu\frac{\partial u_z}{\partial z},$$

$$\sigma_{\theta z} = \mu\left(\frac{1}{r}\frac{\partial u_z}{\partial \theta} + \frac{\partial u_r}{\partial z}\right) = 2\mu\varepsilon_{\theta z} = \mu\varepsilon''_{\theta z},$$

$$\sigma_{zr} = \mu\left(\frac{\partial u_r}{\partial \theta} + \frac{\partial u_z}{\partial r}\right) = 2\mu\varepsilon_{rz} = \mu\varepsilon''_{rz},$$

$$\sigma_{r\theta} = \mu\left(\frac{\partial u_\theta}{\partial r} - \frac{u_\theta}{r} + \frac{1}{r}\frac{\partial u_r}{\partial \theta}\right) = 2\mu\varepsilon_{r\theta} = \mu\varepsilon''_{r\theta},$$

$$e = \varepsilon_{11} + \varepsilon_{22} + \varepsilon_{33}. \qquad \text{(3.36 Landau–(Suzuki)}'')$$

σ_{ij} is the force per unit area in the direction of ith component on a surface where j = constant. For ε_{ij}, the displacement is along the direction of ith component on j = constant surface.

Exercise 3.7 Draw stress (or strain) in Fig. 3.10.

3.5.2 *Spherical coordinate*

$$
\begin{aligned}
\varepsilon_{rr} &= \frac{\partial u_r}{\partial r},\\
\varepsilon_{\theta\theta} &= \frac{1}{r}\frac{\partial u_\theta}{\partial \theta} + \frac{u_r}{r},\\
\varepsilon_{\phi\phi} &= \frac{1}{r\sin\theta}\frac{\partial u_\phi}{\partial \phi} + \frac{u_\theta}{r}\cot\theta + \frac{u_r}{r},\\
2\varepsilon_{\theta\phi} &= \frac{1}{r}\left(\frac{\partial u_\phi}{\partial \theta} - u_\phi \cot\theta\right) + \frac{1}{r\sin\theta}\frac{\partial u_\theta}{\partial \phi} = \varepsilon''_{\theta\phi},\\
2\varepsilon_{\phi r} &= \frac{1}{r\sin\theta}\frac{\partial u_r}{\partial \phi} + \frac{\partial u_\phi}{\partial r} - \frac{u_\phi}{r} = \varepsilon''_{\phi r},\\
2\varepsilon_{r\theta} &= \frac{\partial u_\theta}{\partial r} - \frac{u_\theta}{r} + \frac{1}{r}\frac{\partial u_r}{\partial \theta} = \varepsilon''_{r\theta}.
\end{aligned}
\tag{3.37}
$$

$$
\begin{aligned}
\sigma_{rr} &= (\lambda + 2\mu)\,\varepsilon_{rr} + \lambda\varepsilon_{\theta\theta} + \lambda\varepsilon_{\phi\phi},\\
\sigma_{\theta\theta} &= \lambda\varepsilon_{rr} + (\lambda + 2\mu)\,\varepsilon_{\theta\theta} + \lambda\varepsilon_{\phi\phi},\\
\sigma_{\phi\phi} &= \lambda\varepsilon_{rr} + \lambda\varepsilon_{\theta\theta} + (\lambda + 2\mu)\,\varepsilon_{\phi\phi},\\
\sigma_{\theta\phi} &= 2\mu\varepsilon_{\theta\phi} = \mu\varepsilon''_{\theta\phi}, \sigma_{\phi r} = 2\mu\varepsilon_{\phi r} = \mu\varepsilon''_{\phi r}, \sigma_{r\theta} = 2\mu\varepsilon_{r\theta} = \mu\varepsilon''_{r\theta}.
\end{aligned}
\tag{3.38}
$$

$$
\begin{aligned}
\sigma_{rr} &= \lambda e + 2\mu\frac{\partial u_r}{\partial r},\\
\sigma_{\theta\theta} &= \lambda e + 2\mu\left(\frac{1}{r}\frac{\partial u_\theta}{\partial \theta} + \frac{u_r}{r}\right),\\
\sigma_{\phi\phi} &= \lambda e + 2\mu\left(\frac{1}{r\sin\theta}\frac{\partial u_\phi}{\partial \phi} + \frac{u_\theta}{r}\cot\theta + \frac{u_r}{r}\right),\\
\sigma_{\theta\phi} &= \mu\left\{\frac{1}{r}\left(\frac{\partial u_\phi}{\partial \theta} - u_\phi\cot\theta\right) + \frac{1}{r\sin\theta}\frac{\partial u_\theta}{\partial \phi}\right\} = 2\mu\varepsilon_{\theta\phi} = \mu\varepsilon''_{\theta\phi},\\
\sigma_{\phi r} &= \mu\left(\frac{1}{r\sin\theta}\frac{\partial u_r}{\partial \phi} + \frac{\partial u_\phi}{\partial r} - \frac{u_\phi}{r}\right) = 2\mu\varepsilon_{\phi r} = \mu\varepsilon''_{\phi r},\\
\sigma_{r\theta} &= \mu\left(\frac{\partial u_\theta}{\partial r} - \frac{u_\theta}{r} + \frac{1}{r}\frac{\partial u_r}{\partial \theta}\right) = 2\mu\varepsilon_{r\theta} = \mu\varepsilon''_{r\theta}.
\end{aligned}
\tag{3.39[3]}
$$

Exercise 3.8 Draw stress (or strain) in Fig. 3.10.

[3]Refer to Eqs. (3.2) and (3.2′) for definition of ε and ε''.

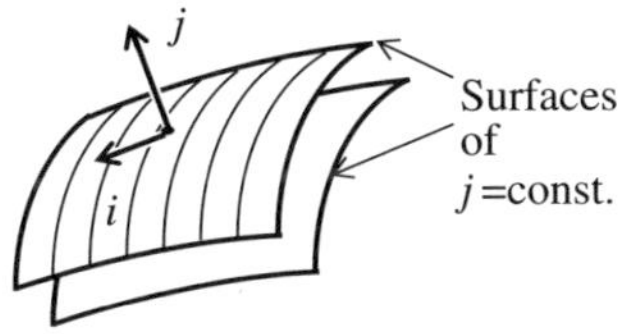

Fig. 3.10 Definition of σ_{ij} in curvilinear coordinate.

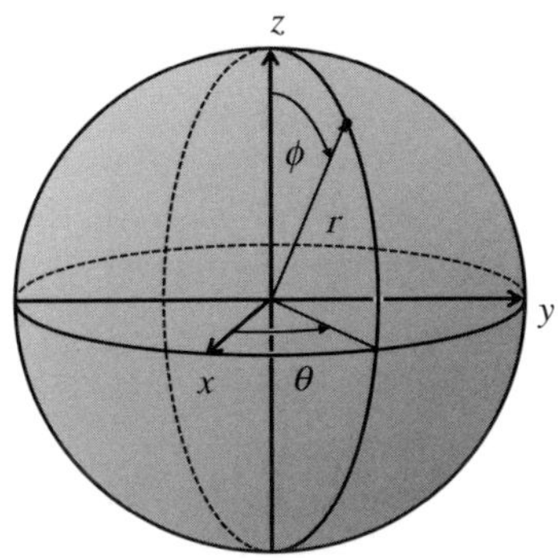

Fig. 3.11 Spherical coordinate.

References

1. Hirth, J. P. and Lothe, J. *Theory of Dislocations*, 1st ed., McGraw-Hill, (1967); 2nd ed, Wiley(1982).
2. Takeuchi, K. Worked *Exampes of Elasticity Theory* (in Japanese) Shokabou, 1969.
3. Landau, L. and Lifshit, E. *Theory of Elasticity* 3rd ed. translated from Russian by U. J. B. Sykes and W. H. Reid (Pergamon Press, 1959).
4. Nye, J. F. *Physical Properties of Crystals*, Oxford Univ. Press, 1985.
5. Timoshenko, S. and Goodier, J. N. *Theory of Elasticity*, 2nd ed. McGraw-Hill, 1955.
6. Suzuki, H. *Introduction to Dislocation Theory* (in Japanese) (Agne, 1967).
7. Love, A. E. H. *A Treatise on the Mathematical Theory of Elasticity* (Dover Pub. 1944).

A tower of the fort in Semur-en Auxios in Bourgognes in France. A vertical crack is a manifest of a radial tensile strain, the amount of which is ν of the compressive strain due to the self-weight.

Chapter 4

Elasticity Theory of Dislocations

4.1 Screw Dislocation

4.1.1 *Displacement, strain and stress*

Figure 4.1(a) shows a screw dislocation sheared by b along the z-axis across x–y plane. It is convenient to use the cylindrical coordinates. Opening-up a surface of a radius r is shown in Fig. 4.1(b). Going around the z-axis counterclockwise completely ($\theta = 2\pi$), the displacement in the direction of the z-axis is b. Therefore, the displacement at a rotation angle of θ from the x-axis is given by

$$\begin{aligned} u_z(r,\theta) &= \frac{b\theta}{2\pi} = \frac{b}{2\pi}\tan^{-1}\frac{y}{x} \\ u_x &= u_y = 0. \end{aligned} \tag{4.1}$$

Strains are given by Eq. (3.2) as follows:

$$\begin{aligned} \varepsilon_{xz} &= \frac{1}{2}\left(\frac{\partial u_z}{\partial x} + \frac{\partial u_x}{\partial z}\right) = \frac{-b}{4\pi}\frac{y}{x^2+y^2}, \\ \varepsilon_{yz} &= \frac{1}{2}\frac{\partial u_z}{\partial y} = \frac{b}{4\pi}\frac{x}{x^2+y^2}. \end{aligned} \tag{4.2}$$

These are the only non-zero strains. The other strains are zero.

Stresses are given by Eq. (3.33) as follows,

$$\begin{aligned} \sigma_{xz} &= 2\mu\varepsilon_{xz} = \frac{-\mu b}{2\pi}\frac{y}{x^2+y^2}, \\ \sigma_{yz} &= 2\mu\varepsilon_{yz} = \frac{\mu b}{2\pi}\frac{x}{x^2+y^2}, \\ \sigma_{xy} &= \sigma_{xx} = \sigma_{yy} = \sigma_{zz} = 0. \end{aligned} \tag{4.3}$$

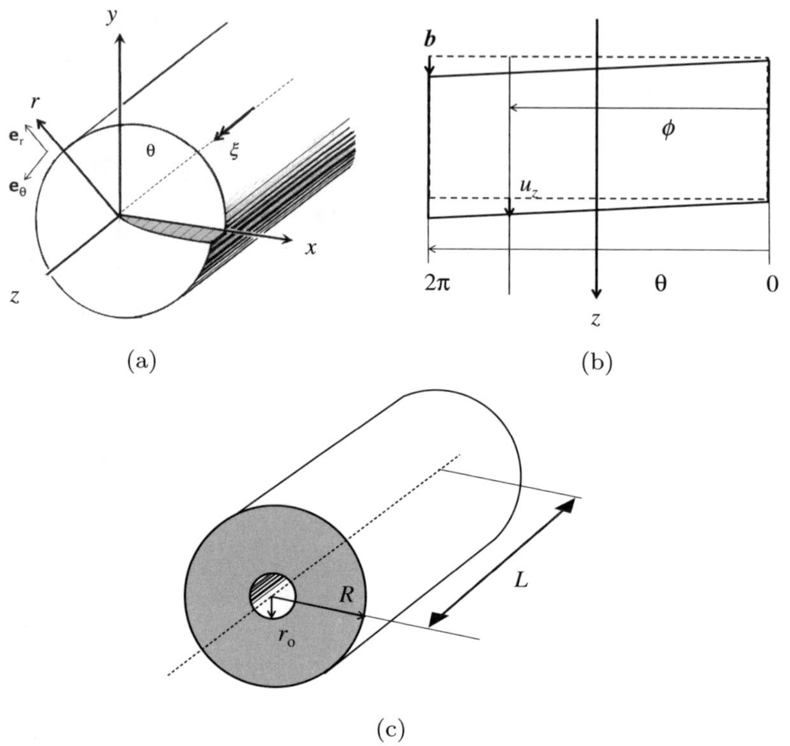

Fig. 4.1 (a) Displacement around a screw dislocation, (b) Opening-up of a surface with a radius r, (c) Dislocation core.

In the polar coordinates stresses are given by

$$\begin{aligned} \sigma_{\theta z} &= 2\mu\varepsilon_{\theta z} = \frac{\mu b}{2\pi r} \\ \sigma_{rz} &= \sigma_{r\theta} = \sigma_{rr} = \sigma_{\theta\theta} = \sigma_{zz} = 0. \end{aligned} \tag{4.4}$$

Exercise 4.1 Derive Eqs. (4.3) and (4.4).

4.1.2 *Strain energy*

Let us calculate the strain energy per unit length around a screw dislocation. As is evident from Eq. (4.4) the stress divers at $r = 0$. In order to overcome the difficulty of diverse, a cylinder with a radius of r_o is cut off from the cylinder with an outer radius R (Fig. 4.1(c)).

Let us indicate the strain energy of a screw dislocation per unit length by W_{el}/L following Hirth-Lothe [1], which is given by

$$\begin{aligned}\frac{W_{el}}{L} &= \int_{r_0}^{R} \frac{1}{2}\sigma_{\theta z}\varepsilon_{\theta z}2\pi r dr \\ &= \frac{\mu b^2}{4\pi}\ln\frac{R}{r_0}. \qquad (4.5)\end{aligned}$$

Exercise 4.2 Derive Eq. (4.5).

It follows from this that:

(1) The strain energy is proportional to b^2. Therefore, a dislocation tries to assume $\boldsymbol{b}$ as small as possible. That is, $\boldsymbol{b}$ is the shortest translational vector of the unit cell, i.e., in other words, the close packed direction.

(2) The strain energy increases with increasing the outer radius R of a crystal. Usually R is taken as follows;

   (a) When only a single dislocation is present in the crystal, R is the shortest distance from the dislocation to the free surface. Beyond this distance contribution to the strain energy decreases rapidly and can be neglected.

   (b) In a crystal containing many dislocations with the average distance l, $R \sim l/2$ is an appropriate approximation.

Exercise 4.3 l is around $100\,\mu$m in a well annealed metal and around $0.01\,\mu$m in a heavily deformed metal. Calculate $\ln(R/r_o)$ for the respective cases taking $r_o = 1\,\text{Å}$.

(3) The energy in the core cannot be calculated by elasticity theory. This energy, referred to as the dislocation core energy, is estimated by a computer simulation. Formally, all contribution is included in Eq. (4.5) by choosing an appropriate αb instead of r_0 to make the following formula give the expected total energy.

$$\begin{aligned}\frac{W_{\text{total}}}{L} &= \frac{W_{\text{el}}}{L} + \frac{W_{\text{core}}}{L} \\ &= \int_{r_0}^{R} \frac{1}{2}\sigma_{\theta z}\varepsilon_{\theta z}2\pi r dr + \frac{W_{\text{core}}}{L}\end{aligned}$$

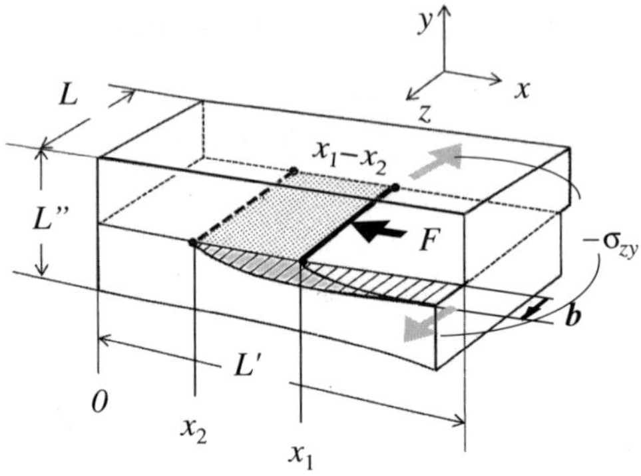

Fig. 4.2 Force acting on a screw dislocation.

$$= \frac{\mu b^2}{4\pi} \ln \frac{R}{r_0} + \frac{W_{\text{core}}}{L}$$

$$= \frac{\mu b^2}{4\pi} \ln \frac{R}{\alpha b}. \tag{4.5'}$$

r_o or αb is referred to as a cut-off parameter. For its experimental evaluation refer to Sec. 8.4.1.

4.1.3 *Force on the screw dislocation*

Consider a screw dislocation moving under the action of an external force. First, let us consider the dislocation as a rod and assume that a force acts on the rod (dislocation). Suppose that, under the applied stress σ_{yz}, the dislocation moves from x_1 to x_2 (Fig. 4.2). The external force acting on the rod (dislocation) per unit length is F_x/L, under which the rod (dislocation) moves $(x_1 - x_2)$, so that the work done by the external force is given by $(F_x/L)L(x_1 - x_2)$.

Now let us consider the actual displacement of atoms. The force applied to the crystal as a whole is $\sigma_{zy} \times (LL')$. Under the action of this force only atoms (material) over the hatched area $(x_1 - x_2) \times L$ are displaced by $\boldsymbol{b}$ along z-axis. Therefore, from the atomic point of view the work done is; $\{\sigma_{yz} \times (x_1 - x_2) \times L\} \times b$.

(In the unhatched area, i.e., $((L' - x_1) \times L + x_2 \times L)$ no atoms are displaced, hence the work done is also zero.)

These expressions are obtained by viewing the identical phenomenon from macroscopic and atomic viewpoints, so that they should be equal, that is,

$$\left(\frac{F_x}{L}\right) L(x_1 - x_2) = \sigma_{yz}(x_1 - x_2)Lb$$

$$\frac{F_x}{L} = \sigma_{yz}b. \tag{4.6}$$

4.1.4 *Image force and Eshelby twist*

Image force

Suppose that a right-handed screw dislocation is present at a distance l away from the free surface (Fig. 4.3(a)). The screw dislocation exerts the following stress on the free surface (at $x = 0$)

$$\sigma_{zx}^{x} = \frac{-\mu b}{2\pi}\frac{y}{l^2 + y^2}.$$

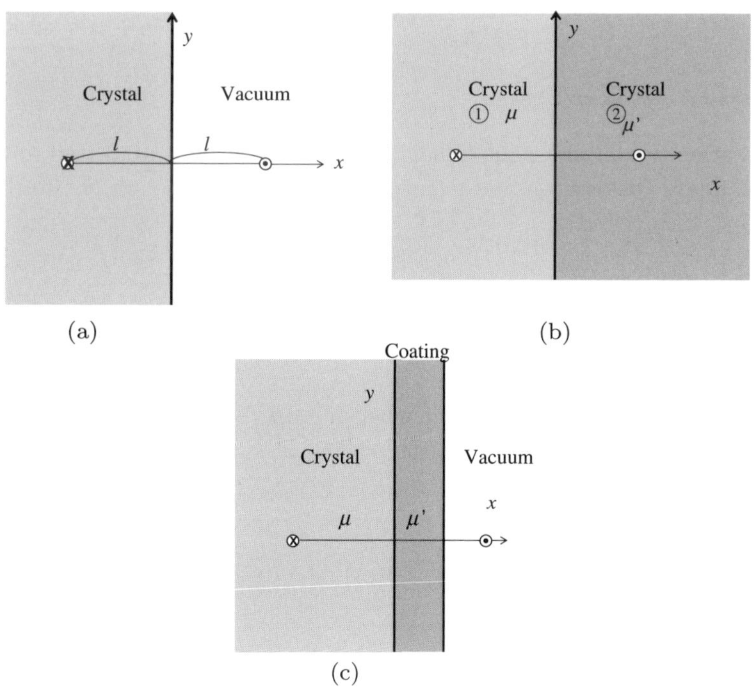

Fig. 4.3 (a)–(c) Imaginary dislocations due to the free surface and the internal interfaces.

The free surface must be free from stress. This boundary condition is met if an imaginary left-handed screw dislocation is located at the mirror position outside the crystal. This imaginary screw dislocation exerts a stress on the free surface ($x = 0$) as follows.

$$\sigma_{zx}^{v} = \frac{\mu b}{2\pi} \frac{y}{l^2 + y^2}.$$

These stresses from the real dislocation and its imaginary one cancel each other, resulting in the zero stress at the surface ($x = 0$).

However, at the same time, this imaginary screw dislocation gives rise to another non-zero stress σ_{yz} according to Eq. (4.4).

$$\sigma_{yz}^{v} = -\frac{\mu b}{2\pi} \frac{x}{x^2 + y^2}.$$

According to Eq. (4.6) this stress gives rise to a force on the real dislocation at $x = -\ell$ (and $y = 0$).

$$\frac{F_x}{L} = b\sigma_{yz}^{v} = \frac{\mu b^2}{4\pi l}. \tag{4.7}$$

Exercise 4.4 Prove this.

In other words, the right-handed (real) dislocation is attracted towards its imaginary dislocation and eventually reaches the free surface to disappear out of the crystal.

Grain boundary

If a crystal is adjacent to a different phase instead of the vacuum (Fig. 4.3(b)), the dislocation is;

(1) repulsed from the free surface when $\mu' > \mu$.
(2) attracted towards the free surface when $\mu' < \mu$.

Surface layer

If a surface layer is present (Fig. 4.3(c)), the dislocation is repulsed when $\mu' > \mu$.

Exercise 4.5 Consider the case where $\mu' < \mu$.

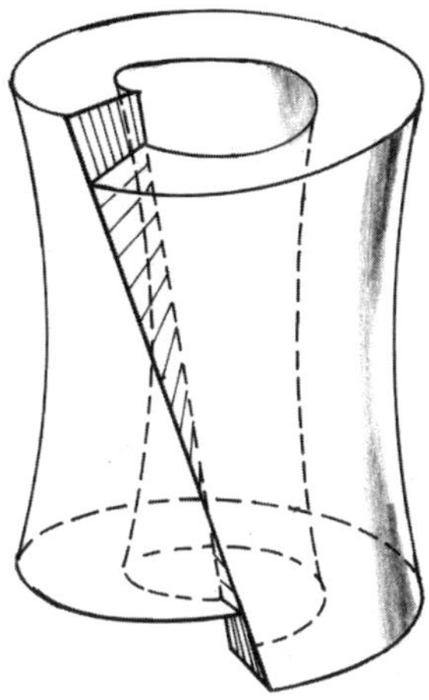

Fig. 4.4 Eshelby twist (Redrawn based on [2]).

Eshelby twist

According to Eq. (4.4) a stress is generated around a screw dislocation as follows:

$$\sigma_{\theta z} = 2\mu\varepsilon_{\theta z} = \frac{\mu b}{2\pi r}. \qquad (4.4\text{ (again)})$$

Apart from a dislocation in an infinitely long crystal, a screw dislocation in a finite long crystal gives rise to a torque $\sigma_{\theta z} = \sigma_{z\theta}$ at the end surfaces $z = 0$ and $z = z$ (Fig. 4.4). This torque gives rise to a twist called Eshelby twist. The Eshelby twist plays an important role in observing a screw dislocation vertical to the foil surface by transmission electron microscope [3].

4.2 Edge Dislocation

4.2.1 *Displacement and stress*

The situation for an edge dislocation is much more complicated than that for a screw dislocation (Fig. 4.5). The stresses are given by

$$\sigma_{xx} = -\frac{\mu b}{2\pi(1-\nu)}\frac{y(3x^2+y^2)}{(x^2+y^2)^2},$$

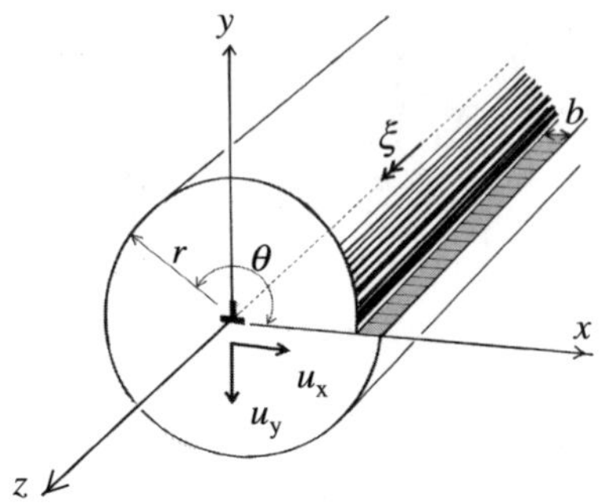

Fig. 4.5 An edge dislocation.

$$\begin{aligned}
\sigma_{yy} &= \frac{\mu b}{2\pi(1-\nu)} \frac{y(x^2-y^2)}{(x^2+y^2)^2}, \\
\sigma_{xy} &= \frac{\mu b}{2\pi(1-\nu)} \frac{x(x^2-y^2)}{(x^2+y^2)^2}, \\
\sigma_{zz} &= \nu(\sigma_{xx}+\sigma_{yy}) = -\frac{\mu b \nu}{\pi(1-\nu)} \frac{y}{x^2+y^2}, \\
\sigma_{xz} &= \sigma_{yz} = 0.
\end{aligned} \tag{4.8}$$

In the polar coordinate

$$\begin{aligned}
\sigma_{rr} &= \sigma_{\theta\theta} = -\frac{\mu b \sin\theta}{2\pi(1-\nu)r}, \\
\sigma_{r\theta} &= \frac{\mu b \cos\theta}{2\pi(1-\nu)r}, \\
\sigma_{zz} &= \nu(\sigma_{rr}+\sigma_{\theta\theta}) = -\frac{\mu b \nu \sin\theta}{\pi(1-\nu)r}, \\
\sigma_{rz} &= \sigma_{\theta z} = 0.
\end{aligned} \tag{4.9}$$

The displacements are given by

$$\begin{aligned}
u_x &= \frac{b}{2\pi}\left[\tan^{-1}\frac{y}{x} + \frac{xy}{2(1-\nu)(x^2+y^2)}\right], \\
u_y &= -\frac{b}{2\pi}\left[\frac{1-2\nu}{4(1-\nu)}\ln(x^2+y^2) + \frac{x^2-y^2}{4(1-\nu)(x^2+y^2)}\right],
\end{aligned} \tag{4.10}$$

where ν is the Poisson ratio.

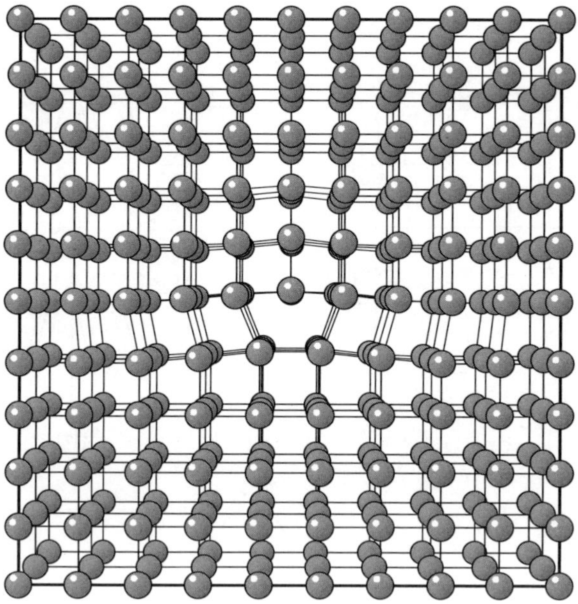

Fig. 4.6 Schematic illustration of lattices about an edge dislocation. Note the horizontal planes are slightly bent near the dislocation.

The displacements and stresses are schematically illustrated in Figs. 4.6 and 4.7(a)–(e), respectively.

4.2.2 *Strain energy*

The strain energy per unit length of an edge dislocation is given by

$$\frac{W}{L} = \int_{r_o}^{R} rdr \int_{0}^{2\pi} d\theta \left[\frac{1}{2\mu}\sigma_{xy}^2 + \frac{1}{2E}(\sigma_{xx}^2 + \sigma_{yy}^2 - 2\nu\sigma_{xx}\sigma_{yy} - \sigma_{zz}^2) \right]$$
$$= \frac{\mu b^2}{4\pi(1-\nu)} \ln \frac{R}{r_o}. \tag{4.11}$$

This is identical to that of a screw dislocation given by Eq. (4.5) except for a factor $1/(1-\nu)$. For an isotropic solid $\nu = 1/3$, so that the strain energy of an edge dislocation is larger than that of a screw dislocation by a factor of 1.5. As a result a dislocation loop is elongated in the direction of $\boldsymbol{b}$ at equilibrium (Fig. 4.8).

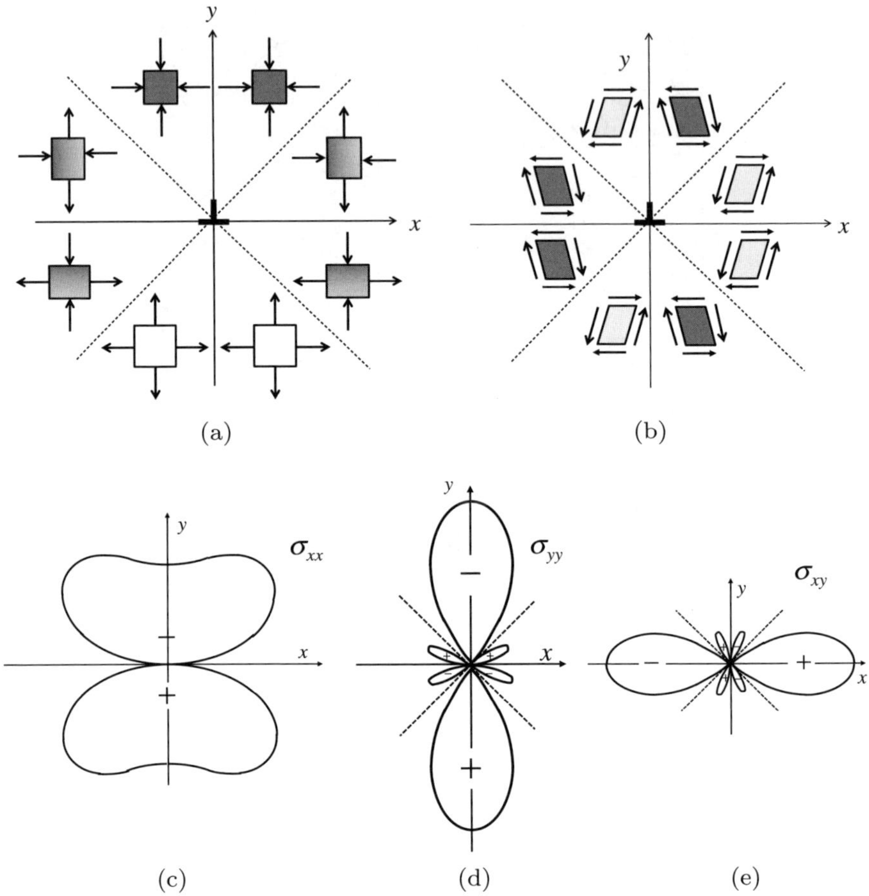

Fig. 4.7 Stress field about an edge dislocation, (a) Tension and compression, (b) Shear. Contours of (c) σ_{xx}, (d) σ_{yy} and (e) σ_{xy}.

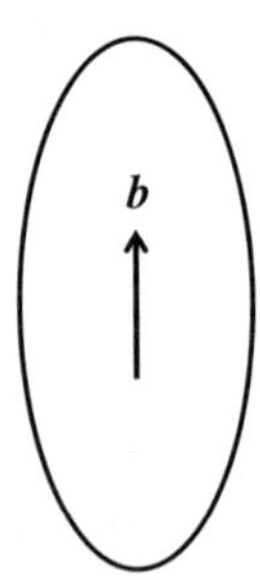

Fig. 4.8

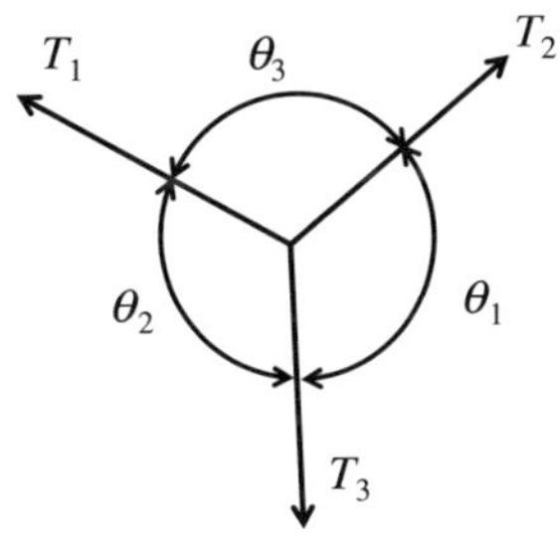

Fig. 4.9

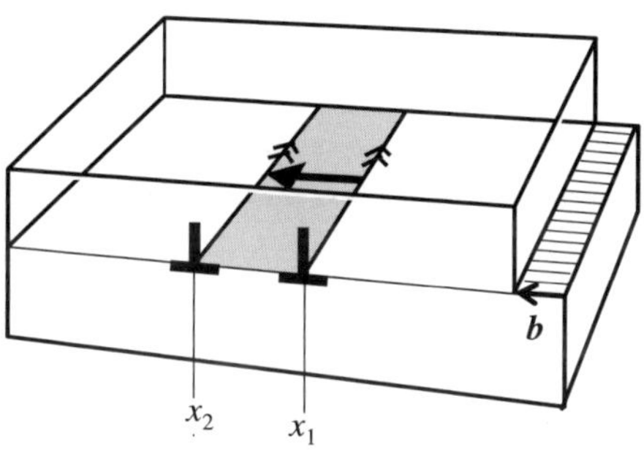

Fig. 4.10 Force acing on an edge dislocation.

As is described in Sec. 2.6, dislocations often form networks. The balance of tensions gives the following relation,

$$\frac{T_1}{\sin\theta_1} = \frac{T_2}{\sin\theta_2} = \frac{T_3}{\sin\theta_3}. \tag{4.12}$$

(see Fig. 4.9).

It is noted here that the orientation dependence of the line tension is neglected, so that Eq. (4.12) is approximate.

4.2.3 *Force on an edge dislocation*

As described in Sec. 4.2, an edge dislocation moves both conservatively (glide) and non-conservatively (climb).

Glide motion

Suppose that an external stress σ_{xy} causes an edge dislocation to move from x_1 to x_2 (Fig. 4.10). Discussion similar to the case of a screw dislocation

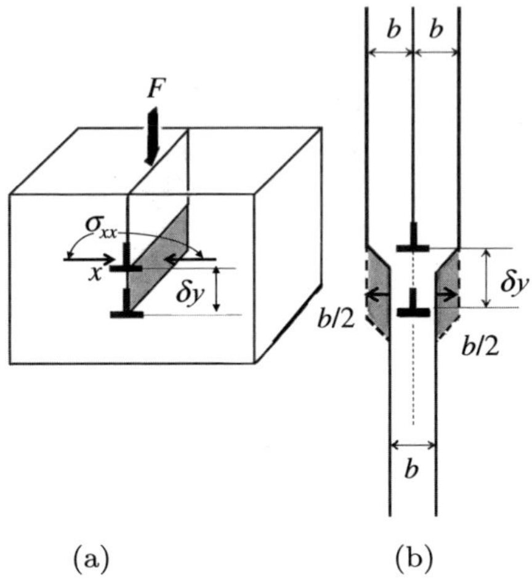

Fig. 4.11 External force required for an edge dislocation to climb.

leads to the following relation:

$$\frac{F_x}{L} = \sigma_{yx} b. \tag{4.13}$$

Exercise 4.6 Derive Eq. (4.13).

Climb motion

An edge dislocation accompanies an extra-half plane (ehp). The ehp can be regarded as a wedge with one atomic thickness. Suppose that the wedge is pushed downward by a force F_y and is displaced by δy (Fig. 4.11(a)). Work done by the external force is $F_y \times \delta y$.

The wedge exerts a stress σ_{xx} acting on the x-plane and displaces atoms in the hatched area by $b/2$ both rightward and leftward. The work done by this stress is $\sigma_{xx}(\delta y \cdot L) \times b$.

These two expressions must be equal.

$$\therefore \quad F_y \times \delta y = \sigma_{xx}(\delta y \times L) \times b, \tag{4.14}$$

$$\frac{F_y}{L} = \sigma_{xx} b.$$

Inversely when a negative stress $-\sigma_{xx}$ acts (i.e., presses the flanks of the crystal) the wedge is displaced upwards. The force acting on the edge dislocation is $F_y/L = \sigma_{xx} b$.

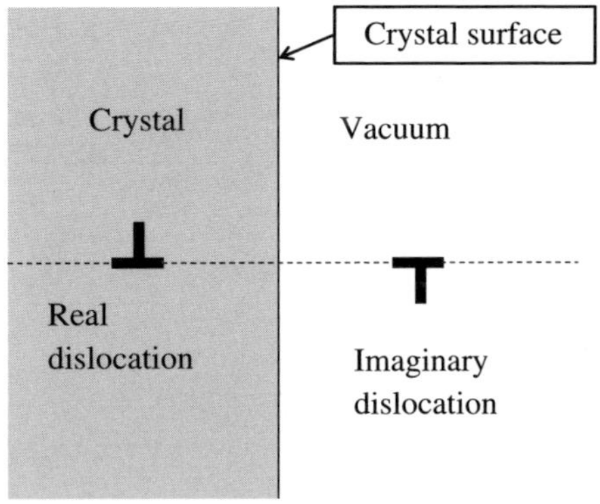

Fig. 4.12 Imaginary dislocation of an edge dislocation.

4.2.4 *Image force*

If an imaginary edge dislocation with $-\boldsymbol{b}$ is located by analogy with a screw dislocation (Fig. 4.12), the stress does not vanish perfectly at the surface. Let us superimpose the stress fields (Figs. 4.6(a) and 4.6(b)) of a real edge dislocation ① and its imaginary dislocation ② near the free surface (Figs. 4.13(a) and 4.13(b)). Obviously the tension and compression stresses of σ_{xx}, and σ_{yy} cancel each other (Fig. 4.13(a)), however, the shear stress $\sigma_{xy}(=\sigma_{yx})$ is not canceled, or rather strengthened (Fig. 4.13(b)). This can be visualized by drawing atomic planes near the edge dislocations (Fig. 4.13(c)). However, an edge dislocation is attracted toward the free surface similarly to a screw dislocation.

4.3 Mixed Dislocation

A dislocation whose $\boldsymbol{b}$ is inclined from ξ by an angle β is referred to as a mixed dislocation (Fig. 4.14). Its Burgers vector $\boldsymbol{b}$ is resolved as follows:

$$\boldsymbol{b} = \boldsymbol{b}_e + \boldsymbol{b}_s,$$

where

$$\begin{aligned} b_e &= b \sin\beta \\ b_s &= b \cos\beta. \end{aligned} \tag{4.15}$$

In other words, a mixed dislocation can be regarded as superposition of an edge dislocation with Burgers vector $\boldsymbol{b}_e$ and a screw dislocation with

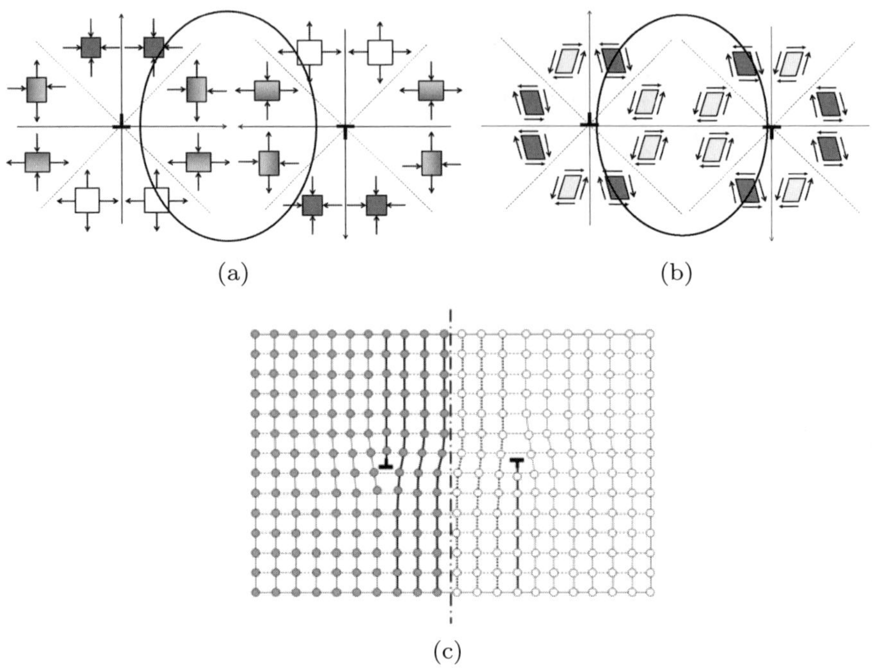

Fig. 4.13 (a),(b) Superimposition of stress fields of a real edge dislocation and its imaginary edge dislocation, (a) σ_{xx} and σ_{yy}, (b)σ_{xy}. (c) Superimposition of strained lattices around the real and imaginary dislocations.

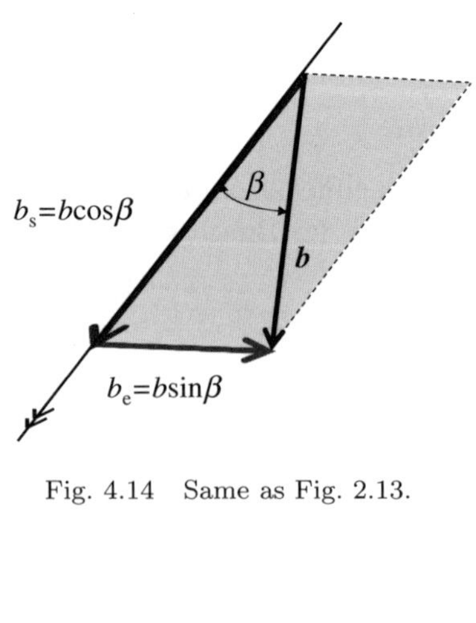

Fig. 4.14 Same as Fig. 2.13.

Burgers vector $\boldsymbol{b}_s$. For instance, the stress field Σ is given by

$$\sum = \overset{e}{\sum} + \overset{s}{\sum}$$

That is,

$$\sum = \overset{e}{\sum} + \overset{s}{\sum} = \begin{pmatrix} \sigma^e_{xx} & \sigma^e_{xy} & 0 \\ \sigma^e_{xy} & \sigma^e_{yy} & 0 \\ 0 & 0 & \sigma^e_{zz} \end{pmatrix} + \begin{pmatrix} 0 & 0 & \sigma^s_{xz} \\ 0 & 0 & \sigma^s_{yz} \\ \sigma^s_{xz} & \sigma^s_{yz} & 0 \end{pmatrix}$$

$$= \begin{pmatrix} \sigma^e_{xx} & \sigma^e_{xy} & \sigma^s_{xz} \\ \sigma^e_{xy} & \sigma^e_{yy} & \sigma^s_{yz} \\ \sigma^s_{xz} & \sigma^s_{yz} & \sigma^e_{zz} \end{pmatrix}, \tag{4.16}$$

where

$$\begin{aligned}
\sigma^e_{xx} &= -\frac{\mu b_e}{2\pi(1-\nu)}\frac{y(3x^2+y^2)}{(x^2+y^2)^2} = -\frac{\mu b \sin\beta}{2\pi(1-\nu)}\frac{y(3x^2+y^2)}{(x^2+y^2)^2} \\
&= -A\frac{y(3x^2+y^2)}{(x^2+y^2)^2}, \\
\sigma^e_{yy} &= \frac{\mu b_e}{2\pi(1-\nu)}\frac{y(x^2-y^2)}{(x^2+y^2)^2} = \frac{\mu b \sin\beta}{2\pi(1-\nu)}\frac{y(x^2-y^2)}{(x^2+y^2)^2} = A\frac{y(x^2-y^2)}{(x^2+y^2)^2}, \\
\sigma^e_{xy} &= \frac{\mu b_e}{2\pi(1-\nu)}\frac{x(x^2-y^2)}{(x^2+y^2)^2} = \frac{\mu b \sin\beta}{2\pi(1-\nu)}\frac{x(x^2-y^2)}{(x^2+y^2)^2} = A\frac{x(x^2-y^2)}{(x^2+y^2)^2}, \\
\sigma^e_{zz} &= \nu(\sigma_{xx}+\sigma_{yy}) = -\frac{\mu b_e \nu}{\pi(1-\nu)}\frac{y}{x^2+y^2} = -\frac{\mu b \nu \sin\beta}{\pi(1-\nu)}\frac{y}{x^2+y^2} \\
&= -2A\nu\frac{y}{x^2+y^2}, \\
\sigma^e_{xz} &= \sigma^e_{yz} = 0,
\end{aligned} \tag{4.17}$$

$$\left(\text{where } A = \frac{\mu b \sin\beta}{2\pi(1-\nu)}\right)$$

are stresses caused by the edge component of $\boldsymbol{b}_e$.

$$\begin{aligned}
\sigma^s_{xz} &= \frac{-\mu b_s}{2\pi}\frac{y}{x^2+y^2} = \frac{-\mu b \cos\beta}{2\pi}\frac{y}{x^2+y^2} = -B\frac{y}{x^2+y^2}, \\
\sigma^s_{yz} &= \frac{\mu b_s}{2\pi}\frac{x}{x^2+y^2} = \frac{\mu b \cos\beta}{2\pi}\frac{x}{x^2+y^2} = B\frac{x}{x^2+y^2}, \\
\sigma^s_{xy} &= \sigma^s_{xx} = \sigma^s_{yy} = \sigma^s_{zz} = 0,
\end{aligned} \tag{4.18}$$

$$\left(\text{where } B = \frac{\mu b \cos\beta}{2\pi}\right)$$

are stresses caused by the screw component of $\boldsymbol{b}_s$.

The strain energy is given by Eqs. (4.5) and (4.11) as follows:

$$\begin{aligned}\frac{W}{L} &= \frac{W_s}{L} + \frac{W_e}{L},\\ &= \frac{\mu b_s^2}{4\pi}\ln\frac{R}{r_o} + \frac{\mu b_e^2}{4\pi(1-\nu)}\ln\frac{R}{r_o},\\ &= \frac{\mu b^2}{4\pi}\left(\cos^2\beta + \frac{\sin^2\beta}{1-\nu}\right)\ln\frac{R}{r_o}.\end{aligned} \tag{4.19}$$

The force acting on the mixed dislocation is given by

$$\frac{F_x}{L} = \sigma_{yz}^s b\cos\beta \quad \text{for the glide motion of the screw component,}$$

$$\frac{F_y}{L} = \sigma_{xy}^e b\sin\beta \quad \text{for the glide motion of the edge component,}$$

$$\frac{F_y}{L} = \sigma_{xx}^e b\sin\beta \quad \text{for the climb motion of the edge component.}$$

In summary,

$$\frac{\boldsymbol{F}}{L} = \left(\boldsymbol{b}\cdot\sum\right)\times\boldsymbol{\xi}, \tag{4.20}$$

where ξ is the unit vector indicating the direction (and the sense) of the dislocation line. Equation (4.26) is referred to as Peach–Koehler formula.

4.4 Application of Peach–Koehler Formula

4.4.1 *Parallel screw dislocations*

A screw dislocation ① lies along the z-axis and a second screw dislocation ② lies parallel to the z-axis passing through a coordinate (x, y) (Fig. 4.15(a)). Let the stress field of dislocation ① (in Eq. (4.20)) be given by $\Sigma_{①}$. ξ is the unit vector representing the direction of dislocation ②, i.e., $\xi_2 = (0\boldsymbol{i}, 0\boldsymbol{j}, 1\boldsymbol{k})$. $\boldsymbol{b}$ is for dislocation ②, i.e., $\boldsymbol{b}_2 = b(0\boldsymbol{i}, 0\boldsymbol{j}, 1\boldsymbol{k})$. $\boldsymbol{F}$ is the force that dislocation ② receives in the stress field of dislocation ①, and hence let it be indicated by $\boldsymbol{F}_{1\to 2}$.

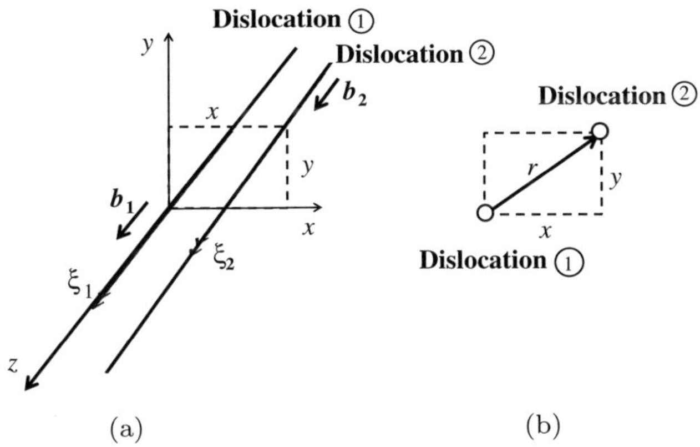

Fig. 4.15 (a),(b)

$$
\begin{aligned}
\frac{\boldsymbol{F}_{1\to 2}}{L} &= \left(\boldsymbol{b}_2 \cdot \sum_1\right) \times \xi_2, \\
&= b\begin{pmatrix} 0\boldsymbol{i} & 0\boldsymbol{j} & 1\boldsymbol{k} \end{pmatrix} \begin{pmatrix} 0\boldsymbol{ii} & 0\boldsymbol{ij} & \sigma_{13}\boldsymbol{ik} \\ 0\boldsymbol{ji} & 0\boldsymbol{jj} & \sigma_{23}\boldsymbol{jk} \\ \sigma_{13}\boldsymbol{ki} & \sigma_{23}\boldsymbol{kj} & 0\boldsymbol{kk} \end{pmatrix} \times \boldsymbol{\xi}_2, \\
&= b\left(\sigma_{31}\boldsymbol{i}, \sigma_{23}\boldsymbol{j}, 0\boldsymbol{k}\right) \times \left(0\boldsymbol{i}, 0\boldsymbol{j}, 1\boldsymbol{k}\right) \\
&= b\begin{bmatrix} \boldsymbol{i} & \boldsymbol{j} & \boldsymbol{k} \\ \sigma_{31} & \sigma_{23} & 0 \\ 0 & 0 & 1 \end{bmatrix} \qquad (4.21) \\
&= b(\boldsymbol{i}\sigma_{23} - \boldsymbol{j}\sigma_{31} + \boldsymbol{k}0) \\
\frac{F_x}{L} &= b\sigma_{23} = \frac{\mu b^2}{2\pi}\frac{x}{x^2+y^2} \\
\frac{F_y}{L} &= -b\sigma_{31} = \frac{\mu b^2}{2\pi}\frac{y}{x^2+y^2}
\end{aligned}
$$

In calculating Eq. (4.21) Gibbs's dyadic[1] is used (see Eq. (3.24)).

[1]While Gibbs's dyadic is mathematically elegant, substituting σ_{ij} is tedious and quite often leads to errors in practical calculation. In order to avoid such errors in calculation the following methods are recommended.

Exercise 4.7 Show that the force acting between two parallel screw dislocations is in the direction of $\boldsymbol{r}$ connecting these two dislocations. (Hint) Use Eq. (4.21) (Fig. 4.15(b)).

4.4.2 *Parallel edge dislocations*

With parallel Burgers vectors

An edge dislocation ① lies along the z-axis and a second edge dislocation ② lies parallel to the z-axis passing through a coordinate (x, y) (Fig. 4.16).

In calculation of

$$\begin{pmatrix} a_1 & a_2 & a_3 \end{pmatrix} \begin{pmatrix} b_{11} & b_{12} & b_{13} \\ b_{21} & b_{22} & b_{23} \\ b_{31} & b_{32} & b_{33} \end{pmatrix}$$
$$= \begin{pmatrix} a_1 b_{11} + a_2 b_{21} + a_3 b_{31} & a_1 b_{12} + a_2 b_{22} + a_3 b_{32} & a_1 b_{13} + a_2 b_{23} + a_3 b_{33} \end{pmatrix}$$

converting $(a_1 \ \ a_2 \ \ a_3)$ to a row vector gives

$$\begin{pmatrix} a_1 \\ a_2 \\ a_3 \end{pmatrix} \begin{pmatrix} \boldsymbol{i} & \boldsymbol{j} & \boldsymbol{k} \\ b_{11} & b_{12} & b_{13} \\ b_{21} & b_{22} & b_{23} \\ b_{31} & b_{32} & b_{33} \end{pmatrix}$$
$$= ((a_1 b_{11} + a_2 b_{21} + a_3 b_{31})\boldsymbol{i}, (a_1 b_{12} + a_2 b_{22} + a_3 b_{32})\boldsymbol{j},$$
$$(a_1 b_{13} + a_2 b_{23} + a_3 b_{33})\boldsymbol{k}).$$

Similarly, calculation of

$$\begin{pmatrix} b_{11} & b_{12} & b_{13} \\ b_{21} & b_{22} & b_{23} \\ b_{31} & b_{32} & b_{33} \end{pmatrix} \begin{pmatrix} a_1 \\ a_2 \\ a_3 \end{pmatrix} = \begin{pmatrix} b_{11} a_1 + b_{12} a_2 + b_{13} a_3 \\ b_{21} a_1 + b_{22} a_2 + b_{23} a_3 \\ b_{31} a_1 + b_{31} a_2 + b_{33} a_3 \end{pmatrix},$$

is converted as follows:

$$\begin{pmatrix} \boldsymbol{i} & b_{11} & b_{12} & b_{13} \\ \boldsymbol{j} & b_{21} & b_{22} & b_{23} \\ \boldsymbol{k} & b_{31} & b_{32} & b_{33} \end{pmatrix} \atop \begin{pmatrix} a_1 & a_2 & a_3 \end{pmatrix} = \begin{pmatrix} (b_{11} a_1 + b_{12} a_2 + b_{13} a_3)\boldsymbol{i} \\ (b_{21} a_1 + b_{22} a_2 + b_{23} a_3)\boldsymbol{j} \\ (b_{31} a_1 + b_{32} a_2 + b_{33} a_3)\boldsymbol{k} \end{pmatrix}$$

The author owes much to Dr Takashi Saka for this discussion.

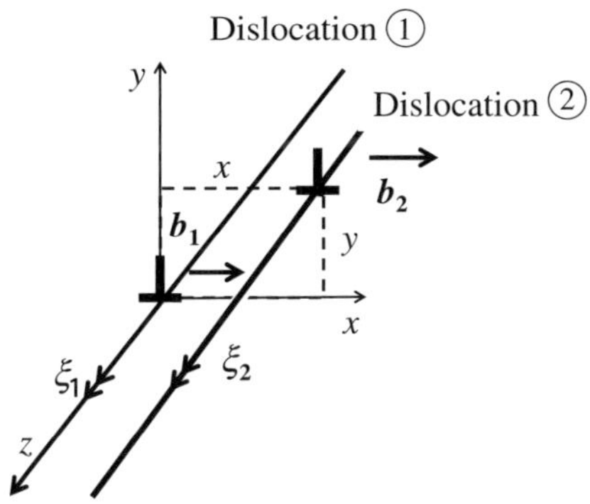

Fig. 4.16

Their Burgers vectors are $\boldsymbol{b}_1 = \boldsymbol{b}_2 = (b\boldsymbol{i}, 0\boldsymbol{j}, 0\boldsymbol{k})$. $\xi_2 = (0\boldsymbol{i}, 0\boldsymbol{j}, 1\boldsymbol{k})$.

$$\begin{aligned}
\frac{\boldsymbol{F}_{1\to 2}}{L} &= (\boldsymbol{b}_2 \cdot \boldsymbol{\Sigma}_1) \times \xi_2,\\
&= b\begin{pmatrix} 1\boldsymbol{i} & 0\boldsymbol{j} & 0\boldsymbol{k}\end{pmatrix}\begin{pmatrix} \sigma_{11}\boldsymbol{ii} & \sigma_{12}\boldsymbol{ij} & 0\boldsymbol{ik}\\ \sigma_{12}\boldsymbol{ji} & \sigma_{22}\boldsymbol{jj} & 0\boldsymbol{jk}\\ 0\boldsymbol{ki} & 0\boldsymbol{kj} & \sigma_{33}\boldsymbol{kk}\end{pmatrix} \times \xi_2,\\
&= b(\sigma_{11}\boldsymbol{i}, \sigma_{12}\boldsymbol{j}, 0\boldsymbol{k}) \times (0\boldsymbol{i}, 0\boldsymbol{j}, 1\boldsymbol{k}),\\
&= b\begin{bmatrix} \boldsymbol{i} & \boldsymbol{j} & \boldsymbol{k}\\ \sigma_{11} & \sigma_{12} & 0\\ 0 & 0 & 1\end{bmatrix}, \qquad (4.22)\\
&= b(\boldsymbol{i}\sigma_{12} - \boldsymbol{j}\sigma_{11} + \boldsymbol{k}0),\\
\frac{F_x}{L} &= b\sigma_{12} = \frac{\mu b^2}{2\pi(1-\nu)}\frac{x(x^2-y^2)}{(x^2+y^2)^2},\\
\frac{F_y}{L} &= -b\sigma_{11} = \frac{\mu b^2}{2\pi(1-\nu)}\frac{y(3x^2+y^2)}{(x^2+y^2)^2},
\end{aligned}$$

Exercise 4.8 There are three solutions to satisfy $F_x = 0$ in Eq. (4.22). They are $x = 0, x = y, x = -y$. These positions are at equilibrium. Among them, at $x = \pm y$ the two edge dislocations are at unstable equilibrium, while at $x = 0$ they are at stable equilibrium. Prove this. (Hint) F_x is given by

$$F_x = -\frac{\partial G}{\partial x},$$

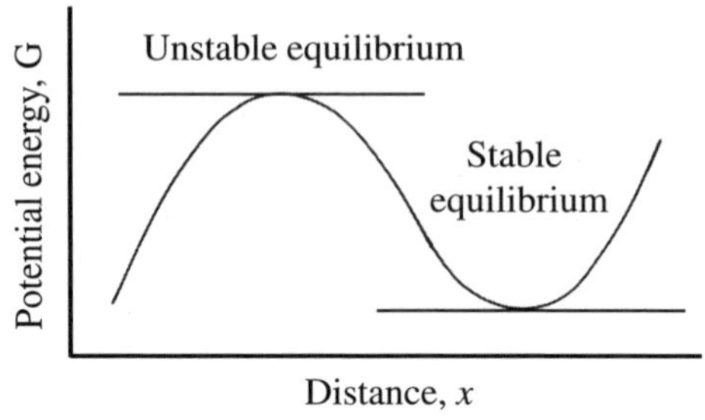

Fig. 4.17

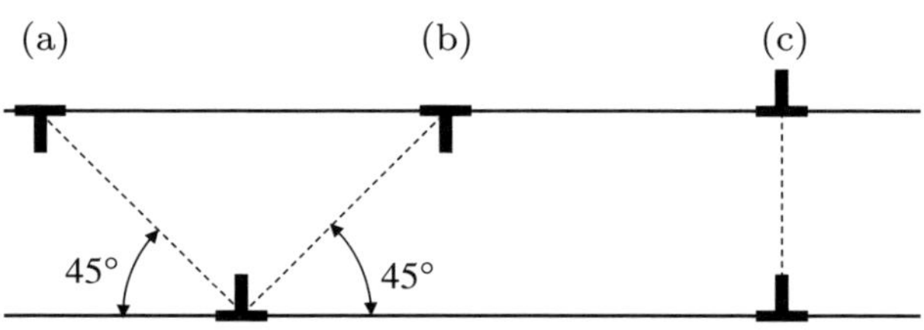

Fig. 4.18

where G is the potential energy. At $x = 0, x = \pm y$, G shows either maximum or minimum (Fig. 4.17). The maximum corresponds to unstable equilibrium and the minimum corresponds to stable equilibrium.

Exercise 4.9 Perform similar calculation to Fig. 4.16 but with $\boldsymbol{b}_1 = -\boldsymbol{b}_2$ and discuss which of $x = 0$ or $x = \pm y$ is the stable equilibrium.

It is obvious that a pair of edge dislocations with opposite signs makes an angle of 45°. Such a pair of edge dislocations with $\pm\boldsymbol{b}$ is referred to as a dipole (Figs. 4.18(a) and 4.18(b)). On the other hand, a pair of edge dislocations with the same sign aligns themselves vertically (Fig. 2.18(c)).

With perpendicular Burgers vectors

As is shown in Fig. 4.19, $\boldsymbol{b}_1 = (b\boldsymbol{i}, 0\boldsymbol{j}, 0\boldsymbol{k}), \boldsymbol{b}_2 = (0\boldsymbol{i}, b\boldsymbol{j}, 0\boldsymbol{k})$ and $\xi_2 = (0\boldsymbol{i}, 0\boldsymbol{j}, 1\boldsymbol{k})$.

$$\frac{\boldsymbol{F}_{1\to 2}}{L} = (\boldsymbol{b}_2 \cdot \boldsymbol{\Sigma}_1) \times \xi_2,$$

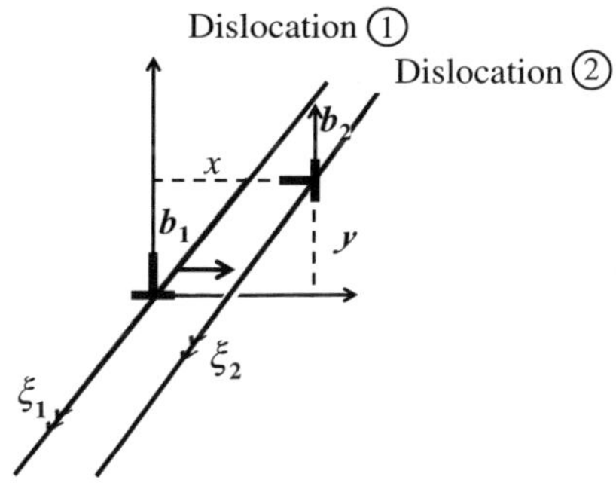

Fig. 4.19

$$
\begin{aligned}
&= b\begin{pmatrix}0\boldsymbol{i} & 1\boldsymbol{j} & 0\boldsymbol{k}\end{pmatrix}\begin{pmatrix}\sigma_{11}\boldsymbol{ii} & \sigma_{12}\boldsymbol{ij} & 0\boldsymbol{ik}\\ \sigma_{12}\boldsymbol{ji} & \sigma_{22}\boldsymbol{jj} & 0\boldsymbol{jk}\\ 0\boldsymbol{ki} & 0\boldsymbol{kj} & \sigma_{33}\boldsymbol{kk}\end{pmatrix}\times\xi_2,\\
&= b(\sigma_{12}\boldsymbol{i},\sigma_{22}\boldsymbol{j},0\boldsymbol{k})\times(0\boldsymbol{i},0\boldsymbol{j},1\boldsymbol{k}),\\
&= b\begin{bmatrix}\boldsymbol{i} & \boldsymbol{j} & \boldsymbol{k}\\ \sigma_{12} & \sigma_{22} & 0\\ 0 & 0 & 1\end{bmatrix}, \qquad (4.23)\\
&= b(\boldsymbol{i}\sigma_{22}-\boldsymbol{j}\sigma_{12}+\boldsymbol{k}0).
\end{aligned}
$$

$$\frac{F_x}{L} = b\sigma_{22} = \frac{\mu b^2}{2\pi(1-\nu)}\frac{y(x^2-y^2)}{(x^2+y^2)^2},$$

$$\frac{F_y}{L} = -b\sigma_{12} = -\frac{\mu b^2}{2\pi(1-\nu)}\frac{x(x^2-y^2)}{(x^2+y^2)^2}.$$

4.4.3 *Perpendicular screw dislocations*

As is shown in Fig. 4.20, $\boldsymbol{b}_1 = (0\boldsymbol{i},0\boldsymbol{j},b\boldsymbol{k}), \boldsymbol{b}_2 = (b\boldsymbol{i},0\boldsymbol{j},0\boldsymbol{k})$ and $\xi_2 = (1\boldsymbol{i},0\boldsymbol{j},0\boldsymbol{k})$.

$$
\begin{aligned}
\frac{\boldsymbol{F}_{1\to 2}}{L} &= (\boldsymbol{b}_2\cdot\boldsymbol{\Sigma}_1)\times\xi_2,\\
&= b\begin{pmatrix}1\boldsymbol{i} & 0\boldsymbol{j} & 0\boldsymbol{k}\end{pmatrix}\begin{pmatrix}0\boldsymbol{ii} & 0\boldsymbol{ij} & \sigma_{13}\boldsymbol{ik}\\ 0\boldsymbol{ji} & 0\boldsymbol{jj} & \sigma_{23}\boldsymbol{jk}\\ \sigma_{13}\boldsymbol{ki} & \sigma_{23}\boldsymbol{kj} & 0\boldsymbol{kk}\end{pmatrix}\times\xi_2,
\end{aligned}
$$

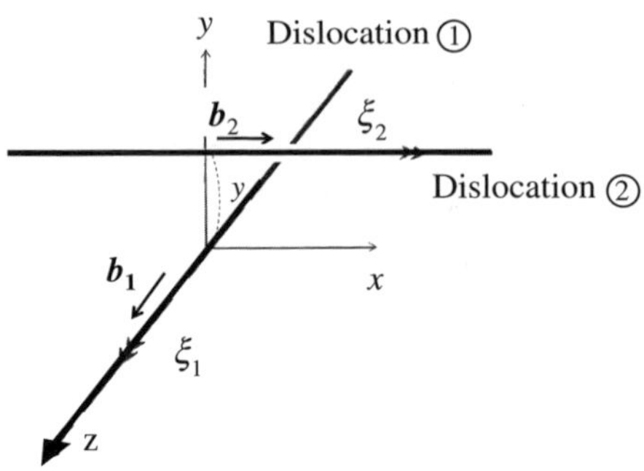

Fig. 4.20

$$
\begin{aligned}
&= b(0\boldsymbol{i}, 0\boldsymbol{j}, \sigma_{13}\boldsymbol{k}) \times (1\boldsymbol{i}, 0\boldsymbol{j}, 0\boldsymbol{k}), \\
&= b\begin{bmatrix} \boldsymbol{i} & \boldsymbol{j} & \boldsymbol{k} \\ 0 & 0 & \sigma_{13} \\ 1 & 0 & 0 \end{bmatrix}, \qquad (4.24) \\
&= b(\boldsymbol{i}0 - \boldsymbol{j}\sigma_{13} + \boldsymbol{k}0).
\end{aligned}
$$

$$
\frac{F_y}{L} = -b\sigma_{13} = -\frac{\mu b^2}{2\pi}\frac{y}{x^2 + y^2},
$$

$$
\frac{F_x}{L} = \frac{F_z}{L} = 0.
$$

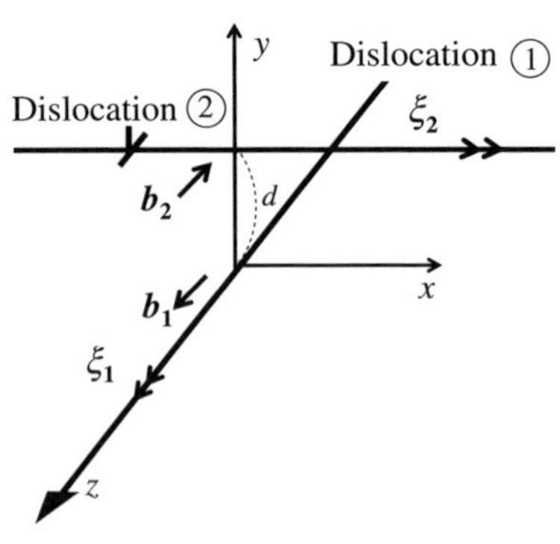

Fig. 4.21

4.4.4 *Edge and screw perpendicular with each other (Fig. 4.21)*

$\boldsymbol{b}_1 = (0\boldsymbol{i}, 0\boldsymbol{j}, b\boldsymbol{k}), \boldsymbol{b}_2 = (0\boldsymbol{i}, 0\boldsymbol{j}, -b\boldsymbol{k}), \xi_2 = (1\boldsymbol{i}, 0\boldsymbol{j}, 0\boldsymbol{k})$

$$\begin{aligned}
\frac{\boldsymbol{F}_{1\to2}}{L} &= (\boldsymbol{b}_2 \cdot \boldsymbol{\Sigma}_1) \times \xi_2 \\
&= b\begin{pmatrix}0\boldsymbol{i} & 0\boldsymbol{j} & -1\boldsymbol{k}\end{pmatrix}\begin{pmatrix}0\boldsymbol{ii} & 0\boldsymbol{ij} & \sigma_{13}\boldsymbol{ik} \\ 0\boldsymbol{ji} & 0\boldsymbol{jj} & \sigma_{23}\boldsymbol{jk} \\ \sigma_{13}\boldsymbol{ki} & \sigma_{23}\boldsymbol{kj} & 0\boldsymbol{kk}\end{pmatrix} \times \xi_2 \\
&= b(-\sigma_{13}\boldsymbol{i}, -\sigma_{23}\boldsymbol{j}, 0\boldsymbol{k}) \times (1\boldsymbol{i}, 0\boldsymbol{j}, 0\boldsymbol{k}) \\
&= -b\begin{bmatrix}\boldsymbol{i} & \boldsymbol{j} & \boldsymbol{k} \\ \sigma_{13} & \sigma_{23} & 0 \\ 1 & 0 & 0\end{bmatrix} \\
&= -b(\boldsymbol{i}0 - \boldsymbol{j}0 + \boldsymbol{k}(-\sigma_{23})) \\
\frac{F_z}{L} &= b\sigma_{23} = \frac{\mu b^2}{2\pi}\frac{x}{x^2+d^2} \\
\frac{F_x}{L} &= \frac{F_y}{L} = 0.
\end{aligned} \tag{4.25}$$

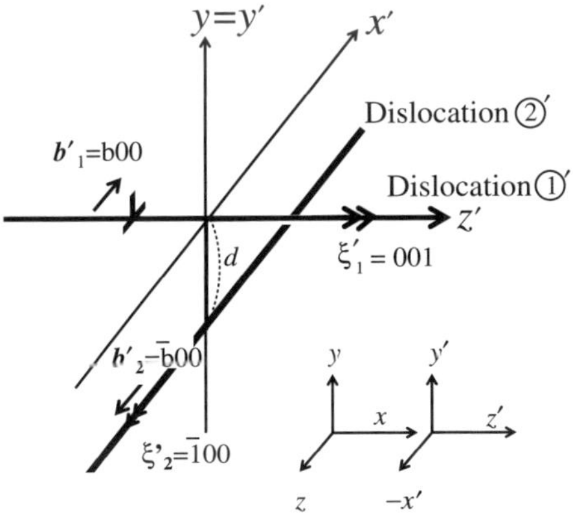

Fig. 4.22

Exercise 4.10 Edge dislocation ② exerts a force on screw dislocation ① in a manner shown in Fig. 4.23(a). Prove this.

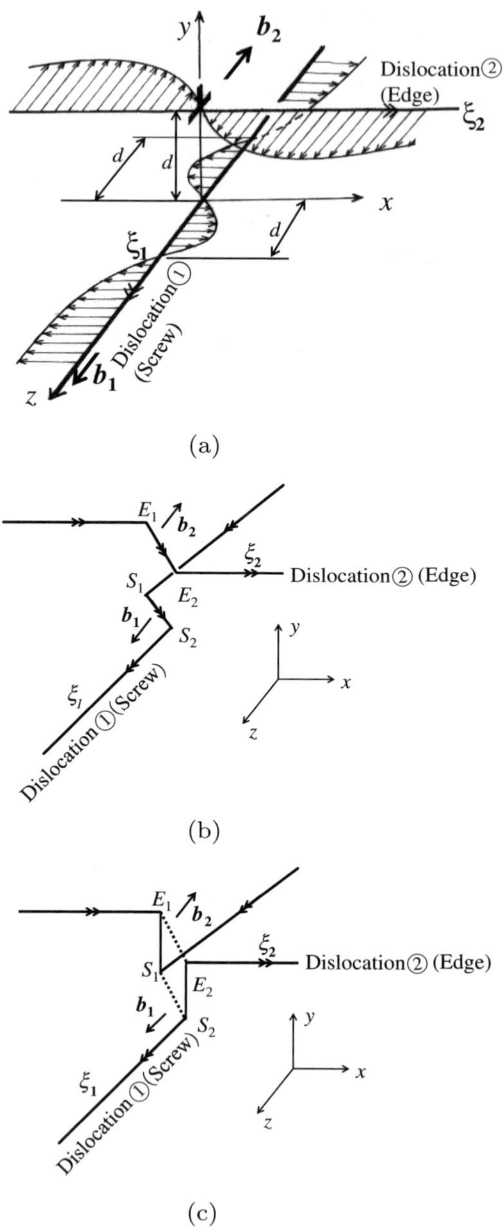
y
b2
Dislocation②
(Edge)
ξ2
d
d
d
x
ξ1
Dislocation①
(Screw)
b1
z
(a)
E1
b2
ξ2
Dislocation② (Edge)
S1
E2
b1
S2
ξ1
Dislocation①(Screw)
y
x
z
(b)
E1
b2
ξ2
Dislocation② (Edge)
S1
E2
b1
S2
ξ1
Dislocation①(Screw)
y
x
z
(c)

Fig. 4.23 (a)–(c)

(Hint) First transform the coordinates as shown in Fig. (4.22), and then apply Eq. (4.20). Finally transform the coordinates back to the original coordinates.

$$\begin{aligned} \frac{F_{2\to 1}}{L} &= (b_1 \cdot \Sigma_2) \times \xi_1 \\ \frac{F_x}{L} &= -\frac{\mu b^2}{2\pi(1-\nu)} \frac{z\left(z^2-d^2\right)}{\left(d^2+z^2\right)^2} \\ \frac{F_y}{L} &= \frac{F_z}{L} = 0. \end{aligned} \tag{4.26}$$

Screw dislocation ① and edge dislocation ② rotate as shown in Fig. 4.23.

Let us suppose that in Fig. 4.23(a) an edge dislocation ② is in the lower part of a prismatic loop of interstitial type lying on the $x-y$ plane (Fig. 4.24(a)). Dislocation ② rotates counterclockwise when viewed downwards in the direction of the y-axis. Screw dislocation ① rotates near the origin in the $x-y$ plane. As a result, segment S_1S_2 of dislocation ① and segment of E_1E_2 of dislocation ② become nearly parallel (Fig. 4.23(b)). Aligning their directions ξ's of S_1S_2 and E_1E_2 to the same sign, their Burgers vectors become anti-parallel. Thus, they attract each other and collapse to annihilate (indicated by broken lines). The result is shown in Fig. 4.23(c). That is, the screw dislocation ① transforms itself into a helical dislocation.

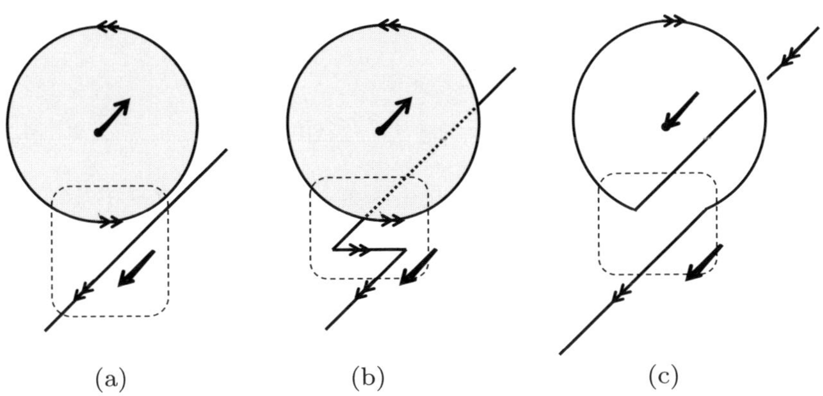

Fig. 4.24 (a)–(c)

Exercise 4.11 Judge whether this helical dislocation is right-handed or left-handed. The result is in agreement with that described in Sec. 2-9-2. (Hint) Fig. 4.24.

4.5 Dislocations in Anisotropic Crystals

The number of the independent elastic constants is two in isotropic crystals. However few real crystals are perfectly isotropic, although they are often approximated as isotropic crystals. As a measure to indicate the deviation from isotropy,

$$A = \frac{2c_{44}}{c_{11} - c_{12}}, \qquad \text{(3.27 (again))}$$

is used. Crystals with $A = 1$ are perfectly isotropic crystals. As A deviates from unity, the crystal becomes more anisotropic.

To what degree the approximation using isotropic elasticity theory can be used depends on the individual case, however, there are two cases where the application of anisotropic elasticity is crucial. One is in the calculation of the dissociation width of a dissociated dislocation (see Chapter 8) and the other is concerned with the equilibrium shape of a dislocation loop.

The equilibrium shape of a dislocation loop can be given by the inverse Wulff construction. This method was originally developed to analyze the three-dimensional (3D) outer shape of a crystal, but can be used to analyze the two-dimensional (2D) shape of a dislocation loop. In anisotropic elasticity the strain energy of a dislocation with $\boldsymbol{b}$ inclined by an angle β with respect to ξ is given by (corresponing to Eq. (4.20) in isotropic elasticity)

$$\frac{W}{L} = (K_s \cos^2 \beta + K_e \sin^2 \beta), \qquad (4.27)$$

where K_s and K_e are the energy factors for screw and edge dislocations, respectively. Draw L/W as a function of β (Fig. 4.25(a)). If one can draw a tangential line $\beta_1\beta_2$ as shown in Fig. 4.25(a), the dislocation is unstable between β_1 and $\beta_2(\beta_1 < \beta < \beta_2)$. Figure 4.25(b) shows a dislocation of $\boldsymbol{b} = [001]$ in β'-CuZn, and shows a prominent anisotropy. Indeed, β'-CuZn is quite anisotropic and its anisotropy factor $A = 10$ (Eq. (3.34)).

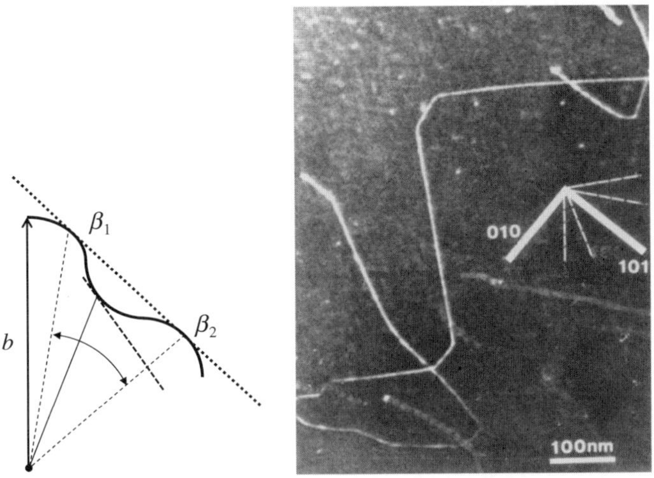

(a) (b)

Fig. 4.25 (a)(b) (Reproduced with permission from [4]).

References

1. Hirth, J. P. and Lothe, J. *Theory of Dislocations*, 1st ed., McGraw-Hill, (1967); 2nd ed., Wiley, (1982).
2. Friedel, J. *Dislocations*, Pergamon Press, Oxford, 1964.
3. Tunstall, W. J., Hirsch, P. B. and Steeds, J. W. *Phil. Mag.* **9** (1964), 99.
4. Saka, H. *Phil. Mag. A*, **49** (1984), 327.

Chapter 5

Elastic Interaction between Dislocations and Solute Atoms

5.1 Isotropic strain (Cottrell effect)

5.1.1 *General theory*

Figure 5.1 shows the schematic representation of a solid solution. Solid solutions are classified into a substitutional solid solution and an interstitial solid solution. Needless to say, radius of a solute atom r_1 in a solid solution is different from that of the solvent r_o (S and s in Fig. 5.1). In an interstitial solid solution, solute atoms enter interstitial positions between the lattices, so that the solute atoms are limited to those with very small radius such as B, C, N and O. In any case a strain is generated around a solute atom.

solid solution (ss)

$$\begin{cases} \text{substitutional ss.} \quad r_1 > r_o \quad (\varepsilon > 0) \quad \text{(S in Fig. 5.1)} \\ \qquad\qquad\qquad\qquad r_1 < r_o \quad (\varepsilon < 0) \quad \text{(s in Fig. 5.1)} \\ \text{interstitial ss (B,C,N,O)} \quad (\varepsilon \gg 0) \quad \text{(I in Fig. 5.1)} \end{cases}$$

In a substitutional solid solution, a solvent atom is replaced by a solute atom. Suppose that a solvent with radius r_o is replaced with a solute atom with radius r_1 (Figs. 5.2(a), 5.2(b)).

Take x-, y- and z-axes as the principal axes of strain. The principal axes x-, y- and z-axes are selected in such a way that only $\varepsilon_{\mathrm{ii}}$, $\varepsilon_{\mathrm{jj}}$, and $\varepsilon_{\mathrm{kk}}$ (that is, normal (compressive or tensile strains)) are non-zero ($\varepsilon_{\mathrm{ii}}$, $\varepsilon_{\mathrm{jj}}, \varepsilon_{\mathrm{kk}} \neq 0$)

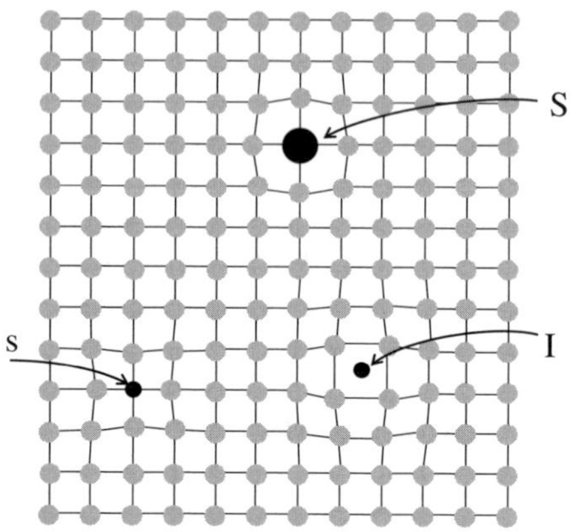

Fig. 5.1

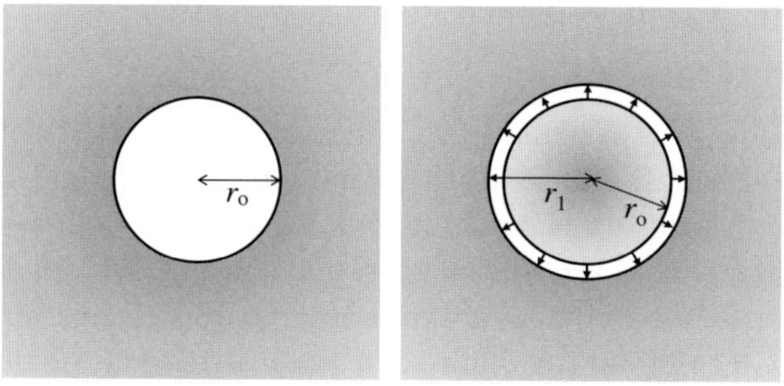

(a) (b)

Fig. 5.2 (a),(b)

with $\varepsilon_{ij} = 0$ (i $\neq$ j)(shear strain) = 0. In other words,

$$\varepsilon = \begin{pmatrix} \varepsilon_{11} & 0 & 0 \\ 0 & \varepsilon_{22} & 0 \\ 0 & 0 & \varepsilon_{33} \end{pmatrix} = \begin{pmatrix} \varepsilon_1 & 0 & 0 \\ 0 & \varepsilon_2 & 0 \\ 0 & 0 & \varepsilon_3 \end{pmatrix}. \tag{5.1}$$

An elastic interaction W between this strain field of the solute atom and the stress field of a dislocation is given by,

$$\begin{aligned} W &= -\int \varepsilon\sigma dS \\ &= -\int \begin{pmatrix} \varepsilon_1 & 0 & 0 \\ 0 & \varepsilon_2 & 0 \\ 0 & 0 & \varepsilon_3 \end{pmatrix} \begin{pmatrix} \sigma_{11} & \sigma_{12} & \sigma_{13} \\ \sigma_{21} & \sigma_{22} & \sigma_{23} \\ \sigma_{31} & \sigma_{32} & \sigma_{33} \end{pmatrix} dS, \end{aligned} \tag{5.2}$$

where the strain field of the solute atom extends over the volume of S, which is approximated by the volume of the solute atom.

When the strain is isotropic,

$$\varepsilon_1 = \varepsilon_2 = \varepsilon_3 = \varepsilon$$

$$W = -\int \varepsilon \sigma dS$$

$$= -\int \begin{pmatrix} \varepsilon & 0 & 0 \\ 0 & \varepsilon & 0 \\ 0 & 0 & \varepsilon \end{pmatrix} \begin{pmatrix} \sigma_{11} & 0 & 0 \\ 0 & \sigma_{22} & 0 \\ 0 & 0 & \sigma_{33} \end{pmatrix} dS$$

$$= -\varepsilon(\sigma_{11} + \sigma_{22} + \sigma_{33})\Omega \tag{5.2$'$}$$

Here $\Omega = \frac{4\pi r_o^3}{3}$ and r_o is the radius of the solute atom.

5.1.2 *Edge dislocation*

Now let us calculate the interaction energy in a geometry shown in Fig. 5.3. Substituting σ_{11}, σ_{22} and σ_{33} into Eq. 4.8,

$$W = -\Omega\varepsilon A\left\{-\frac{y(3x^2+y^2)}{(x^2+y^2)^2} + \frac{y(x^2-y^2)}{(x^2+y^2)^2} - \frac{2y\nu}{x^2+y^2}\right\} = \frac{2A(1+\nu)y\varepsilon\Omega}{r^2}$$

$$= \frac{4}{3}\left(\frac{1+\nu}{1-\nu}\right)\frac{\mu b r_o^3 \varepsilon y}{x^2+y^2}, \tag{5.3}$$

where A $= \mu$b$/2\pi(1-\nu)$.

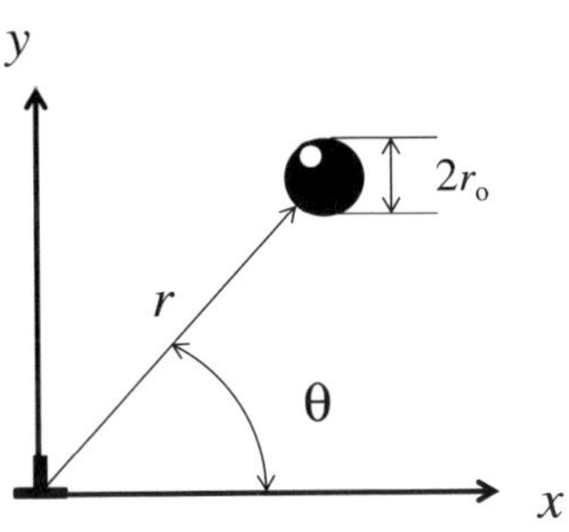

Fig. 5.3

In the cylindrical coordinates

$$W = \frac{4}{3}\left(\frac{1+\nu}{1-\nu}\right)\frac{\mu b r_o^3 \varepsilon \sin\theta}{r}, \tag{5.4}$$

Exercise 5.1 Derive Eqs. (5.3) and (5.4).

Exercise 5.2 Show that contours of W drawn from Eq. 5.4 are as shown in Fig. 5.4. When $\varepsilon > 0$, i.e., a solute is larger than solvent, contours at $y > 0$ are positive, i.e., hill and those at $y < 0$ are valley. When $\varepsilon < 0$, i.e., a solute is smaller than solvents, the reverse is true. Solute atoms constituting a Cottrell atmosphere migrate along a path for which the gradient of W is maximum. In other words, the path is at right angle with the contour (for instance along ① in Fig. 5.4).

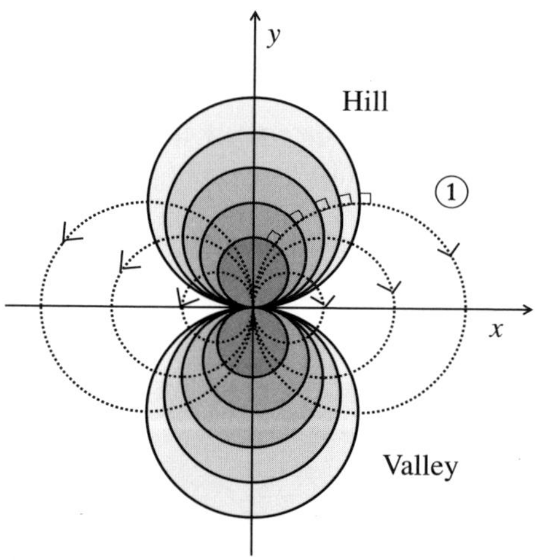

Fig. 5.4

Here, it is assumed that a solute atom is larger than solvent atoms. The absolute value increases with increasing tint. That is, the hill is the higher and the valley is the deeper.

According to Eq. (5.4) the interaction energy shows a maximum when $\theta = 90°$ and $270°$. When $\varepsilon > 0$, solute atoms segregate just beneath the edge dislocation and when $\varepsilon < 0$, the reverse is true. In a substitutional solid solution, solute atoms larger than the matrix segregate below an edge dislocation and those smaller above an edge dislocation (Fig. 5.5(a)). In an

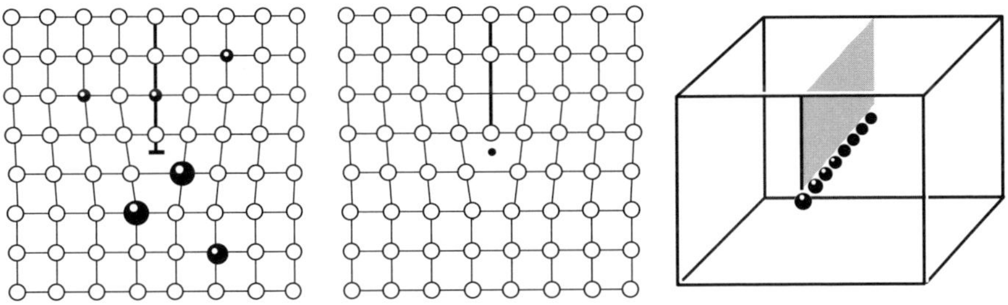

(a) (b) (c)

Fig. 5.5 Cottrell effect.

interstitial solid solution, $\varepsilon \gg 0$, so that interstitial atoms always segregate beneath an edge dislocation (Fig. 5.5(b)). Since the strain (ε) of interstitial atoms such as carbon (C) and nitrogen (N) atoms in iron (Fe) is quite large, and also the diffusion of interstitials is very fast (faster than substitutional solutes by an order of three), they segregate just beneath an edge dislocation line and pin it over its total length (Fig. 5.5(c)). This phenomenon is referred to as Cottrell effect. State where solute atoms are segregated over a rather wider region around an edge dislocation is referred to as Cottrell atmosphere.

5.1.3 *Screw dislocation*

For a screw dislocation,

$$\sigma_{11} = \sigma_{22} = \sigma_{33} = 0,$$

so that it is obvious from Eq. (5.2′) that:

$$W = 0.$$

In other words, there is no interaction between a screw dislocation and an isotropic strain. Neither screw dislocations nor solute atoms notice the existence of each other.

5.2 Anisotropic Strain

Steels are alloys of iron (Fe) and carbon (C). Figure 5.6 shows the phase diagram of the Fe–C binary system. They crystallize in two structures, i.e., face centered cubic (FCC) (γ-phase, austenite) at high temperatures and BCC (α-phase, ferrite) at room temperature. In the high-temperature

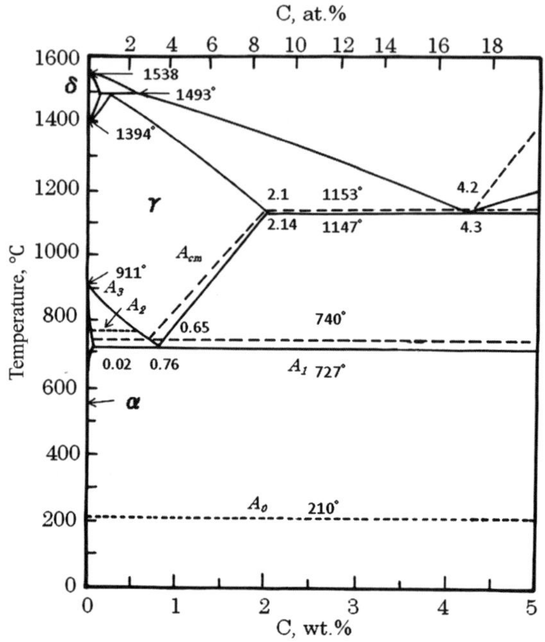

Fig. 5.6 The phase diagram of the Fe–C system(Reproduced with permission from [1]).

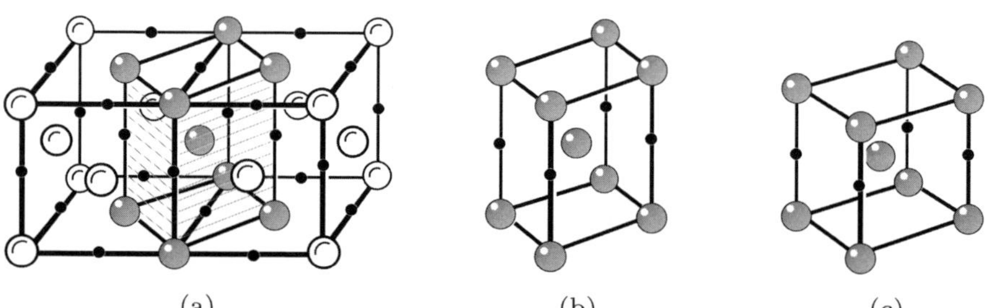

Fig. 5.7 (a)–(c) Bain's relationship.

γ-phase, i.e., a solid solution with the FCC structure, C atoms are distributed randomly at positions shown by small solid circles (•) in the FCC lattice of Fe (Fig. 5.7(a)). The solubility limit is ~8 wt.%. That is to say, up to ~8 wt.% C atoms are accommodated in the matrix of Fe. However, in the low-temperature α-phase (body centered cubic (BCC)) the solubility limit is as small as 0.02 wt.%. Suppose that a steel containing, say, 1 wt.% C is cooled to room temperature, only 0.02 wt.% C out of 1 wt.% C can stay in α-Fe matrix and the rest are expelled from α-Fe. Those C atoms which are expelled from α-Fe matrix either exist as graphite (pure C) or react with Fe to form Fe_3C (cementite), both of which are different phases from α-Fe. The former is present in a cast iron and the latter in a mild steel.

However, when γ-phase is cooled rapidly (quenched) from high temperatures, Fe transforms from FCC to BCC so quickly that C atoms cannot escape from the Fe matrix, i.e., trapped in the Fe matrix down to room temperature. This is referred to as a supersaturated solid solution,

The sites where C atoms occupy in the high-temperature γ (FCC) phase are shown in Fig. 5.7(a). They are located at the centres of the x-, y- and z-axes of the FCC unit cell. As shown in Fig. 5.7(b), only one of the x-, y- and z-axes is contained in a newly formed BCC structure (z-axis in Fig. 5.7(b)). This is referred to as Bain's relationship. As a result, in the supersaturated α-solid solution, C atoms are inserted only along the z-axis. The z-axis is elongated, resulting in the c/a ratio larger than unity (c/a>1). Such a strain, where a strain along one axis is larger than those along the other two axes, is referred to as a tetragonal strain, and the supersaturated α-solid solution containing C is referred to as martensite.

Because of symmetry of FCC structure, x-, y- and z-axes (a-, b- and c-axes) are crystallographically equivalent. Therefore, as shown in Figs. 5.8(a)–5.8(c), martensite is formed with tetragonality along x-, y- and z-axes (a-, b- and c-axes) with the same frequency, however, it is customary to refer the axis along which C is inserted to the c-axis.

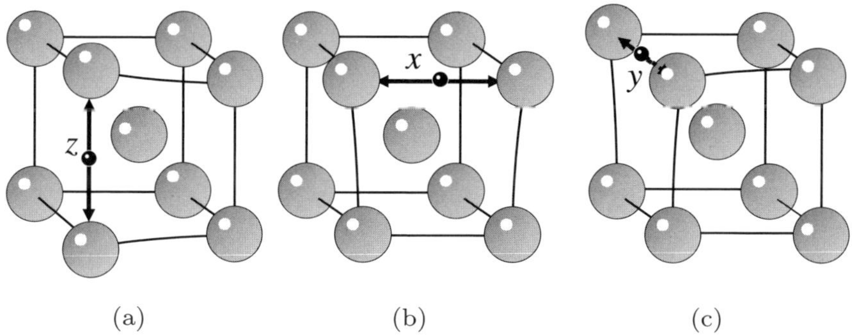

Fig. 5.8 (a)–(c) Sites occupied by C in martensite.

The strain tensor is, as is obvious from Fig. 5.8, written as follows:

$$\varepsilon = \begin{pmatrix} \varepsilon_1 & 0 & 0 \\ 0 & \varepsilon_1 & 0 \\ 0 & 0 & \varepsilon_2 \end{pmatrix}.$$

Therefore,

$$W = -\{\varepsilon(\sigma_{11} + \sigma_{22}) + \varepsilon_2\sigma_{33}\}, \tag{5.5}$$

where experiment shows,

$$\varepsilon_1 = -0.026,$$
$$\varepsilon_2 = 0.38.$$

The coordinates referred to here are the usual orthogonal coordinates used to represent the unit cell. That is, the usual x-, y- and z-axes are the principal axes. Equation 5.2 is expressed in this coordinates. However, the Burgers vector $\boldsymbol{b}$ of a dislocation in the BCC structure is $1/2 < 111 >$. It is customary to express its stress (see Eq. (4.3)) by taking $< 111 >$ direction as the z-axis (in order to distinguish the principal axis of strain let us write it as z' axis). In other words, while the principal axes for a strain are expressed in the usual (x, y, z) coordinates, the stress field of a screw dislocation is expressed in different coordinates (x', y', z'). Equation 4.3 gives only σ'_{11}, σ'_{22}, and σ'_{33}, not σ_{11}, σ_{22}, and σ_{33}, which are to be used to calculate the interaction energy W using Eq. (5.2).

Here, we transform the old coordinate system (x, y, z) into a new (x', y', z') coordinate system in such a way that the [111] axis, referred to the old coordinates, becomes (is parallel to) the new z'-axis (Fig. 5.9(a)).

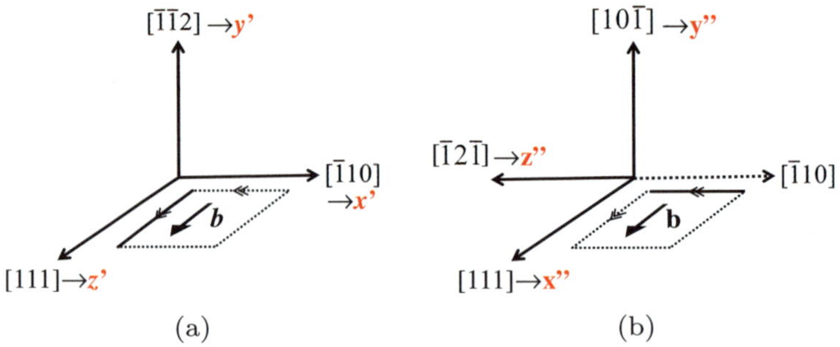

Fig. 5.9 (a) Screw dislocation (b) Edge dislocation.

The new z'-axis is parallel to [111], but the choice of x'- and y'-axes is not unique. If we choose the new coordinate as shown in Fig. 5.9(a),

$$
\begin{aligned}
x' &= \frac{1}{\sqrt{2}}[\bar{1}10] \\
y' &= \frac{1}{\sqrt{6}}[\bar{1}\bar{1}2]. \\
z' &= \frac{1}{\sqrt{3}}[111]
\end{aligned}
\qquad (3.16\ (\text{again})^{1})
$$

Here, the indices in the right side are (x', y', z') referred to the old (x, y, z) coordinate system.[2]

The transformation given by Eqs. (3.18) and (3.20) is as follows:

$$\begin{pmatrix} x' \\ y' \\ z' \end{pmatrix} = \{T_{ij}\} \begin{pmatrix} x \\ y \\ z \end{pmatrix}.$$

([UVW]//[$\bar{1}$10], $[uvw]$//[$\bar{1}\bar{1}$2], $[UVW]$//[111])

$$\{T_{ij}\} = \begin{bmatrix} T_{11} & T_{12} & T_{13} \\ T_{21} & T_{22} & T_{22} \\ T_{31} & T_{32} & T_{33} \end{bmatrix} = \begin{bmatrix} -1 & 1 & 0 \\ -1 & -1 & 2 \\ 1 & 1 & 1 \end{bmatrix} \begin{matrix} \div\sqrt{2} \\ \div\sqrt{6} \\ \div\sqrt{3} \end{matrix} = \begin{pmatrix} \frac{-1}{\sqrt{2}} & \frac{1}{\sqrt{2}} & 0 \\ \frac{-1}{\sqrt{6}} & \frac{-1}{\sqrt{6}} & \frac{2}{\sqrt{6}} \\ \frac{1}{\sqrt{3}} & \frac{1}{\sqrt{3}} & \frac{1}{\sqrt{3}} \end{pmatrix}.$$

Transformation from (x', y', z') system to (x, y, z) system is given by Eq. 3.15

$$\begin{pmatrix} x \\ y \\ z \end{pmatrix} = \begin{pmatrix} T_{11} & T_{21} & T_{31} \\ T_{12} & T_{22} & T_{32} \\ T_{13} & T_{23} & T_{33} \end{pmatrix} \begin{pmatrix} x' \\ y' \\ z' \end{pmatrix}.$$

Equation 5.2 indicates that only σ_{11}, σ_{22} and σ_{33} are required.

---

[1] $x'//[\bar{1}10>, y'//[\bar{1}\bar{1}2>, z'//[111>$.

[2] The ordinary coordinates can be indicated by (x', y', z'), and coordinates containing the dislocation can be indicated by (x, y, z). Needless to say, this is equivalent to what is used in the text.

Equation 3.21

$$\sigma_{ij} = \sum_{l,m} T_{li} T_{mj} \sigma'_{lm}, \tag{3.21}$$

gives σ_{11} as follows:

$$\begin{aligned}
\sigma_{11} &= \sum_{l,m} T_{l1} T_{m1} \sigma'_{lm} \\
&= T_{11}T_{11}\sigma'_{11} + T_{11}T_{21}\sigma'_{12} + T_{11}T_{31}\sigma'_{13} \\
&\quad + T_{21}T_{11}\sigma'_{21} + T_{21}T_{21}\sigma'_{22} + T_{21}T_{31}\sigma'_{23} \\
&\quad + T_{31}T_{11}\sigma'_{31} + T_{31}T_{21}\sigma'_{32} + T_{31}T_{31}\sigma'_{33} \\
&= T_{11}T_{11}0 + T_{11}T_{21}0 + T_{11}T_{31}\sigma'_{13} \\
&\quad + T_{21}T_{11}0 + T_{21}T_{21}0 + T_{21}T_{31}\sigma'_{23} \\
&\quad + T_{31}T_{11}\sigma'_{31} + T_{31}T_{21}\sigma'_{32} + T_{31}T_{31}0 \\
&= T_{11}\left(= -\frac{1}{\sqrt{2}}\right) T_{31}\left(= \frac{1}{\sqrt{3}}\right)\sigma'_{13} + T_{21}\left(= \frac{-1}{\sqrt{6}}\right) T_{31}\left(= \frac{1}{\sqrt{3}}\right)\sigma'_{23} \\
&\quad + T_{31}\left(= \frac{1}{\sqrt{3}}\right) T_{11}\left(= -\frac{1}{\sqrt{2}}\right)\sigma'_{31} + T_{31}\left(= \frac{1}{\sqrt{3}}\right) T_{21}\left(= \frac{-1}{\sqrt{6}}\right)\sigma'_{32} \\
&= 2\left(-\frac{1}{\sqrt{2}}\right)\left(\frac{1}{\sqrt{3}}\right)\sigma'_{13} + 2\left(\frac{-1}{\sqrt{6}}\right)\left(\frac{1}{\sqrt{3}}\right)\sigma'_{23} \\
&= -\frac{2}{\sqrt{6}}\left(\sigma'_{13} + \frac{1}{\sqrt{3}}\sigma'_{23}\right) \\
&= \frac{\mu b}{\sqrt{6}\pi} \frac{1}{x'^2 + y'^2}\left(y' - \frac{x'}{\sqrt{3}}\right) \\
&= \frac{-\mu b}{\sqrt{6}\pi r}\left(\frac{\cos\theta}{\sqrt{3}} - \sin\theta\right)
\end{aligned} \tag{5.6}$$

(See Eq. 4.3).

Here, θ is referred to the coordinates (x', y', z') for the screw dislocation, corresponding to θ in the cylindrical coordinates shown in Fig. 4.1.

Similarly,

$$\begin{aligned}\sigma_{22} &= \sum_{l,m} T_{l2}T_{m2}\sigma'_{lm} \\ &= \frac{-\mu b}{\sqrt{6}\pi}\frac{1}{x'^2+y'^2}\left(\frac{x'}{\sqrt{3}}-y'\right) \\ &= \frac{-\mu b}{\sqrt{6}\pi r}\left(\frac{\cos\theta}{\sqrt{3}}-\sin\theta\right),\end{aligned} \tag{5.7}$$

$$\begin{aligned}\sigma_{33} &= \sum_{l,m} T_{l3}T_{m3}\sigma'_{lm} \\ &= \frac{\sqrt{2}\mu b}{3\pi}\frac{x'}{x'^2+y'^2} \\ &= \frac{\sqrt{2}\mu b}{3\pi r}\cos\theta.\end{aligned} \tag{5.8}$$

Exercise 5.3 Prove Eqs. (5.7) and (5.8).

Therefore, the interaction energy is given by,

$$\begin{aligned}W_s &= -(\sigma_{11}\varepsilon_1 + \sigma_{22}\varepsilon_1 + \sigma_{33}\varepsilon_2)\Omega \\ &= -\frac{\mu b\Omega}{\sqrt{6}\pi r}\Delta\varepsilon\left(\frac{\cos\theta}{\sqrt{3}}+\sin\theta\right),\end{aligned} \tag{5.9}$$

where $\Delta\varepsilon = \varepsilon_2 - \varepsilon_1$. W_s is a function of r and θ and is shown in Fig. 5.10.

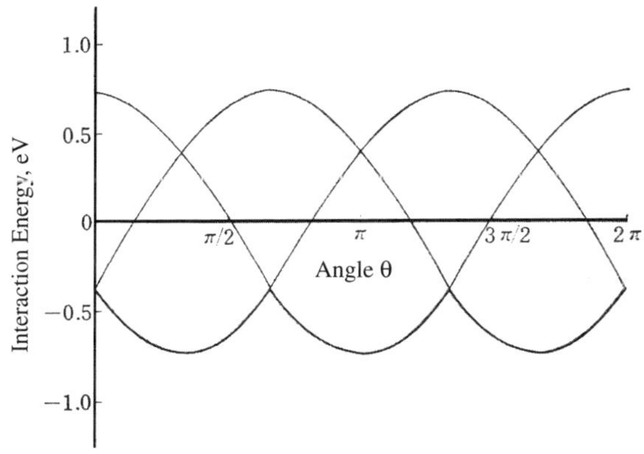

Fig. 5.10 Interaction energy between a screw dislocation and a C atom. Three curves correspond to the three sites of C atom shown in Fig. 5.8. It is assumed that r=b (Reproduced with permission from [2]).

A tetragonal strain interacts with a screw dislocation. Needless to say, when $\varepsilon_1 = \varepsilon_2$, i.e., for an isotropic strain W_s is zero.

Exercise 5.4 Show that interaction of an edge dislocation with a tetragonal strain is expressed as follows. Use transformation of coordinates shown in Fig. 5.9(b).

$$W_{e_3} = \frac{A\Omega}{r}\left[2\varepsilon(1+\nu)\sin\theta + \frac{1}{6}\{2(2+\nu)\sin\theta\right.$$
$$\left. + \cos 2\theta(2\sqrt{2}\cos\theta + \sin\theta)\}\Delta\varepsilon\right]$$
$$W_{e_1} = \frac{A\Omega}{r}\left[2\varepsilon(1+\nu)\sin\theta + \frac{1}{6}\{2(2+\nu)\sin\theta\right.$$
$$\left. - \cos 2\theta(2\sqrt{2}\cos\theta + \sin\theta)\}\Delta\varepsilon\right]$$
$$W_{e_2} = \frac{A\Omega}{r}\left[2\varepsilon(1+\nu)\sin\theta - \sin\theta\{\cos 2\theta + 2(1+2\nu)\}\frac{\Delta\varepsilon}{3}\right], \qquad (5.10)$$

where $\mathrm{A} = \mu\mathrm{b}/2\pi(1-\nu)$.

Here subscripts 1,2,3 of W_e indicate that the respective tetragonality is along x-, y- and z-axes.

References

1. Saka, H. *Introduction to Phase Diagrams for Materials Scientists* (in Japanese), Asakura, (2012) pp. 201.
2. Cochardt, A.W., Schoeck, G. and Wiedersich, H. *Acta Met.* **3**, (1955), 533.

Chapter 6

Motion (Peierls Force) and Multiplication (Frank-Read Source, Bardeen-Herring Source) of Dislocations

6.1 Peierls Force

6.1.1 *General theory*

In Chapter 4, it is shown that the total energy of a dislocation = core energy $(E_c)+$ elastic energy outside the core(E_e).

Let us consider the motion of a dislocation in a crystal. Suppose that a dislocation moves from a position A to an adjacent position B by an atomic distance ($\boldsymbol{b}$) (Fig. 6.1(a)). E_e is independent of the position of the dislocation core, so that it remains unchanged during the motion of the dislocation from A to B by $\boldsymbol{b}$. By contrast, E_c should change periodically with a period of $\boldsymbol{b}$ (Fig. 6.1(b)). This periodical potential with the period of $\boldsymbol{b}$ is referred to as Peierls–Nabarro potential (in contracted form as P–N potential), and the force required to overcome the potential is referred to as the Peierls–Nabarro force (P–N force) or simply the Peierls force.

The Peierls force (τ_p) is given by

$$\tau_p = \frac{2\mu}{1-\nu}\exp\left\{-\frac{2\pi d}{b(1-\nu)}\right\}, \tag{6.1}$$

where d is the spacing of the slip plane. On increasing d, τ_p decreases rapidly. This is easily rationalized by considering an orthogonal unit cell with dimension of $d \times b$ being deformed in shearing by b. The longer is d, the easier is the shearing (Fig. 6.2).

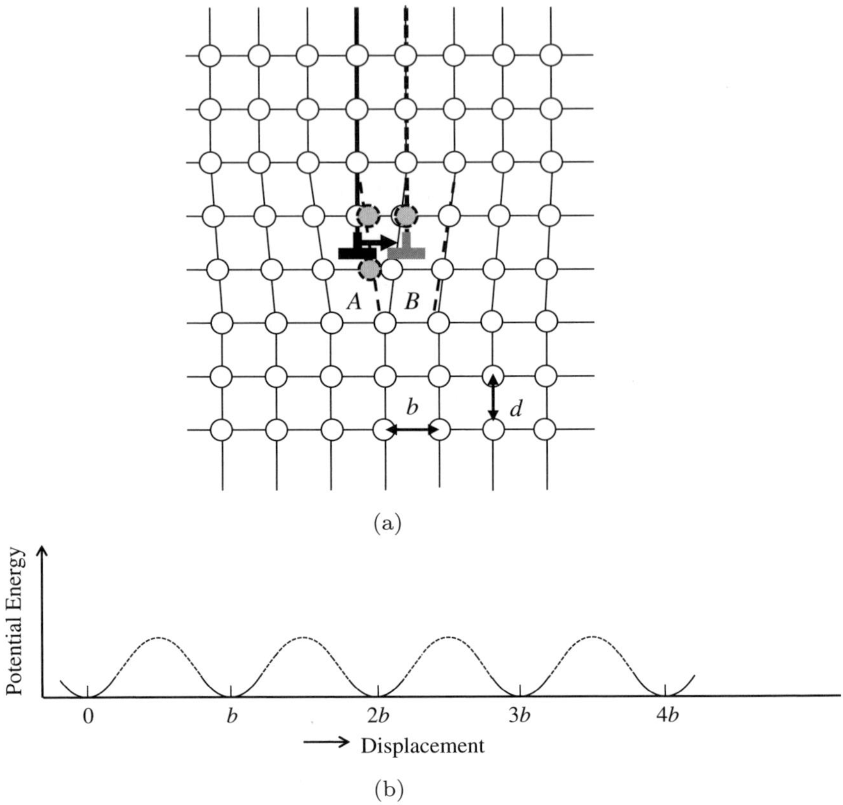

Fig. 6.1 Variation in potential energy accompanied with a dislocation motion.

Exercise 6.1 The spacing of $\{hkl\}$ planes in a cubic crystal, d_{hkl}, is given by

$$d_{hkl} = \frac{a}{\sqrt{h^2 + k^2 + l^2}}.$$

This indicates that the Peierls force on low-index planes (with small digits of h, k, l) is low. In FCC crystals $\boldsymbol{b} = 1/2\langle 110\rangle$, that is, the slip planes belong to the $\langle 110\rangle$ zone axis. Show that among the planes belonging to the $\langle 110\rangle$ zone axis the most closely packed plane, i.e., $\{111\}$ has the largest value of d. This indicates that, in FCC crystals, Peierls force on the most closely packed $\{111\}$ plane is a minimum.

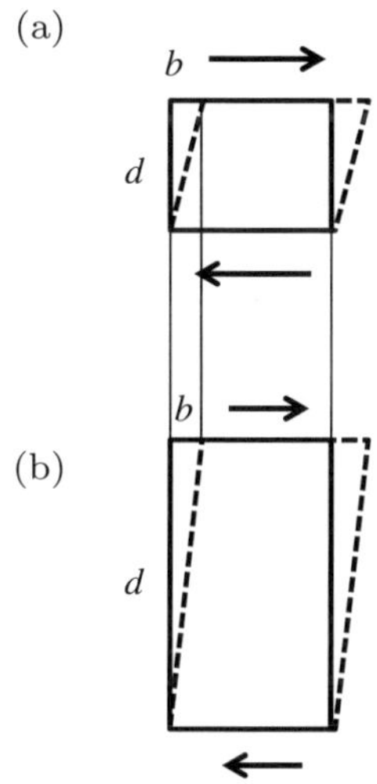

Fig. 6.2 (a), (b) A qualitative explanation of difference in a large Peierls force (a) and a small one (b).

6.1.2 *The Peierls force in BCC metals*

The P–N forces on the closed packed slip planes in FCC crystals and on the basal plane of hexagonal close packed (HCP) crystals are very small and can be neglected practically. That is, a dislocation in face-centered cubic (FCC) crystals and HCP crystals (basal plane slip) can move at a very small force. On the other hand, the P–N force in body-centered cubic (BCC) metals is large. This will be described in detail in Chapter 12.

6.1.3 *The Peierls force in Si, Ge, GaAs*

The P–N forces in covalent crystals such as Si, Ge and GaAs are large. This is not attributed to Eq. (6.1) but to the high directionality of their bonds and will be described in detail in Chapter 11.

6.1.4 *Overcoming of P–N potential*

Consider a process by which a dislocation overcomes the P–N potential. A dislocation does not overcome the P–N potential while keeping the straight shape as a whole. Part of the dislocation overcomes the hill of P–N potential and reaches the adjacent valley (Fig. 6.3). This is referred to as a double–kink mechanism. A kink is a bend of dislocation on the slip plane. A pair of kinks is referred to as a double-kink. As is obvious from Fig. 6.4(b), the

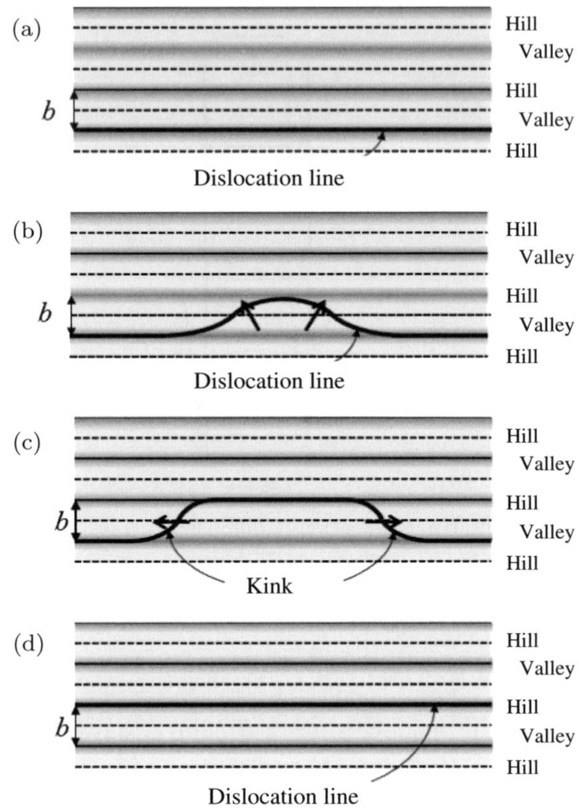

Fig. 6.3 Escape from the P–N valley by formation of a double kink.

kink ① and ②′ are anti-parallel with respect to $\boldsymbol{b}$. Therefore, they attract each other and when an external stress to separate the kinks away is absent or very low, the double kinks collapse to annihilate. It is not until the kinks are separated by a critical distance l_c that the external stress succeeds in separating them further. The process from the initial straight state to this critical state is a thermally activated process in which a thermal energy assists the motion of atoms.

6.2 Dislocation Sources

6.2.1 *Frank-Read source*

When a single dislocation completely sweeps the slip plane, the step formed at the crystal surface is an order of Å(1Å=0.1 nm). Macroscopic plastic

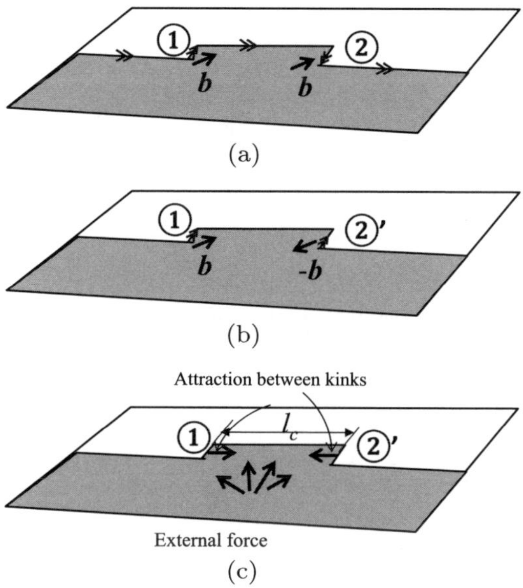

Fig. 6.4 A double kink. In (a) the directions ξ of kink ① and ② are anti-parallel but their $\boldsymbol{b}$'s are parallel. In (b) ξ of ② is reversed to ②′, so that ξ's of ① and ②′ are now parallel. Burgers vector of ②′ is reversed to $-\boldsymbol{b}$.

deformation requires motion of a vast amount of dislocations. Dislocation density of a well annealed crystal is in the order of $10^4\,\text{cm}^{-2}$. But a plastic strain of 1–2% increases the dislocation density up to 10^{8-9}cm^{-2} (10,000–100,000 times larger).

A mechanism by which a dislocation multiplies itself is the famous Frank–Read source. In an annealed crystal, dislocations comprise a network (Fig. 6.5(a)). Now, a stress is applied externally to a crystal to deform. Suppose that the shear stress acting on a segment of a dislocation network happens to be high enough for this segment to move while all the other segments of the dislocation network are blocked. The dislocation segment will bow out (Fig. 6.5(b)). The dislocations that anchor the bowing-out dislocation are referred to as pole dislocations. The bowing-out dislocation surmounts the pole dislocations (Fig. 6.5(c)), and annihilate each other (Fig. 6.5(d)). Then, the bowing-out dislocation is now unanchored from the nodes of the pole dislocations and continues to move, expanding the slipped area. On the other hand, the dislocation segment connecting the two pole dislocations returns to the original position and repeats the same process. This process can therefore be repeated indefinitely. This is referred

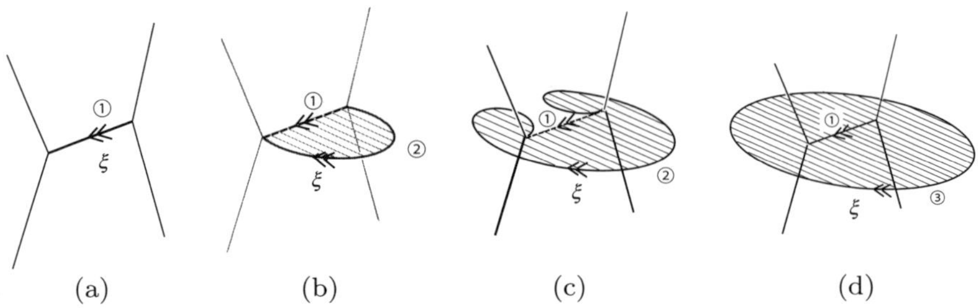

Fig. 6.5 Frank–Read source.

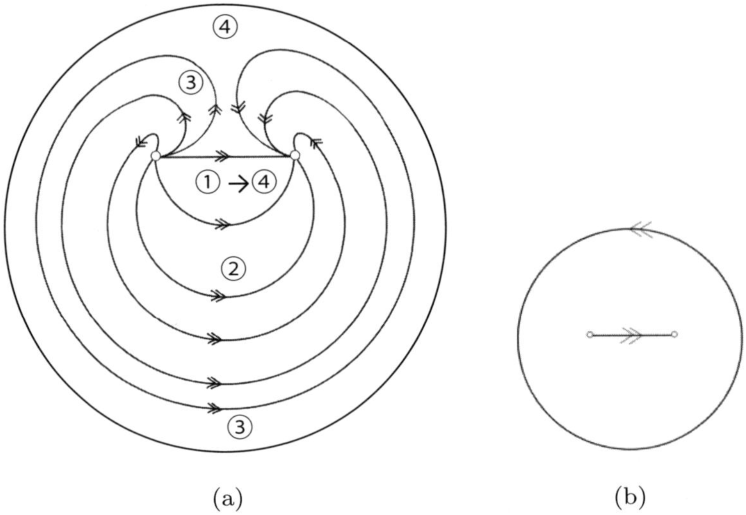

Fig. 6.6 Repetition of the operation of a Frank–Read source.

to as the Frank–Read source. Figure 6.7 shows an infrared photograph of a Frank–Read source in Si [1].

Exercise 6.2 In Fig. 6.5(c) when dislocation segments bowing from right and left sides meet, they annihilate. Explain this using Kirchhoff's law. (Hint) Figures 6.6(a) and 6.6(b).

A back stress acts on a bowing-out dislocation which is emitted from a F–R source due to the line tension, and straightens the bowing-out dislocation. When an external stress $\tau = F/b$ acting on the bowing-out segment is larger than this back stress, the dislocation continues to expand. However,

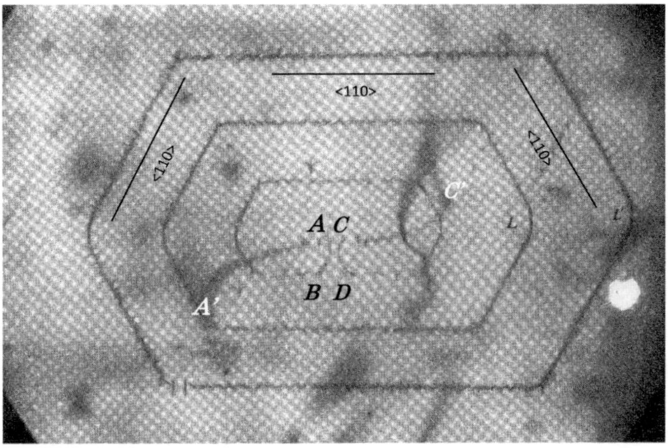

Fig. 6.7 Frank–Read source in Si with a slip plane $\{111\}$. Dislocation lines are decorated by precipitates of Cu and observed with an infrared optical microscope. Blurred images of AA′ and CC′ are pole dislocations. The focal length of an infrared microscope is short, so that pole dislocations, which lie nearly vertically in the crystal, are blurred. Hexagonal shapes of dislocations are attributed to the P–N force being high along $\langle 110 \rangle$ directions in Si (see Chapter 11) (Reproduced from [1]).

when the external stress is smaller than the back stress, the bowing-out dislocation shrinks. This is analogous to expanding a balloon. When the blowing force is larger than the surface tension of the balloon, the balloon continues to expand, but when smaller it deflates.

Now consider the minimum stress required to activate a F–R source. What is analogous to the surface tension of a balloon is the line tension (T) of a dislocation. The line tension is an increment of the elastic energy of a dislocation caused by an incremental increase in its length. Let the distance between two anchoring points of a F–R source be l (Fig. 6.8(a)). An initially straight dislocation bows out rightward to reach the radius R. On the central segment δl of the bowing-out dislocation two forces act, one is an external force to expand the bowing-out dislocation and the other is the back stress due to the line tension T; they balance each other. As is obvious from Fig. 6.8(b) the back stress due to the line tension is given by

$$\begin{aligned} F_{\text{back}} &= 2T \cos\left(\frac{\pi}{2} - \frac{\delta\theta}{2}\right) \\ &= 2T \sin\left(\frac{\delta\theta}{2}\right) \end{aligned}$$

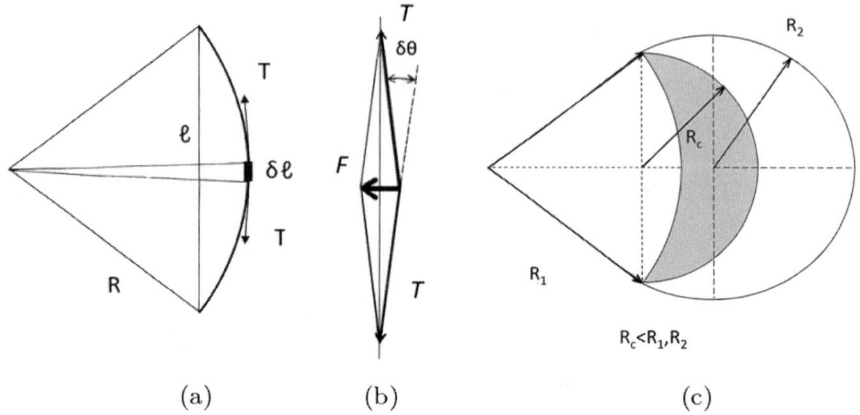

Fig. 6.8 Back stress exerting on a bowing-out dislocation.

$$\approx T\delta\theta$$
$$= T\frac{\delta l}{R}.$$

On the other hand, the external force F_{ext} acting on δl is given by Eqs. (4.6) and (4.13) as follows:

$$F_{ext} = \left(\frac{F}{L}\right)\delta l = (\sigma b)\delta l.$$

These two forces balance, i.e.,

$$T\frac{\delta l}{R} = (\sigma b)\,\delta l$$
$$\therefore \quad \sigma = \frac{T}{Rb}.$$

The line tension T is the self-energy of a dislocation per unit length, and given by Eq. (4.19) as follows to the first approximation:

$$T \approx \frac{\mu b^2}{2}. \tag{6.2}$$

$$\therefore \quad \sigma = \frac{T}{Rb} = \frac{\mu b}{2R}. \tag{6.3}$$

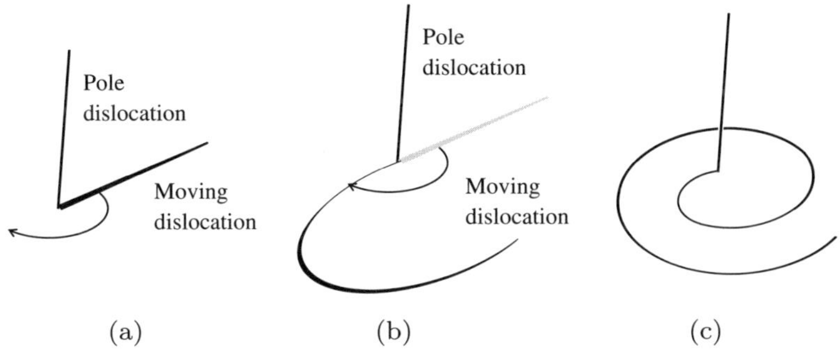

Fig. 6.9 Single-ended Frank–Read source.

Exercise 6.3 Calculate T putting $\beta = 45°$ in Eq. (4.19).

Exercise 6.4 When the dislocation bows out further, the radius of curvature becomes larger. In other words, the maximum stress is reached when $R = l/2$.

$$\sigma_{\max} = \frac{\mu b}{l}. \tag{6.4}$$

Show that this is the stress required to operate the F–R source.

6.2.2 *Modifications of Frank-Read source*

Single-ended F-R source

When only one pole-dislocation is present in a F–R source, a moving dislocation continues to rotate like a propeller as shown in Figs. 6.9(a)–(c). Figure 6.10 shows an example of a single-ended F–R source in Si.

Giant jog and double cross-slip

Two dislocations above and below a giant jog rotate like two propellers. Figure 6.11 shows an example of such a F–R source in pure Fe.

6.2.3 *Bardeen-Herring source*

Geometry of Bardeen–Herring source is quite similar to that of Frank–Read source except that it operates by climb motion. An edge segment connects two pole-dislocations (Fig. 6.12). Under supersaturation of point defects, the edge segment climbs to form a dislocation loop in a manner

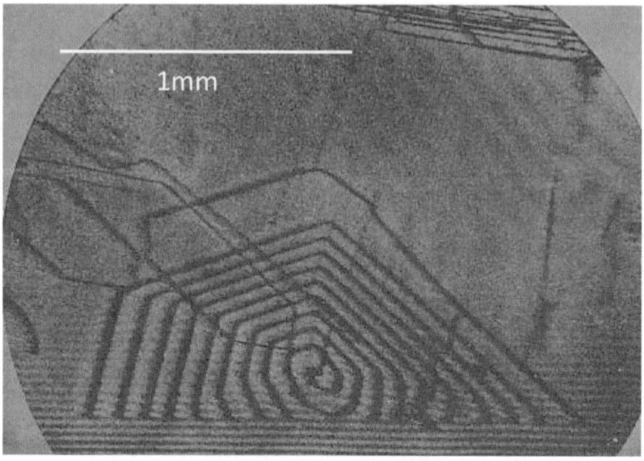

Fig. 6.10 A single-ended F–R source in Si. X-ray Lang topograph. Note the magnification (Reproduced with permission from [2]).

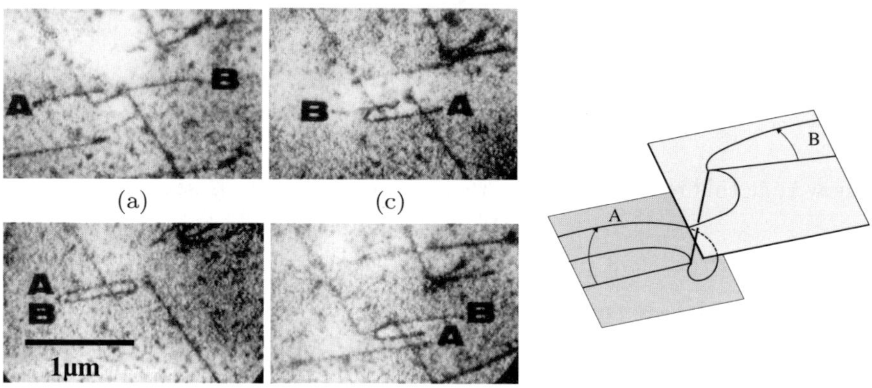

(b) (d) (e)

Fig. 6.11 (a)–(e) Dislocation source at a giant jog (Reproduced with permission from [3]).

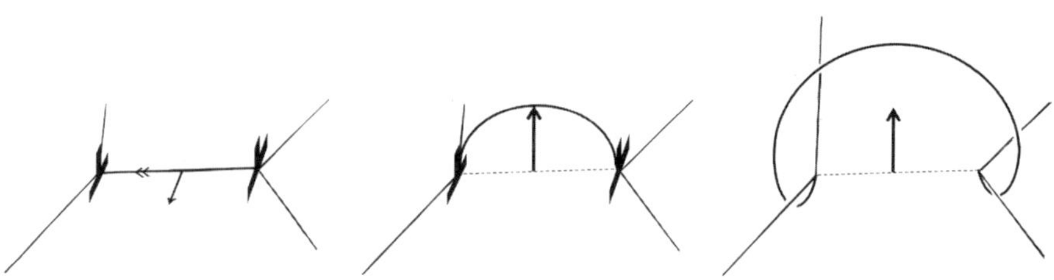

(a) (b) (c)

Fig. 6.12 Bardeen–Herring source.

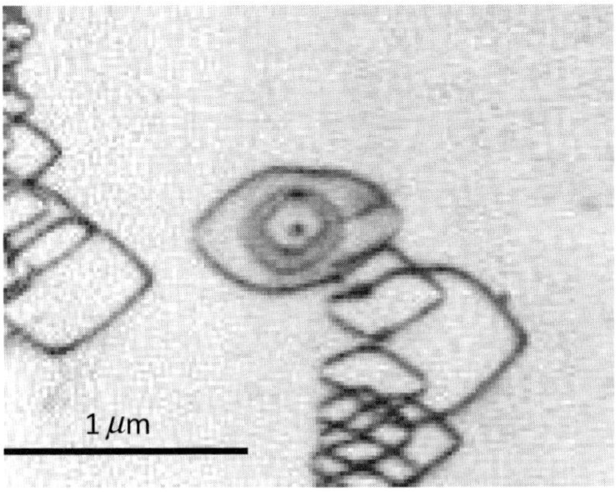

Fig. 6.13 A Bardeen–Herring source (after T. Imura and T. Hashimoto).

quite similar to F–R source. However, unlike the case of F–R source, this loop is a prismatic loop whose ***b*** is perpendicular (or nearly perpendicular) to the plane on which the dislocation lies. Figure 6.13 shows a Bardeen–Herring source in an Al–Mg alloy.

References

1. Dash, W. C., cited in Friedel, J. *Dislocations*, Pergamon Press, Oxford, 1964.
2. Authier, A. and Lang, A. *J. appl. Phys.* **35** (1964), 1956.
3. Saka, H., Noda, K. and Imura, T. *Trans* ISIJ, **13** (1973), 318.

Chapter 7

Dislocation Groups

7.1 Pileup of Dislocations

When a group of dislocations meet a very strong barrier (Fig. 7.1) such as a hard surface oxide layer (Fig. 7.1(a)), a second phase (inclusion or precipitate Fig. 7.1(b)) and a grain boundary) (Fig. 7.1(c)), the leading dislocation comes to a halt by the barrier, and the trailing dislocations emitted from a F–R source are jammed. This is referred to as a pileup of dislocations. Figure 7.2 shows an example of such a pileup against a grain boundary.

As can be noted by intuition, the spacing between the neighbouring dislocations in a pileup is narrower near the head of the pileup (Fig. 7.2). Consider a dislocation pileup under the action of an external stress σ. Let the coordinates of dislocation ① be x_1, the coordinate of dislocation ② be x_2, etc. The repulsive forces between the dislocations produce an interaction energy which depends only on the relative spacing (Fig. 7.3),

$$W_{\text{int}} = W(x_2 - x_1, x_3 - x_1, \ldots, x_N - x_1)$$

At equilibrium, the net force on the individual dislocations ②–Ⓝ is zero. Suppose that the leading dislocation moves δx, then the rest of the dislocations will move δx. The work done by the external stress is $(\sigma b \cdot \delta x)N$. The work, done by the leading dislocation against the internal stress σ_i, is $\sigma_i b \cdot \delta x$. Since these works should be equal,

$$bN = \sigma_i.$$

The stress concentration factor on the leading dislocation is the same as the number of dislocations held up in a pileup. In other words, a pileup of N

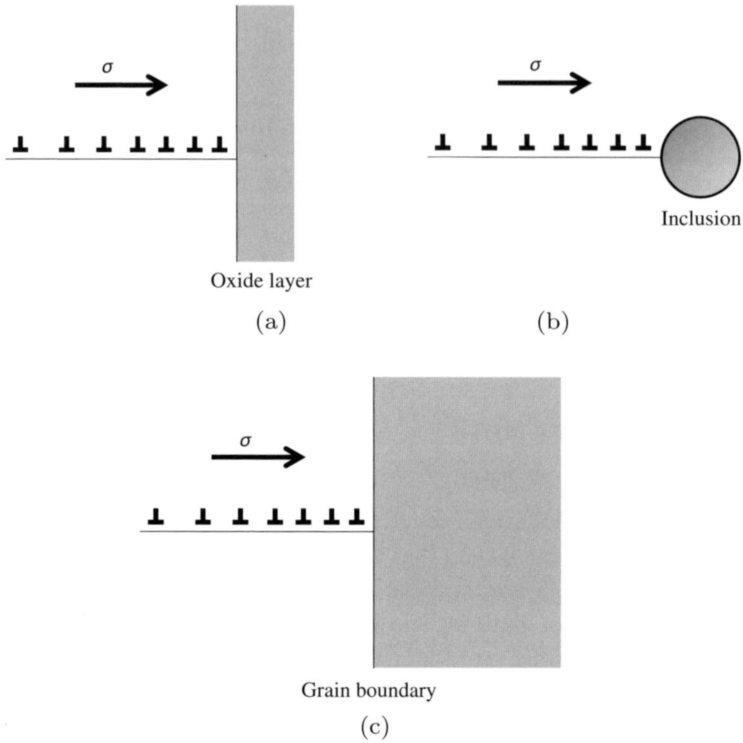

Fig. 7.1 Dislocation pileups at a surface oxide layer (a), an inclusion (b) and a grain boundary (c).

dislocations will exert a stress that a superdislocation with $\boldsymbol{b} = N\boldsymbol{b}$ would do against the barrier. This stress concentration can be strong enough to nucleate a crack (Figs. 7.4(a) and 7.4(b)).

Exercise 7.1. Discuss the validity of this discussion qualitatively based on Kirchhoff's law.

The piled-up dislocations exert a back stress to a F–R source. If an applied stress on the F–R source given by Eq. (6.4)[1] is lower than the back stress, the F–R source stops operating.

1

$$\sigma_{\max} = \frac{\mu b}{l} \tag{6.4 (again)}$$

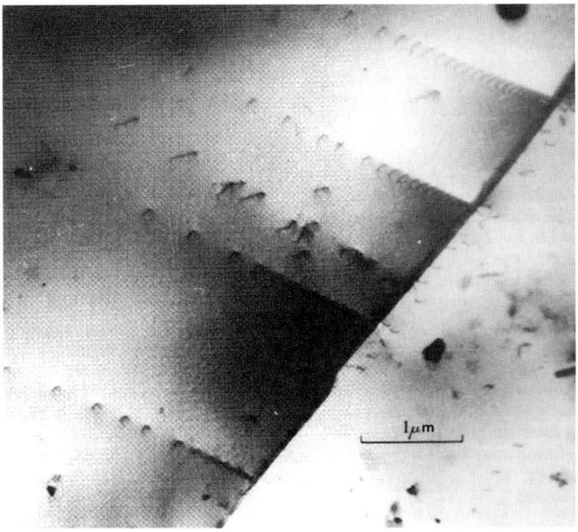

Fig. 7.2 Dislocation pileups at a grain boundary (Reproduced with permission from [1]).

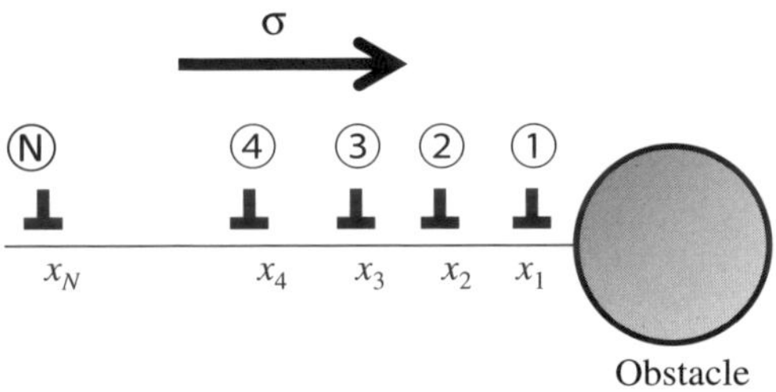

Fig. 7.3 Pileup of dislocations.

7.2 Polygonization and Small Angle Tilt Boundary

When a single crystal is bent, an excess of positive (+) dislocations will be introduced as shown in Fig. 7.5(a). On annealing, the edge dislocations assume stable positions. In Sec. 4.4.2, it is shown that edge dislocations with same signs align themselves just above each other as shown in Fig. 7.5(b). This is referred to as polygonization, and the resultant sub-grain boundary is referred to as a small angle tilt boundary (Fig. 7.6). The number of dislocations in a small angle tilt boundary is given by

$$n_p = \frac{2h}{b} \sin \frac{\theta}{2}.$$

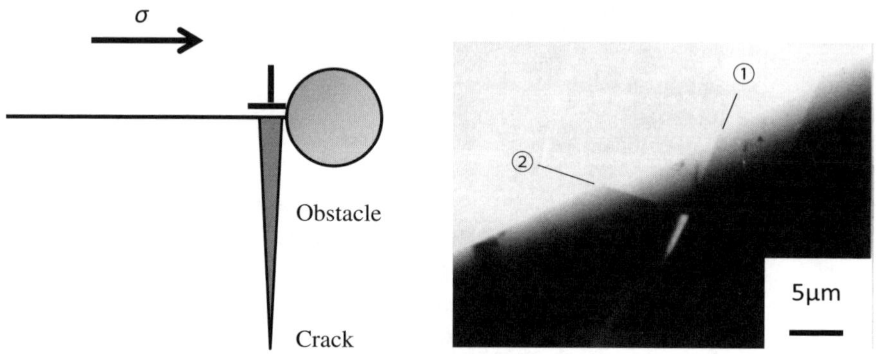

(a) (b)

Fig. 7.4 Nucleation of a crack at a dislocation pileup. (a) Schematic diagram. In (b), a pre-existing slip band ① hinders the motion of slip band ②, and at the cross a crack is nucleated. Note that the width of the crack and surface step associated with the slip band ② are the same (Reproduced with permission from [2]).

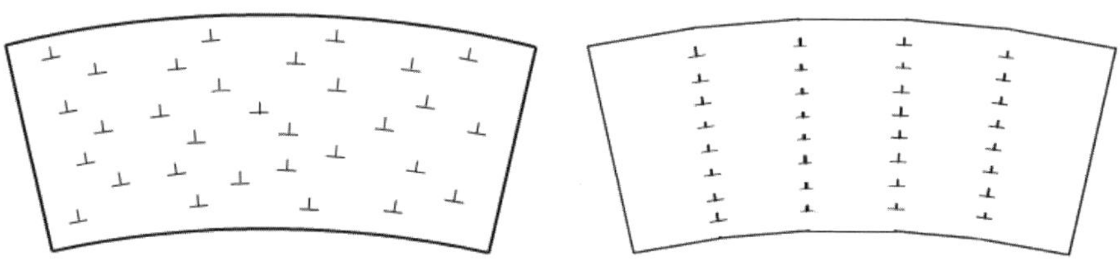

(a) (b)

Fig. 7.5 Polygonization.

The spacing between dislocations is given by

$$D = \frac{h}{n_p} = \frac{b}{2\sin(\theta/2)}. \tag{7.1}$$

When θ is very small,

$$D \cong \frac{b}{\theta}. \tag{7.2}$$

Figure 7.7 shows a TEM micrograph (a) and the corresponding diffraction pattern (b) of a small angle tilt boundary in YAG.[2] In (a) contrasts of

[2]Yittrium aluminium garnet, $Y_3Al_5O_{12}$.

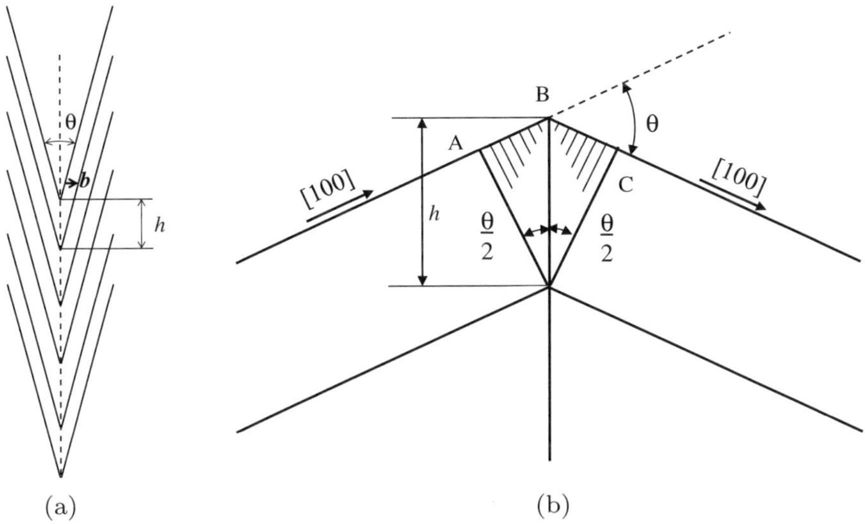

Fig. 7.6 A tilt boundary (Reproduced with permission from [3]).

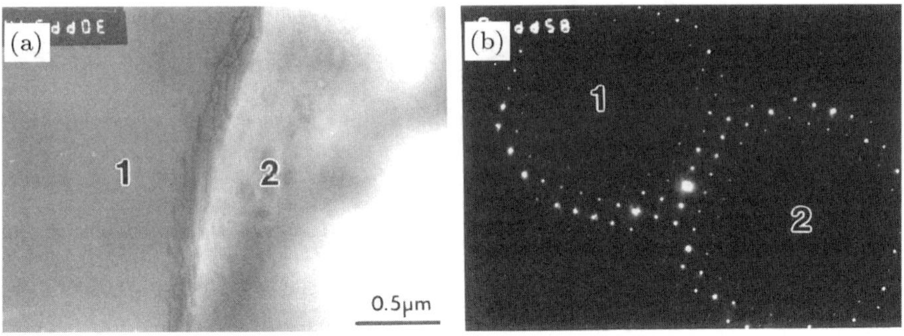

Fig. 7.7 A tilt boundary in YAG, (a) TEM, (b) The corresponding diffraction pattern (Reproduced with permission from [4]).

sub-grains 1 and 2 are slightly different. This is because their orientations are slightly different. Sub-grain 1 satisfies the Bragg condition while sub-grain 2 does not. Diffraction pattern in (b) indicates that the misorientation between 1 and 2 is $\sim 1^\circ$. In TEM micrograph (a), a group of dislocations is observed at the boundary between 1 and 2, indicating that the boundary is a small angle tilt boundary.

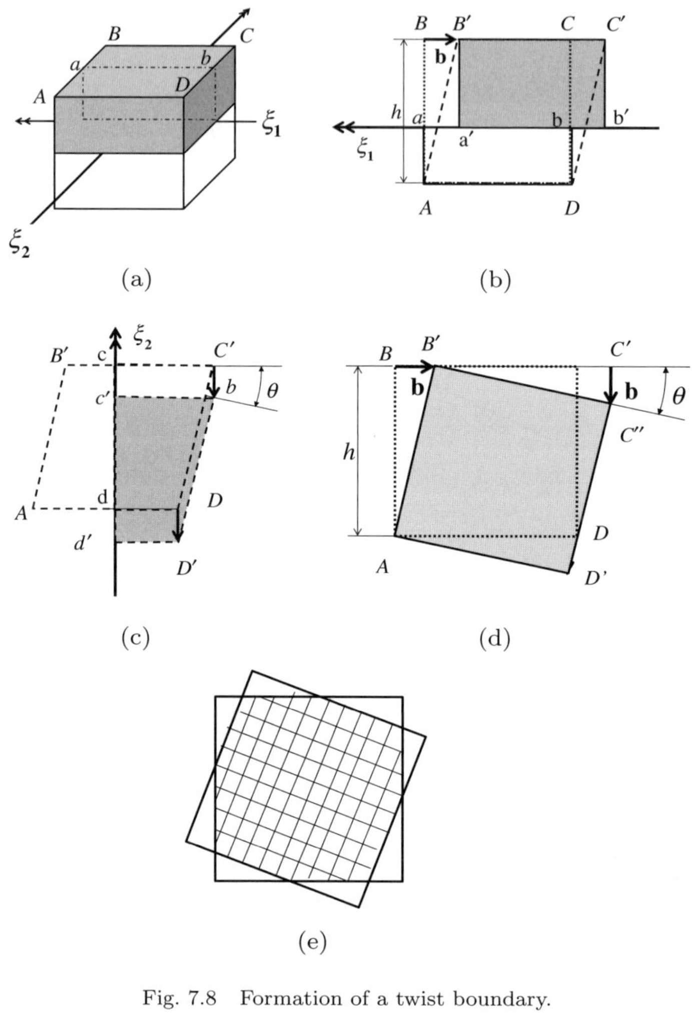

Fig. 7.8 Formation of a twist boundary.

7.3 Twist Boundary

A simple twist boundary is formed by joining two crystals with a rotating by θ along an axis perpendicular to the boundary plane (Fig. 7.8(a)). For the simple twist boundary to be at equilibrium at least two sets of screw dislocations are required. Suppose that only one screw dislocation is introduced. Two parts of the crystal are joined with cross-section elastically deformed in shear, and therefore it is not a stress-free boundary (Fig. 7.8(b)). However, if another screw dislocation is introduced orthogonally to the first dislocation as shown in Fig. 7.8(c), two parts of the crystal meet without elastic deformation (Fig. 7.8(d)). That is, two sets of screw dislocations are required (Fig. 7.8(e)).

Exercise 7.2. Show that the average spacing of two sets of screw dislocation in Fig. 7.8 is given by Eq. (7.1) or (7.2).

References

1. Whelan, M. J., Hirsch, P. B., Horne, R. W., and Bollmann, W. *Proc. Roy. Soc.* London, A **240** (1957), 524.
2. Saka, H., Imura, T., and Yukawa, N. *Jpn. J. Appl. Phys.*, **19** (1971), 1.
3. Hirth, J. P. and Lothe, J. *Theory of Dislocations*, 1st ed., McGraw-Hill, (1967); 2nd ed, Wiley, (1982).
4. Moon, W. J., Saka, H., and Waku, Y. *Materi. Trans. JIM*, **40** (1999), 627.

Chapter 8

Dissociated Dislocations in FCC Structure

8.1 Dissociated Dislocations

By definition, a dislocation is the boundary between slipped and unslipped regions. However, in some crystal structures the slip does not take place non-stop but takes place in two stages or more. A typical example of this is dissociated dislocations[1] in the face-centered cubic (FCC) structure. The concept of the dissociated dislocation is the most important in dislocation theory. Its understanding necessitates comprehensive knowledge on the Burgers vector (including Burgers circuit, Kirchhoff's law), elasticity theory and crystallography. This chapter is the most crucial in dislocation theory. This chapter is closely related with Chapter 9 (the hexagonal close packed (HCP) structure) and Chapter 11 (the diamond, the zincblende, the wurtzite structures).

8.1.1 *Thompson tetrahedron*

The FCC structure is shown in Fig. 1.1. From the viewpoint of dislocation theory, its slip planes, i.e., $\{111\}$ planes are most important. As shown in Fig. 8.1(a) four $\{111\}$ planes are present in the unit cell of the FCC structure, that is, DAB, DAC, DBC, ABC. They are labelled (c), (b), (a) and (d) on the outside surface, respectively, and $(\bar{c})$, $(\bar{b})$, $(\bar{a})$ and $(\bar{d})$ on the inside surface. Figure 8.1(b) displays (d) plane, i.e., ΔABC. The midpoint of (d) is labelled as δ. For the other planes their midpoints are labelled as α, β and γ. These four planes (a), (b), (c) and (d) planes comprise a tetrahedron $ABCD$, which is referred to as Thompson tetrahedron. Figure 8.1(c) shows

[1] Also referred to as extended dislocations or imperfect dislocations.

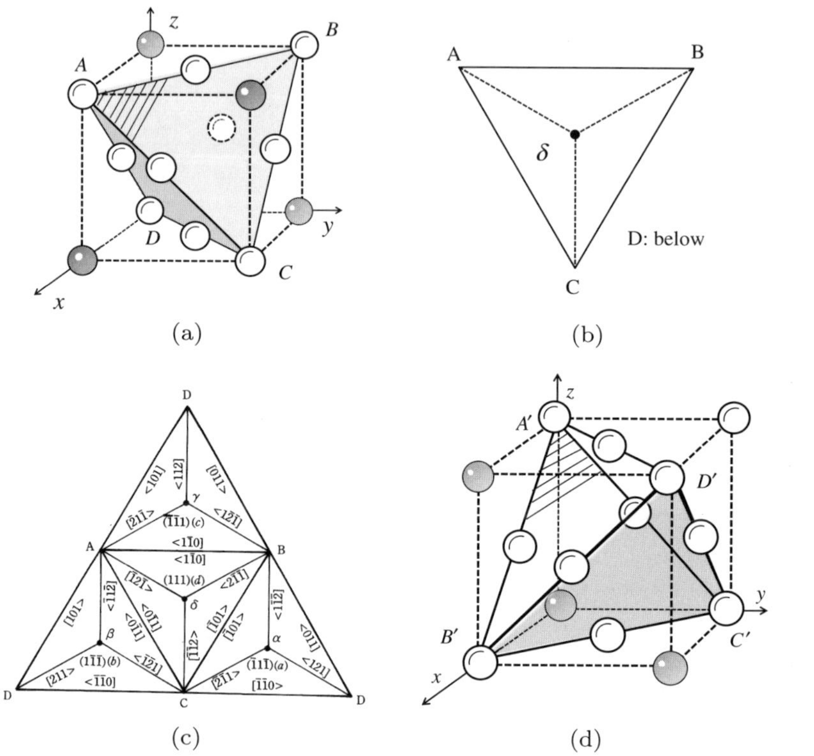

Fig. 8.1 (a) $ABCD$ is Thompson tetrahedron, composed of atoms ○, (b) δ is the midpoint of (d) plane, (c) Thompson tetrahedron opened up at corner D (Reproduced with permission from [1]), (d) The other Thompson tetrahedron.

the Thompson tetrahedron opened-up at the corner D. The reader is recommended to build the Thompson tetrahedron and master how to use it. An enlarged version of Fig. 8.1(c) is attached in Appendix A6. Whether one understands the Thompson tetrahedron or not determines whether one can master this chapter and hence dislocation theory.

In Fig. 8.1(a) BA^{*}[2] represents $\langle 1\bar{1}0]$. This is twice the Burgers vector $\boldsymbol{b}(1/2\langle 110\rangle)$.[3] That is,

$$BA = <1\bar{1}0] = 2b.$$

[2]Vectors defined using a Thompson tetrahedron are referred to as Thompson vectors.

[3]Originally a Thompson tetrahedron is referred to as 1/8 unit cell of the edge length $1/2a_o$. However, in order to make correlation between the Thompson tetrahedron and stacking of {111} planes clear, the Thompson tetrahedron is referred to as one unit cell with the edge length a_o throughout this book.

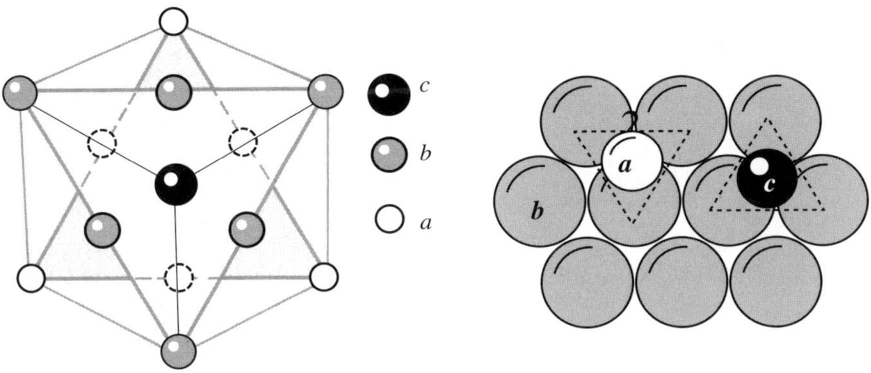

(a) (b)

Fig. 8.2 (a),(b) Stacking of (111) planes in the FCC structure. There are three equivalent positions ***a***, ***b*** and ***c***. In the HCP structure only two of them are present (see Fig. 1.1(f)).

In the Thompson tetrahedron BA is written $<1\bar{1}0]$, but strictly speaking, this stands for $2\boldsymbol{b} = <1\bar{1}0]$. However, for the reasoning written in footnote 3), let BA represent $\boldsymbol{b}(1/2\langle 110\rangle)$ in this book.

$\{111\}$ planes are closed packed planes where atoms are packed together most tightly. Among them (d) plane (ΔABC) has a Miller index (111). Viewing this plane along [111] direction towards the origin (Fig. 8.2(a)), there are three possible (equivalent) positions of atoms indicated by ○, ◍ and ● (Fig. 8.2(b). Let us label ◍ by ***b***, ○ by ***a***, and ● by ***c***. The stacking sequence of the (111) planes is, from the origin (D), ***cabcabc*** (Fig. 8.2(a)). As is shown in Fig. 8.1(d) there is one more Thompson tetrahedron.

Exercise 8.1 Confirm using Fig. A.6 in Appendix that there are only two Thompson tetrahedra.

Exercise 8.2 Show that basal planes ABC and $A'B'C'$ (hatched in Figs. 8.1(a) and (d), respectively) correspond to ***b*** and ***a*** in Fig. 8.2(a).

We employ the Thompson tetrahedron shown in Fig. 8.1(a).[4]

[4]This rigidity in the choice is important in discussing the polarity of dislocations in the zincblende structure, the wurtzite structure and SiC, which have the polarity (see for details Chapter 11).

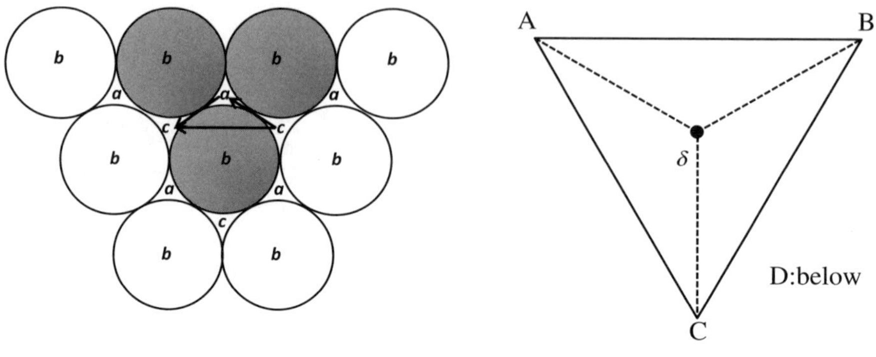

Fig. 8.3 Motion of atoms over {111} plane. (a) Actual motion, (b) represented by Thompson vectors (on the Thompson tetrahedron).

If one of ***abcabc*** is missing, the stacking order becomes ***abababab***, ***bcbcbcb***, or ***cacacaca***. This corresponds to the stacking of (0001) plane in the HCP structure (Fig. 1.1(f)).

8.1.2 *Dissociated dislocations and partial dislocations*

The Burgers vector $\boldsymbol{b}$ in FCC crystals is the minimum translational vector of the unit cell, i.e., $1/2 < 110 >$ and represented by AB, BC, CA, AD, BD and CD on the Thompson tetrahedron.

Suppose that an atom ($\boldsymbol{c}$ ●) on the (b) plane moves leftward ($\leftarrow$) to the adjacent $\boldsymbol{c}$ position (Fig. 8.3(a)). When the atom moves from $\boldsymbol{c}$ to $\boldsymbol{c}$, it must move on the flank of atom $\boldsymbol{b}$. However, in between $\boldsymbol{c}$ and $\boldsymbol{c}$, a stable position $\boldsymbol{a}$ is present, albeit in the side road. Therefore, instead of the non-stop translation of $\boldsymbol{c}\rightarrow\boldsymbol{c}$, it may be easier for the atom to translate via $\boldsymbol{a}$ like $\boldsymbol{c}\rightarrow\boldsymbol{a}\rightarrow\boldsymbol{c}$, taking a rest at $\boldsymbol{a}$. In other words, in the first stage it translates $\boldsymbol{c}\rightarrow\boldsymbol{a}$, followed by translation $\boldsymbol{a}\rightarrow\boldsymbol{c}$.

The dislocation associated with the non-stop translation $\boldsymbol{c}\rightarrow\boldsymbol{c}$ is referred to as a perfect dislocation and let the Burgers vector be denoted by $\boldsymbol{b}_{\mathbf{T}}$. Those associated with 2-stage translation, $\boldsymbol{c}\rightarrow\boldsymbol{a}$ and $\boldsymbol{a}\rightarrow\boldsymbol{c}$, are referred to as Shockley partial dislocations (Strictly speaking, Heidenreich-Shockley partial dislocations [2]) or imperfect dislocations and let them be denoted by p_1, p_2 (p_1 is referred to the leading partial and p_2 the trailing partial), with the respective Burgers vectors $\boldsymbol{b}_{\boldsymbol{p1}}$ and $\boldsymbol{b}_{\boldsymbol{p2}}$. Which of the perfect and the imperfect dislocations are energetically favourable can be judged by comparing their elastic self-energies.

Let us assume that the dissociation is given as follows.

$$\boldsymbol{c} \to \boldsymbol{a} = \boldsymbol{b_{p1}}, \quad \boldsymbol{a} \to \boldsymbol{c} = \boldsymbol{b_{p2}}$$

$$\boldsymbol{b_T} = \boldsymbol{b_{p1}} + \boldsymbol{b_{p2}}. \tag{8.1}$$

Exercise 8.3 Confirm that Eq. (8.1) is expressed in Thompson vectors by

$$BA = B\delta + \delta A \tag{8.2}$$

In summary

$$\text{Slip in stage } 1 = \boldsymbol{c} \to \boldsymbol{a} = \boldsymbol{b_{p1}} = \delta A \text{ (leading partial)}$$
$$\text{Slip in stage } 2 = \boldsymbol{a} \to \boldsymbol{c} = \boldsymbol{b_{p2}} = B\delta \text{ (trailing partial)} \tag{8.3}$$

The elastic energy is proportional to b^2. Neglecting natures of dislocations (edge, screw or mixed) (see Eqs. (4.5) and (4.11)) the first approximation is to compare b_T^2 with $b_{p1}^2 + b_{p2}^2$. As shown in Fig. 8.3(b), the angle between $\boldsymbol{b}_{p1}$ and $\boldsymbol{b}_{p2}$ is 120°, so that obviously

$$b_T^2 > b_{p1}^2 + b_{p2}^2. \tag{8.4}$$

(When $\boldsymbol{b}_{p1}$ and $\boldsymbol{b}_{p2}$ are at right angles, i.e., Pythagoras theorem holds, the right and left sides of Eq. (8.4) are equal.) According to this criterion, i.e., Frank criterion, the imperfect dislocations can be energetically favourable.

8.1.3 *Stacking fault*

Now consider the process by which a dissociated dislocation is generated in more detail on atomic level, using the Thompson tetrahedron shown in Fig. 8.3(b). (111) planes are stacked from below in the order of ***abcabc***. Figure 8.4(a) shows the stacking of $\boldsymbol{c}$ above $\boldsymbol{b}$. The atom $\boldsymbol{c}$ lies at the midpoint of a triangle constructed by the close packing of $\boldsymbol{b}$ with its apex upward like Δ. On the other hand, $\boldsymbol{a}$ lies in a triangle constructed by $\boldsymbol{b}$ with its apex downward like ∇ (Fig. 8.2(b)).

First, consider the slip in stage 1. The dislocation line corresponding to the slip in stage 1 is denoted by a straight line attached with a double-headed arrow. Let the direction of the dislocation line be ξ_1. By definition, if the lower layer of the crystal (i.e., plane $\boldsymbol{b}$) at the right side of ξ_T is fixed,

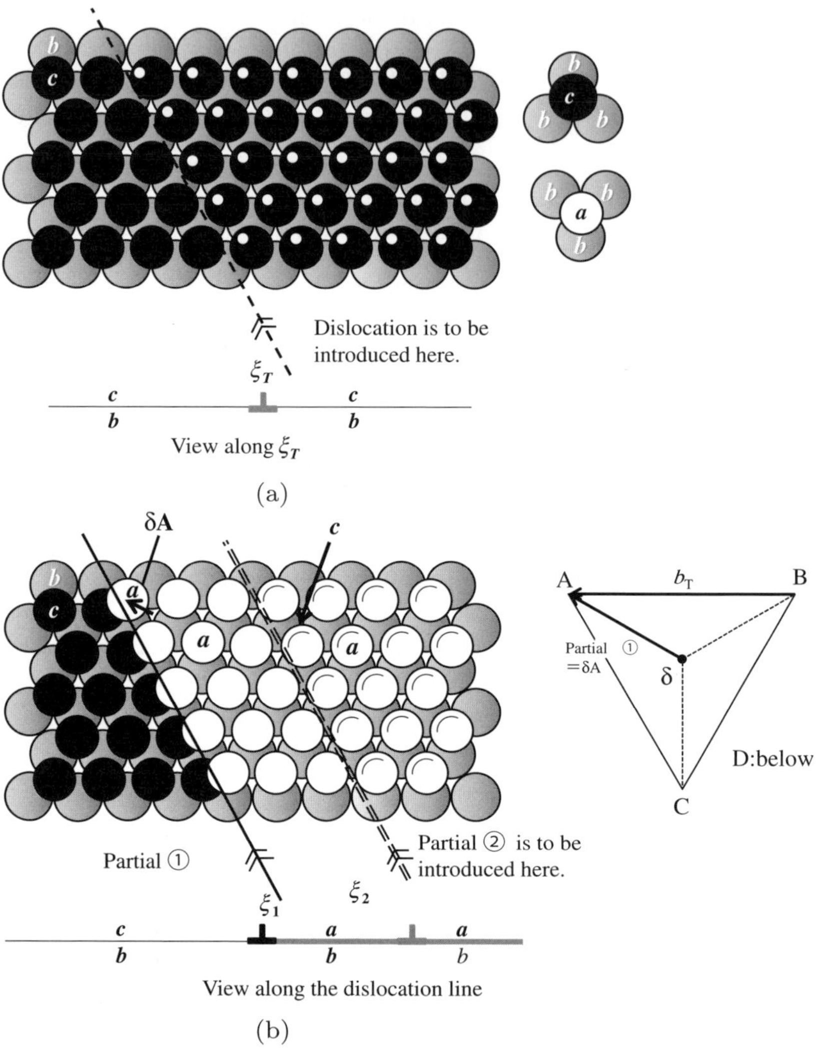

Fig. 8.4 (a)–(d) Motion of a dissociated dislocation. (a) is just before motion. Atoms $\boldsymbol{c}$ (shown in black) at the right side of ξ_T are going to move (illuminated by light from the left), (b) Atoms $\boldsymbol{c}$ in the right side of ξ_1 have moved from $\boldsymbol{c}$ to $\boldsymbol{a}$. Now, atoms in the right side of ξ_2 are going to move, (c) Atoms in the right side of ξ_2 have moved from $\boldsymbol{a}$ to $\boldsymbol{c}$. Between ξ_1 and ξ_2 atoms $\boldsymbol{a}$ sit on $\boldsymbol{b}$. (d) Summary of Burgers vectors of a dissociated dislocation.

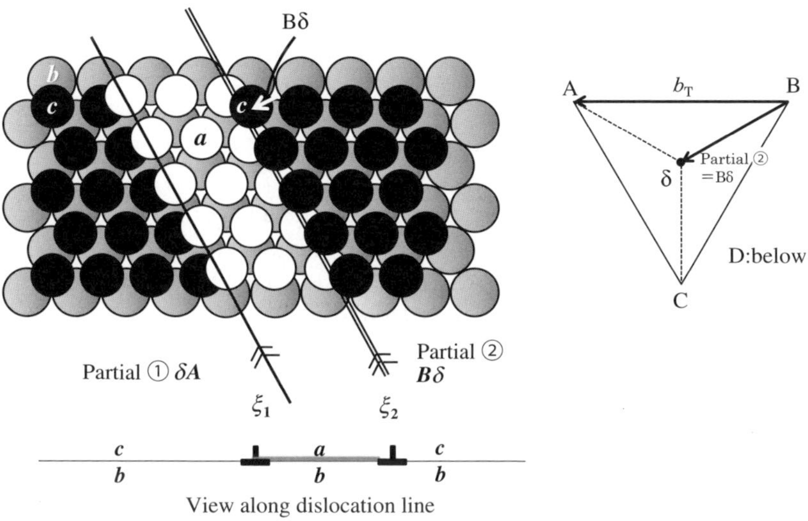

(c)

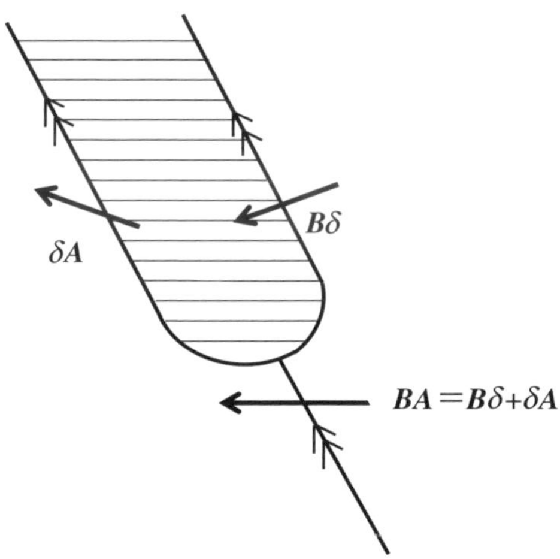

(d)

Fig. 8.4 (*Continued*)

then the upper layer (i.e., ***c*** plane) is displaced by $\boldsymbol{b_{1p}}$ (Fig. 8.4(b)). After the slip, atoms ***b*** at the right side of the dislocation are sitting on ***a***.

Next, let the dislocation line corresponding to the slip in stage 2 be defined by a broken line ξ_2. At its right side, atoms ***a*** (translated from ***c*** in the slip of stage 1) are displaced by $\boldsymbol{b_{p2}}$ back to ***c*** (Fig. 8.4(c)).

As shown in Fig. 8.4(d), in the hatched area bounded by the two partials atoms lie on $\boldsymbol{a}$ instead of its right position $\boldsymbol{c}$. The stacking order of (111) plane is disturbed as follows:

$$\boldsymbol{abc\underline{ab\overset{\downarrow}{a}b}cabc} \tag{8.5}$$

where the stacking indicated by ↓ is disturbed. This is referred to as the stacking fault. The underlined portion has the stacking order of $\boldsymbol{a\underline{ba}b}$... which has the HCP structure. In other words, a very thin layer of HCP is inserted into the region between the two partials comprising the dissociated dislocation. This results in an increase in free energy. This increment in energy is referred to as the stacking-fault energy (SFE), and expressed in energy per unit area (usually indicated by γ or γ_{SFE} with the unit of erg/cm^2 or mJ/m^2).

Thus, the energy of a dissociated dislocation per unit length is given by

$$b_{p1}^2 + b_{p2}^2 + E_{\mathrm{int}} + \gamma\Delta, \tag{8.6}$$

where E_{int} is the elastic interaction energy between partials p_1 and p_2, Δ is the dissociation distance between p_1 and p_2. $b_{p1}^2 + b_{p2}^2 + E_{\mathrm{int}}$ depends on the shear modulus and Poisson ratio but only slightly; it does not differ very much from one crystal to another, so that the SFE γ determines whether the dislocation is dissociated or not.

The same stacking fault as that given by Eq. (8.5) is created by removing one (111) plane from the stacking order $\boldsymbol{abcabcabc}$ (in this case $\boldsymbol{c}$ plane). This type of stacking fault is referred to as the intrinsic stacking fault.

Therefore, if vacancies precipitate on (111) plane in a manner described in Sec. 2.8.1, the intrinsic stacking fault given by Eq. (8.5) is created. This stacking fault is not of glide type. Similarly, if interstitials precipitate on (111) plane, say between $\boldsymbol{b}$ and $\boldsymbol{c}$, the only position available for the interstitials is $\boldsymbol{a}$. Therefore, the stacking order is given by

$$\boldsymbol{abca\underline{b\overset{\downarrow}{\boxed{a}}c}abcabc} \tag{8.7}$$

This type of the stacking fault is referred to as the extrinsic stacking fault. The stacking fault formed by removing or inserting one layer of (111) plane is referred to as the Frank stacking fault. The dislocation loop bounding the Frank SF is referred to as a Frank partial dislocation. The Burgers vector of the Frank partial dislocation is out of the plane on which the dislocation

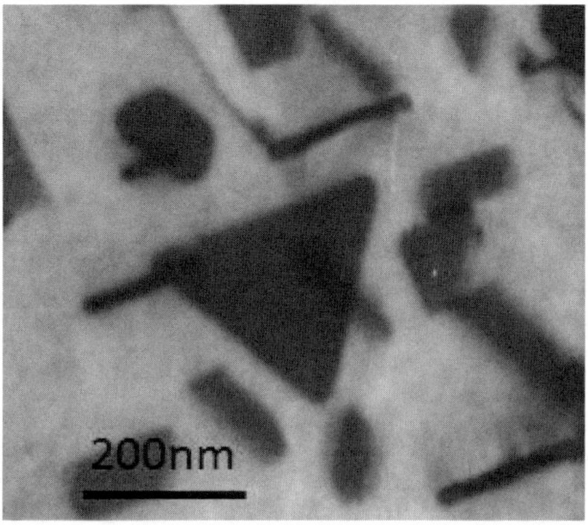

Fig. 8.5 Frank stacking fault formed in Ag by precipitation of interstitials. It lies on (111) plane and viewed from above. Frank partials bounding the stacking fault lie along $\langle 110 \rangle$ (for details see [3]).

lies, so that it cannot glide: It is a sessile dislocation. Figure 8.5 shows a Frank SF formed by precipitation of interstitials.

Stacking orders used in describing the HCP structure are ***abababab***, ***bcbcbcbcb*** and ***cacacacac***. They are equivalent but they may give the impression that they are not equivalent. In order to avoid such a confusion Frank proposed to express the stacking order ***ab bc ca*** by Δ and ***ba cb ac*** by ∇ (Frank symbol).

According to this notation the stacking order of the FCC structure

$$\boldsymbol{abcabcabcabc}$$

is expressed by

$$\Delta\Delta\Delta\Delta\Delta\Delta\Delta\Delta\Delta.$$

For the HCP structure all of the stacking orders ***abababab, bcbcbcbcb*** and ***cacacacaca*** are expressed simply by

$$\Delta\nabla\Delta\nabla\Delta\nabla\Delta \quad \Delta\nabla\Delta\nabla\Delta\nabla\Delta \quad \Delta\nabla\Delta\nabla\Delta\nabla\Delta.$$

The intrinsic SF given by Eq. (8.5) is expressed by

$$\begin{array}{c} \downarrow \\ \boldsymbol{abc\underline{ab\ ab}cabc} \\ \Delta\Delta\Delta\Delta\nabla\Delta\Delta\Delta. \end{array} \tag{8.5'}$$

The extrinsic SF given by Eq. (8.7) is expressed by

$$\downarrow$$

$$\boldsymbol{abca\underline{bac}abcabc} \qquad (8.7')$$

$$\Delta\Delta\Delta\Delta\nabla\nabla\Delta\Delta\Delta.$$

Exercise 8.4 Discuss why the extrinsic stacking fault has a higher energy than the intrinsic one.

8.1.4 *Burgers vectors of dissociated dislocations: Thompson vectors*

It is convenient and useful to use the Thompson vectors for expressing Burgers vectors of a dissociated dislocation. Let us consider the Thompson tetrahedron again in more detail (Fig. 8.1).

① Four equilateral triangles comprising the Thompson tetrahedron are $\{111\}$ planes, each of which being labelled (a), (b), (c) and (d) on the outside surface, respectively and $(\bar{a})$, $(\bar{b})$, $(\bar{c})$ and $(\bar{d})$ on the inside surface.

The edges of Thompson tetrahedron, AB, BC, CA, AD, BD and CD represent the Burgers vectors of perfect dislocations, i.e., $1/2\langle 110\rangle$.

② The midpoints of the equilateral triangles (a), (b), (c) and (d) are labelled α, β, γ and δ.

Those vectors such as $A\delta$ and $B\delta$ that lie on one $\{111\}$ plane represent Burgers vector of a Shockley partial dislocation, i.e., $1/6\langle 112\rangle$.

③ For instance when AB is dissociated into Shockley partials on (d),

$$AB = A\delta + \delta B(d).$$

δ indicates that this dissociation takes place on (d) plane.

④ $A\alpha$, $B\beta$, $C\gamma$ and $D\delta$ stand for the Burgers vectors of Frank partials, i.e., $\boldsymbol{b} = 1/3\langle 111\rangle$.

Exercise 8.5 Express the dissociation $AB = A\delta + \delta B$ by conventional vectors such as $1/2[110]$.

Using Thompson vectors provides quite a reduction of labor in considering complex arrays of dissociated dislocations. The vector algebra for the addition and subtraction of the Thompson vectors is defined by the set of relations:

$$
\begin{aligned}
&\mathbf{PQ} = -\mathbf{QP} \\
&\frac{\mathbf{PQ}}{\mathbf{RS}} = -\frac{\mathbf{RS}}{\mathbf{PQ}} \\
&\frac{\mathbf{PQ}}{\mathbf{RS}} = \frac{\mathbf{QP}}{\mathbf{RS}} = \frac{\mathbf{PQ}}{\mathbf{SR}} = \frac{\mathbf{QP}}{\mathbf{SR}} \\
&\mathbf{PQ} + \mathbf{QR} = \mathbf{PR} \\
&\mathbf{PQ} + \mathbf{RS} = \frac{\mathbf{PR}}{\mathbf{QS}}
\end{aligned}
\tag{8.8}
$$

where **PQ** represents a vector connecting P to Q (P→Q). **PQ/RS** represents a vector twice the length of the vector connecting the midpoint of PQ to the midpoint of RS (Fig. 8.6).

Exercise 8.6 Indicate AB/CD and $AB/\delta\gamma$ in concrete indices and confirm them on Thompson tetrahedron.

{111} planes comprising the Thompson tetrahedron are crystallographically equivalent. But throughout this book (d) plane, i.e., Δ ABC will be treated as representing {111} plane. It is noted that the rotation of A→B→C is clockwise.

When the Thompson tetrahedron is viewed with the apex D downwards, the Thompson tetrahedron is viewed from the outside and Δ ABC is labelled (d) plane (i.e., (111) plane defined by Miller index) (Fig. 8.7(a)).

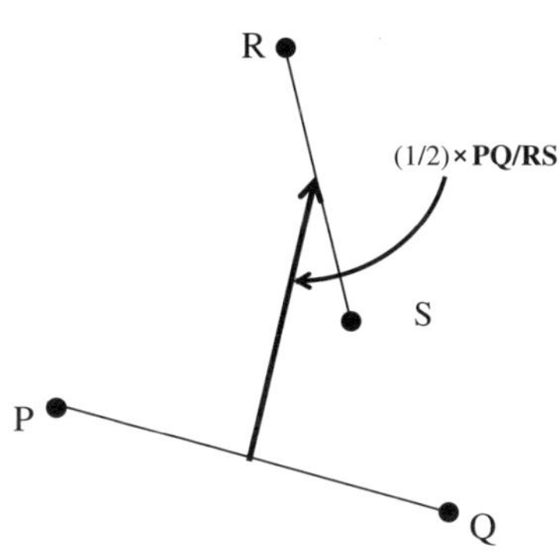

Fig. 8.6 Definition of PQ/RS.

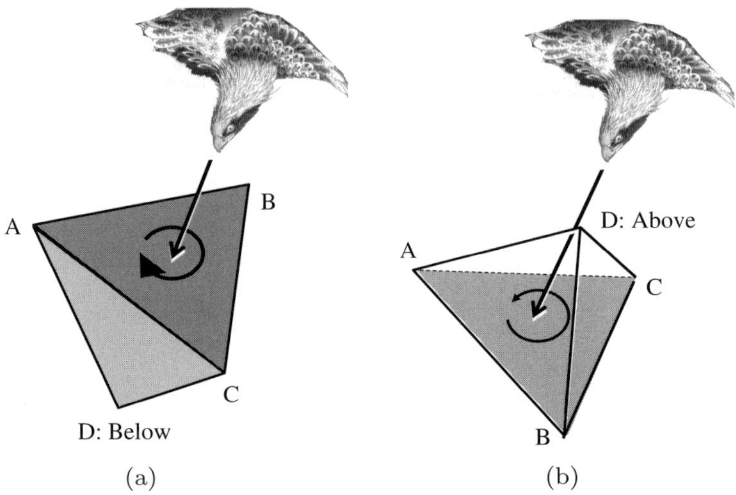

Fig. 8.7 Viewing (d) plane (ΔABC) of the Thompson tetrahedron from the outside (a) and from the inside (b). → indicates the rotating order of ABC, which is right-handed (clockwise) in (a) and left-handed (counterclockwise) in (b).

When the Thompson tetrahedron is viewed with D upwards (Fig. 8.7(b)), it is viewed from the inside. In this case, the order of A→B→C is counterclockwise and ΔABC is labelled ($\bar{d}$) plane (i.e., ($\bar{1}\bar{1}\bar{1}$) plane defined by Miller index). Similar rules are applied to (a), (b) and (c) planes.

In Fig. 8.4 the Thompson tetrahedron is viewed from the outside and $\boldsymbol{b_T} = BA$, where

$$\text{Slip in stage } 1 = \boldsymbol{c} \to \boldsymbol{a} = \boldsymbol{b}_{p1} = \boldsymbol{\delta A}$$

$$\text{Slip in stage } 2 = \boldsymbol{a} \to \boldsymbol{c} = \boldsymbol{b}_{p2} = \boldsymbol{\delta B}$$

As is obvious from Fig. 8.4(d), viewing the dislocations from the outside of the Thompson tetrahedron, the Greeks are located outside and the Romans are inside along the positive sense of the dislocation line.

Exercise 8.7 What about when $\boldsymbol{b}_T = AB$? (Hint) Fig. 8.8.

In Fig. 8.9(b), viewing the Thompson tetrahedron from the outside, the Greeks are located outside along the dislocation line. Therefore, the Burgers vector of partial ①(p_1)$\boldsymbol{b_{p1}} = \boldsymbol{A\delta}$, and that of partial ②($p_2$)$\boldsymbol{b_{p2}} = \delta\boldsymbol{B}$. The Thompson tetrahedron in Fig. 8.9(b) is rotated along AB-axis

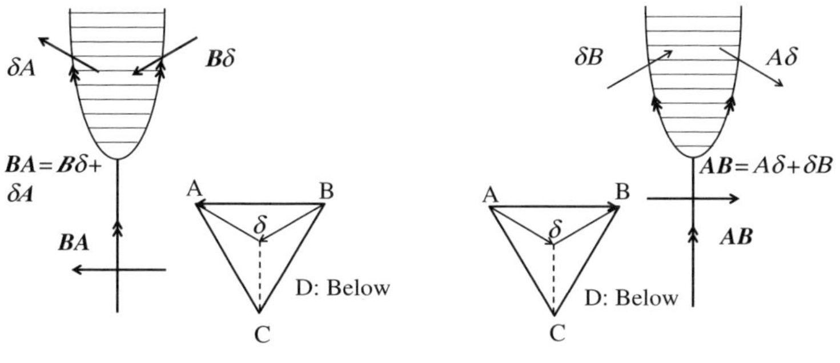

(a) (b)

Fig. 8.8 (a) (b).

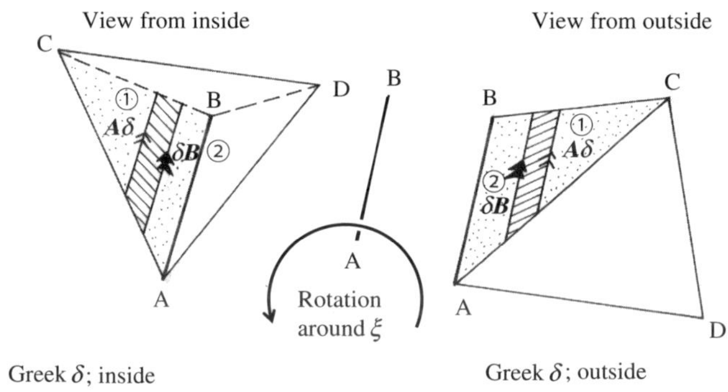

(a) (b)

Fig. 8.9

counterclockwise to Fig. 8.9(a), where the Thompson tetrahedron is now viewed from the inside. Here, partial ①$(p_1)\boldsymbol{b}_{p1} = \boldsymbol{A\delta}$ is at the left side and partial ②$(p_2)\boldsymbol{b}_{p2} = \boldsymbol{\delta B}$ is at the right side. That is, the Greeks are located inside.

> **Hirth–Lothe's Axiom 1**
> Suppose that a dissociated dislocation accompanies **the intrinsic stacking fault**.
> Viewing from the **outside** of the Thompson tetrahedron the **Greeks** are **outside** along the positive sense of the dissociated dislocation (Fig. 8.4(a)).

Viewing from the **inside** of the Thompson tetrahedron the **Greeks** are **inside** along the positive sense of the dissociated dislocation (Fig. 8.4(b)).
When a dissociated dislocation accompanies **the extrinsic stacking fault**, the reverse is true.
Viewing from the **outside** of the Thompson tetrahedron the **Greeks** are **inside** along the positive sense of the dissociated dislocation.
Viewing from the **inside** of the Thompson tetrahedron the **Greeks** are **outside** along the positive sense of the dissociated dislocation.

8.2 Dissociation Distance

The dissociation distance or the extended width is determined by a balance between an elastic repulsive force between the two partials and an attractive force due to the stacking fault. First consider the elastic repulsive force.

Consider two parallel dislocations ① and ② lying in parallel to z-axis in the $x-z$ plane. Let dislocation ① pass through the origin and dislocation ② pass through coordinate (x, y) (Fig. 8.10(a)). Burgers vectors of dislocations ① and ②, i.e., $\boldsymbol{b}_1, \boldsymbol{b}_2$ are inclined by 120° (Fig. 8.10(b)).

$$\boldsymbol{b}_1 = (b \sin \alpha \boldsymbol{i},\ 0\boldsymbol{j}, b \cos \alpha \boldsymbol{k})$$
$$\boldsymbol{b}_2 = (b \sin \beta \boldsymbol{i},\ 0\boldsymbol{j}, b \cos \beta \boldsymbol{k})$$

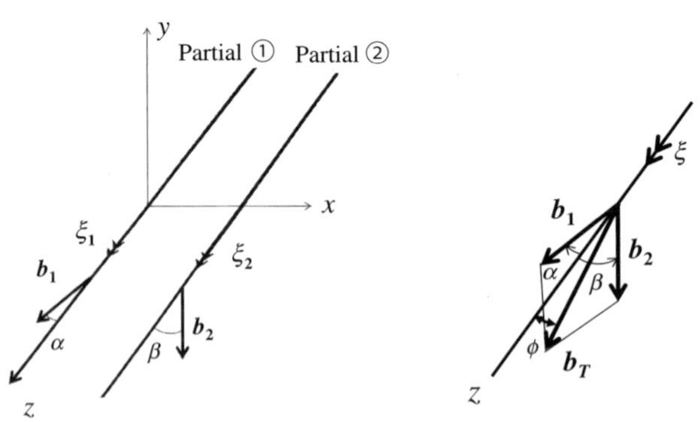

(a) (b)

Fig. 8.10 (a), (b) Calculation of a force acting between partial dislocations.

$$\xi_1 = (0\boldsymbol{i}, 0\boldsymbol{j}, 1\boldsymbol{k}),$$
$$\xi_2 = (0\boldsymbol{i}, 0\boldsymbol{j}, 1\boldsymbol{k}).$$

Substitute these into Peach-Koehler equation (Eq. (4.21))

$$\frac{\boldsymbol{F}_{1\to 2}}{L} = \left(\boldsymbol{b}_2 \cdot \sum_1\right) \times \xi_2 \qquad (4.21)\ \text{(again)}$$

where Σ_1 is the tensor representing the stress field of dislocation ① which lies along z-axis.

$$\sum_1 = \begin{pmatrix} \sigma_{11}^{1e}\boldsymbol{ii} & \sigma_{12}^{1e}\boldsymbol{ij} & \sigma_{13}^{1s}\boldsymbol{ik} \\ \sigma_{21}^{1e}\boldsymbol{ji} & \sigma_{22}^{1e}\boldsymbol{jj} & \sigma_{23}^{1s}\boldsymbol{jk} \\ \sigma_{31}^{1s}\boldsymbol{ki} & \sigma_{32}^{1s}\boldsymbol{kj} & \sigma_{33}^{1e}\boldsymbol{kk} \end{pmatrix}.$$

Here, superscript 1 refers to dislocation ①, and superscript e and s refer to edge and screw components, respectively (see Eq. (4.16)).

$$\left(\boldsymbol{b}_2 \cdot \sum_1\right) = (b\sin\beta\boldsymbol{i}\ \ \boldsymbol{0j}\ \ b\cos\beta\boldsymbol{k}) \begin{pmatrix} \sigma_{11}^{1e}\boldsymbol{ii} & \sigma_{12}^{1e}\boldsymbol{ij} & \sigma_{13}^{1s}\boldsymbol{ik} \\ \sigma_{21}^{1e}\boldsymbol{ji} & \sigma_{22}^{1e}\boldsymbol{jj} & \sigma_{23}^{1s}\boldsymbol{jk} \\ \sigma_{31}^{1s}\boldsymbol{ki} & \sigma_{32}^{1s}\boldsymbol{kj} & \sigma_{33}^{1e}\boldsymbol{kk} \end{pmatrix}$$

$$= (b\sin\beta\sigma_{11}^{1e} + b\cos\beta\sigma_{31}^{1s})\boldsymbol{i} + (b\sin\beta\sigma_{12}^{1e} + b\cos\beta\sigma_{32}^{1s})\boldsymbol{j}$$
$$+ (b\sin\beta\sigma_{13}^{1s} + b\cos\beta\sigma_{33}^{1e})\boldsymbol{k}$$

$\therefore$

$$\frac{\boldsymbol{F}_{1\to 2}}{L} = \begin{pmatrix} \left(b\sin\beta\sigma_{11}^{1e} + b\cos\alpha\sigma_{31}^{1s}\right)\boldsymbol{i}, \left(b\sin\beta\sigma_{12}^{1e} + b\cos\beta\sigma_{32}^{1s}\right)\boldsymbol{j}, \\ \left(b\sin\beta\sigma_{13}^{1s} + b\cos\beta\sigma_{33}^{1e}\right)\boldsymbol{k} \end{pmatrix}$$
$$\times \left(0\boldsymbol{i},\quad 0\boldsymbol{j},\quad 1\boldsymbol{k}\right)$$
$$= \left(b\sin\beta\sigma_{12}^{1e} + b\cos\beta\sigma_{32}^{1s}\right)\boldsymbol{i} - \left(b\sin\beta\sigma_{11}^{1e} + b\cos\beta\sigma_{31}^{1s}\right)\boldsymbol{j}.$$

Both dislocations ① and ② are in the $x - z$ plane, so that $y = 0$.

$$\frac{\boldsymbol{F}_{1\to 2}}{L} = (b\sin\beta\sigma_{12}^{1e} + b\cos\beta\sigma_{32}^{1s})\boldsymbol{i}$$
$$= \frac{b}{2\pi}\left[\sin\beta\frac{\mu b\sin\alpha}{(1-\nu)\,x} + \cos\beta\frac{\mu b\cos\alpha}{x}\right]\boldsymbol{i}$$
$$= \frac{\mu b^2}{2\pi x}\left(\frac{\sin\alpha\sin\beta}{1-\nu} + \cos\alpha\cos\beta\right)\boldsymbol{i}.$$

Let the Burgers vector $\boldsymbol{b}_T$ of the perfect dislocation inclined with respect to the dislocation direction ξ by φ,

$$\alpha = \varphi - \frac{\pi}{6}$$

$$\beta = \varphi + \frac{\pi}{6}$$

so that the force which dislocation ① exerts on dislocation ②, along x direction, is given by

$$\frac{F_x}{L} = \frac{\mu b_p^2 (2-\nu)}{8\pi (1-\nu) x}\left(1 - \frac{2\nu \cos\varphi}{2-\nu}\right). \tag{8.9}$$

When this repulsive force is balanced with an attractive force due to the SFE (γ), x assumes the equilibrium value Δe. That is,

$$\Delta_e = \frac{\mu b_p^2 (2-\nu)}{8\pi(1-\nu)\gamma}\left(1 - \frac{2\nu\cos\varphi}{2-\nu}\right) \tag{8.9$'$}$$

Exercise 8.8 Show that the attractive *force* caused by the SFE is γ, and derive Eq. (8.9).

Exercise 8.9 Show that the dissociation distance in edge orientation is about three times larger than that in screw orientation. Figure 8.11 shows dissociated dislocations in a Cu–13.43at.%Al alloy.

The most reliable method to measure the SFE (γ) is to measure the dissociation width by means of transmission electron microscopy (TEM) and calculate γ using Eq. (8.9′) (Fig. 8.12). Eq. (8.9′) is valid within the framework of isotropic elasticity theory and for more strict discussion, analysis using anisotropic elasticity theory is to be used (Sec. 4.5).

8.3 Jogs on Dissociated Dislocations and Stair-rod Dislocations

8.3.1 *Stair-rod dislocations*

Bends of a dislocation in the slip plane are referred to as kinks (Fig. 8.13(a)). Bends out of the slip plane are referred to as jogs (Fig. 8.13(b)).

Geometry of jogs on a dissociated dislocation is rather complex (see Fig. 8.14). Consider a jog on dislocation of $\boldsymbol{b} = BA$ dissociated on (d) plane. First let the jog be contracted, i.e., a perfect dislocation. Then, the perfect jog is dissociated on one of {111} planes, say, on (c) plane. Taking the

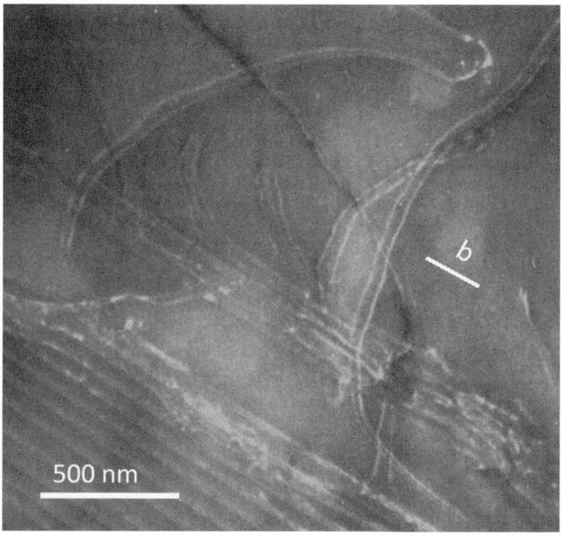

Fig. 8.11 Dissociated dislocations in a Cu–13.43at.%Al alloy (Reproduced with permission from [4]).

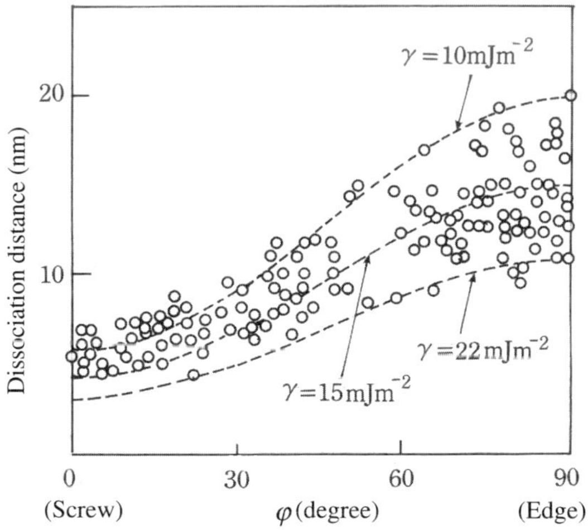

Fig. 8.12 Dissociation distance versus dislocation orientation.

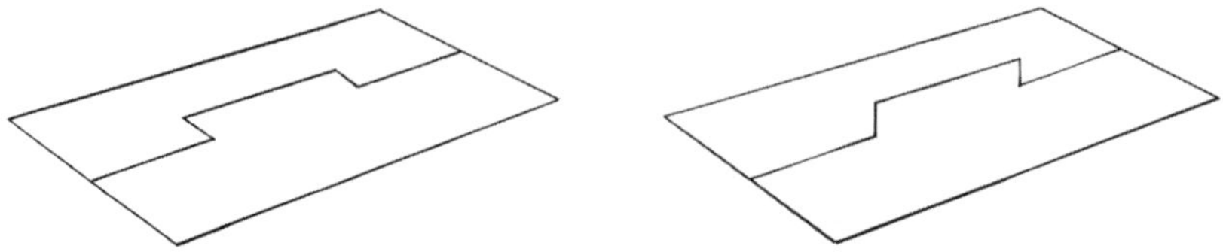

(a) (b)

Fig. 8.13 (a) kink. (b) jog.

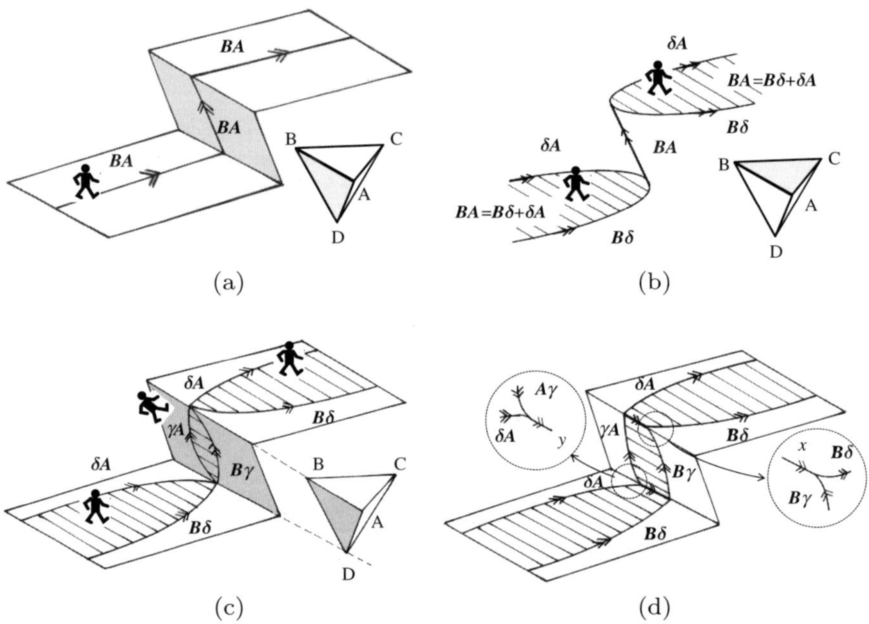

Fig. 8.14 A stair-rod dislocation on the extended jog (I).

direction of the dislocation as shown in Fig. 8.14, there are two possibilities for the direction of the jog, i.e., upward or downward.

First consider the upward case (Fig. 8.14(a)). Place the Thompson tetrahedron in such a way that D is below the tetrahedron. Suppose that you shrink and walk on the stacking fault along the dislocation line. On the (d) plane below the riser (jog) you view the Thompson tetrahedron from the outside with D below ABC, so that according to Axiom 1,

$\boldsymbol{BA} = \boldsymbol{B\delta}+\boldsymbol{\delta A} = \boldsymbol{\delta A}$ (left side of the viewer) $+\boldsymbol{B\delta}$ (right side of the viewer) with the Greeks of the two partials outside (Fig. 8.14(b)).

Now let the dislocation be a 60° dislocation (the angle between ξ and $\boldsymbol{b}$ is 60°). When you reach the jog, you must climb the cliff of (c) plane. In this case you view the (c) plane from the outside of the tetrahedron, so that the jog is dissociated with the Greeks outside as follows:

$$\boldsymbol{BA} = \boldsymbol{B\gamma} + \boldsymbol{\gamma A} = \boldsymbol{\gamma A}(\text{left}) + \boldsymbol{B\gamma}(\text{right})$$
$$(\text{Fig. 8.14(c)}).$$

Now let the nodes of jog be dissociated, then a pair of dislocations is generated along the riser as shown in Fig. 8.14(d). These dislocations are referred to as stair-rod dislocations.[5] Burgers vectors of the stair-rod dislocations are given by applying Kirchhoff's law. Applying the Kirchhoff's law to node $\boldsymbol{x}$ in Fig. 8.14(d),

$$\boldsymbol{x} + \boldsymbol{B\gamma} = \boldsymbol{B\delta}.$$

Therefore,

$$\boldsymbol{x} = \boldsymbol{B\delta} - \boldsymbol{B\gamma} = \boldsymbol{B\delta} + \boldsymbol{\gamma B} = \boldsymbol{\gamma B} + \boldsymbol{B\delta} = \boldsymbol{\gamma\delta}.$$

Exercise 8.10 Apply Kirchhoff's law to node $\boldsymbol{y}$ and show that $\boldsymbol{b}$ is $\boldsymbol{\delta\gamma}$.

If you go down the stair as shown in Fig. 8.15, a pair of stair-rod dislocations shown in Fig. 8.15(c) is created. Irrespective of whether you go down from the left to the right $((a) \to (b) \to (c))$ or you go up from the right to the left $((a') \to (b') \to (c))$, the result shown in Fig. 8.15(c) is reached.

Exercise 8.11 Prove this. (Hint) Reversing the sign of ξ results in reversing the sign of $\boldsymbol{b}$.

Stair rod dislocations shown in Figs. 8.14 and 8.15 are dipoles with positive (+) and negative (−) Burgers vectors. The stress field of a dipole is not long ranged because those of the component dislocations comprising the dipole cancel each other.

8.3.2 *Jogs on a screw dislocation*

Geometry of jogs on a screw dislocation is much more complex. A screw dislocation of $\boldsymbol{BA}$ is dissociated on (d) as follows:

$$\boldsymbol{BA} = \boldsymbol{B\delta} + \boldsymbol{\delta A} \tag{8.10}$$

[5] A stair-rod is a metallic rod holding stair carpet down at each stair angle.

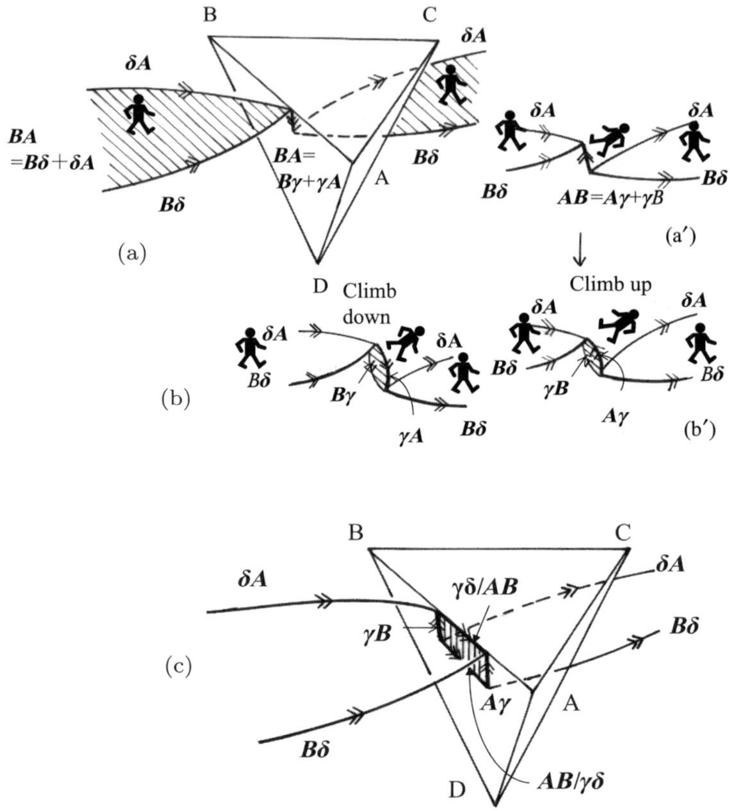

Fig. 8.15 A stair-rod dislocation on the extended jog (II).

Consider a jog shown in Fig. 8.16. You go rightward along the dislocation line. For the sake of simplicity, let the jog be constricted.

A jog on a screw dislocation has an edge component and hence accompanies an ehp. Thus, the jog must move non-conservatively (see 2.9.1) when the screw dislocation moves, either absorbing or emitting point defects.

Exercise 8.12 Suppose that the screw dislocation in Fig. 8.16 moves leftward (backward) emitting point defects. Identify the nature of the point defect (I or V).

Now let the jog be dissociated. Candidates for the riser of the stair are (a), (b) and (c). Dislocation of $\boldsymbol{b} = \boldsymbol{BA}$ is dissociated on the respective

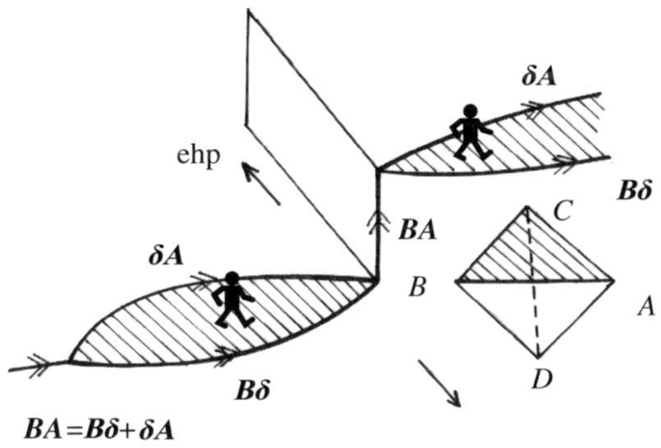

Fig. 8.16 Jog on dissociated screw dislocation.

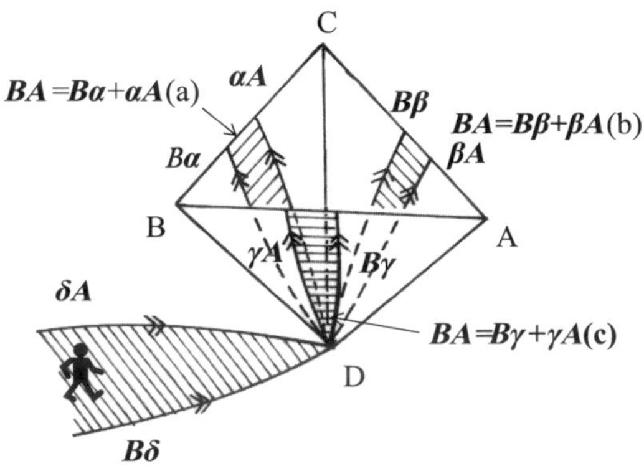

Fig. 8.17 Possible configurations of a jog on a screw dislocation.

planes as follows:

$$(a)\quad \boldsymbol{BA} = \boldsymbol{B\alpha} + \underline{\boldsymbol{\alpha A}} \tag{8.11a}$$

$$(b)\quad \boldsymbol{BA} = \underline{\boldsymbol{B\beta}} + \boldsymbol{\beta A} \tag{8.11b}$$

$$(c)\quad \boldsymbol{BA} = \boldsymbol{B\gamma} + \boldsymbol{\gamma A} \tag{8.11c}$$

as shown in Fig. 8.17. Among them the extended jog on (c) plane is composed of Shockley type alone and essentially the same as that shown in Fig. 8.10(c). However, when dissociated on (a) or (b) planes, Frank partial of $\alpha\boldsymbol{A}$ or $\boldsymbol{B}\beta$ is required to be generated (underlined in Eqs. 8.11(a) and 8.11(b)).

Now consider whether or not Hirth–Lothe's Axiom 1 (see 8.1.3) can be applied to such an extended dislocation consisting of Frank and Shockley partials.

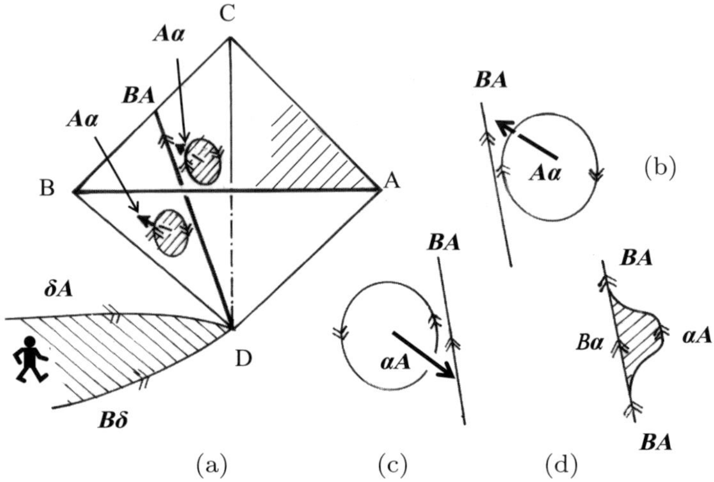

Fig. 8.18 Hirth–Lothe's Axiom II.

Suppose that a dislocation of $\boldsymbol{b} = \boldsymbol{BA}$ climbs on ($\bar{a}$) plane (Fig. 8.18(a)). Let a loop of Frank partial of vacancy type be located just to the right (b) or left (c) side of dislocation $\boldsymbol{BA}$ as shown in Fig. 8.18. Because of vacancy type, the relation of $\boldsymbol{b}$ and ξ of the Frank loop is that of a right handed screw (see Fig. 2.29(b)). If dislocation $\boldsymbol{BA}$ reacts with a Frank loop on its left, the resultant Burgers vector will be $\boldsymbol{BA} + \alpha\boldsymbol{A}$, which is quite large. On the other hand, if $\boldsymbol{BA}$ reacts with a Frank loop at its right side, the resultant Burgers vector will be $\boldsymbol{BA} + \boldsymbol{A\alpha} = \boldsymbol{B\alpha}$ (Fig. 8.18(d)), where $\boldsymbol{B\alpha}$ is a Shockley partial. In other words, reacting with a Frank loop at its right side, $\boldsymbol{BA}$ is dissociated into $\boldsymbol{B\alpha}$(left) $+\boldsymbol{\alpha A}$ (right) with the Greeks inside.

Now let us view (a) plane from the outside, then the Greeks are outside. This implies that Axiom 1 can be applied to a dissociated dislocation consisting of Frank and Shockley partials.

> **Hirth–Lothe's Axiom II**
> Axiom 1 can be applied to the dissociation consisting of Frank and Shockley partials.

Now consider motion of a screw dislocation with an extended jog. Figure 8.19(a) shows an acute jog extended on (a) plane on a screw dislocation $\boldsymbol{b} = \boldsymbol{BA}$. $\boldsymbol{B\alpha}$ is Shockley and $\boldsymbol{\alpha A}$ is Frank partial. When a dissociated dislocation of $\boldsymbol{BA} = \boldsymbol{B\delta} + \boldsymbol{\delta A}$ is gliding on (d) plane, Shockley

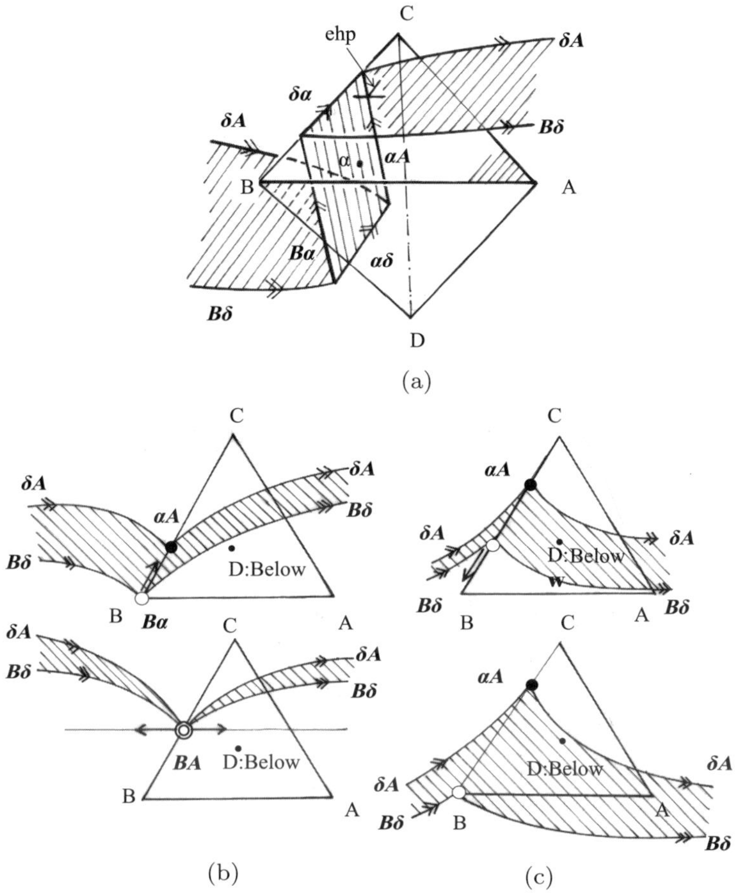

Fig. 8.19 (a)–(c) Motion of a jog along a dissociated screw dislocation.

partial $\boldsymbol{B\alpha}$ can glide on (a) plane, but Frank partial $\boldsymbol{\alpha A}$ cannot glide without absorbing or emitting point defects. When $\boldsymbol{BA}$ moves backwards, $\boldsymbol{B\alpha}$ glides on (a) plane to react with $\boldsymbol{\alpha A}$, and the jog will be constricted as shown in Fig. 8.19(b). The constricted jog now can move conservatively along the dislocation line, during which the jog grows in height when it meets another jog with the same sign, or vanishes when it meets another jog with the opposite sign. On the other hand, when $\boldsymbol{BA}$ moves forwards, $\boldsymbol{B\alpha}$ can glide forwards as well, but $\boldsymbol{\alpha A}$ is left behind, resulting in widening of the extended jog as shown in Fig. 8.19(c). In order for $\boldsymbol{\alpha A}$ to move, either vacancies (V) are emitted or interstitials (I) must be absorbed. Since the concentration of I is low, V must be emitted.

8.3.3 *Stacking fault tetrahedron*

At thermal equilibrium the number of the point defects is given by

$$c = c_o \exp\left(-\frac{E_f}{kT}\right),$$

where C_o is the total number of the lattice sites, k is the Boltzmann constant, T is the absolute temperature and E_f is formation energy of point defect, $\sim 1\,\text{eV}$ for vacancies and $\sim 5\,\text{eV}$ for interstitials.

Exercise 8.13 Compare concentrations of V and I. Calculate the concentration of V at 300 K, 1000 K and 1500 K.

The equilibrium concentration of vacancies increases exponentially with temperature. Therefore, if a crystal is cooled rapidly (quenched) from a high temperature, vacancies are supersaturated and some of them coagulate to form Frank loops (Fig. 8.5 shows an example.).

Suppose that a Frank loop with $\boldsymbol{b} = \boldsymbol{\delta D}$ of vacancy type (with the intrinsic stacking fault) is formed as shown in Fig. 8.20(a). Along the edges CA and AB of the Thompson tetrahedron, $\delta\boldsymbol{D}$ can be dissociated as follows (Fig. 8.20(b)):

$$\boldsymbol{\delta D} = \boldsymbol{\delta\beta} + \boldsymbol{\beta D}$$
$$\boldsymbol{\delta D} = \boldsymbol{\delta\gamma} + \boldsymbol{\gamma D}$$

Here, $\boldsymbol{\delta\beta}$ is a stair-rod dislocation formed along the intersection of (d) and (b) planes and $\boldsymbol{\delta\gamma}$ is a stair-rod formed along the intersection of (d) and (c) planes. $\boldsymbol{\beta D}$ and $\boldsymbol{\gamma D}$ are Shockley partials and can glide on (b) and (c) planes, respectively, and react to form a stair-rod dislocation $\boldsymbol{\gamma\beta}$ (Fig. 8.20 (b′)).

Exercise 8.14 Show that stair-rod dislocations are formed along CD and BD in a similar way.

Eventually a tetrahedron covered completely with four surfaces of the stacking fault is formed (Fig. 8.20(c)). This is referred to as a stacking fault tetrahedron (SFT). Figure 8.21 shows TEM micrographs of a SFT.

A SFT can be formed by deformation. Figure 8.22 shows an example of formation processes of a SFT. A dislocation of $\boldsymbol{b} = \boldsymbol{BD}$ moving rightward on $(\bar{a})$ plane crosses another dislocation of $\boldsymbol{b} = \boldsymbol{AC}$, resulting in formation of a jog on dislocation $\boldsymbol{AC}$ (Fig. 8.22(a)). The jog is dissociated on (b) plane as $\boldsymbol{BD} = \boldsymbol{B\beta} + \boldsymbol{\beta D}$. Concurrently dislocation segment $\boldsymbol{BD}$ at the

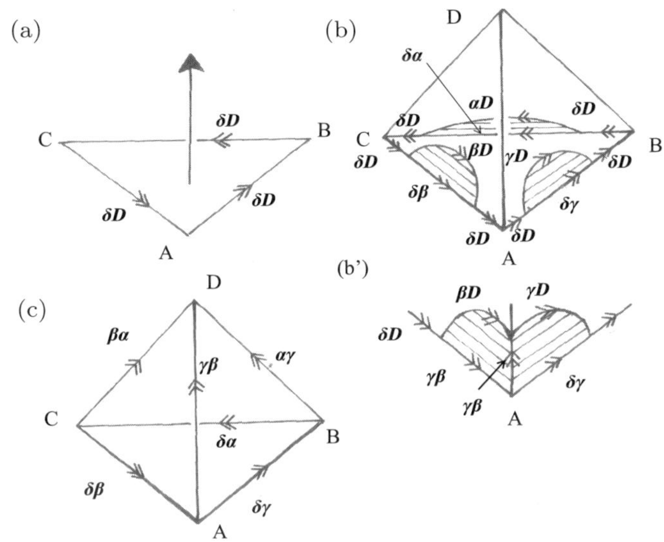

Fig. 8.20 (a)–(c) Formation of a SFT.

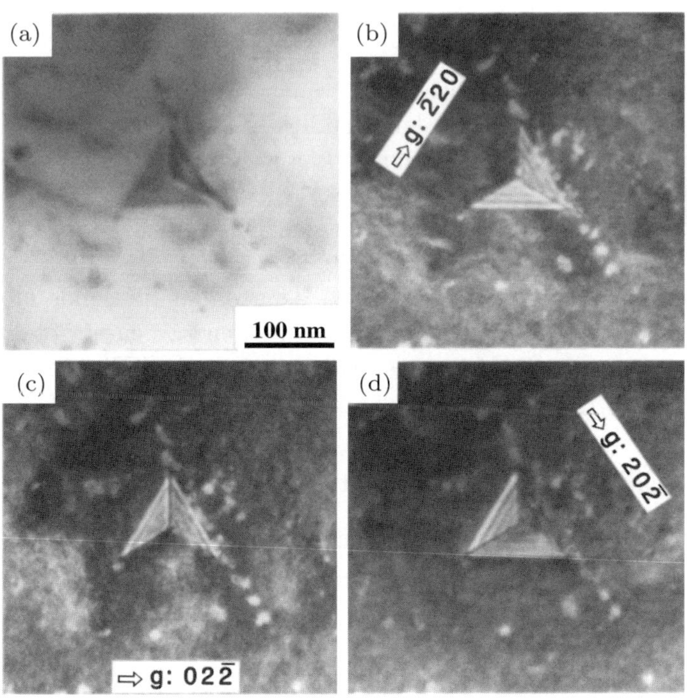

Fig. 8.21 SFT in Si quenched from the liquid, imaged in different reflection vectors **g**, viewed from above (Reproduced with permission from [5]).

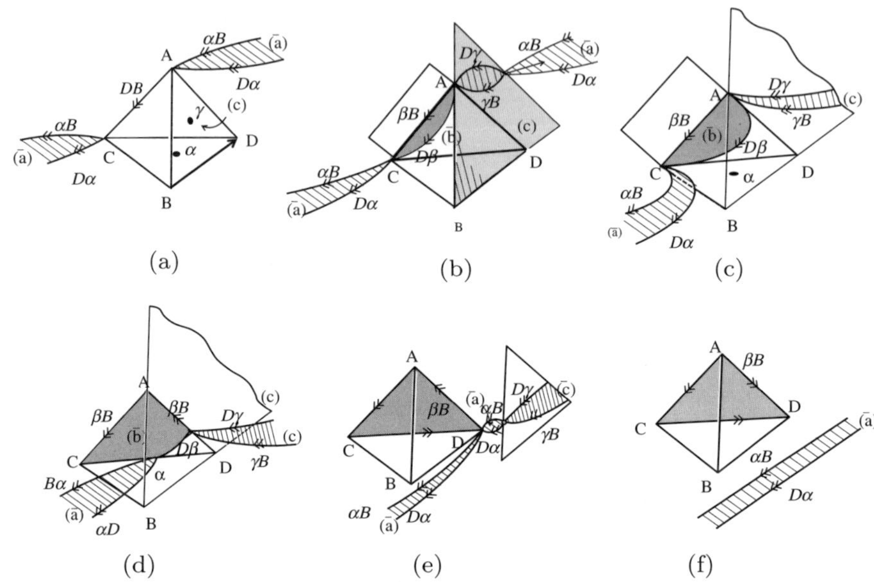

Fig. 8.22 (a)–(f). Formation of a SFT by the motion of a dislocation (Reproduced with permission from [6]). (b)–(e) are added with permission of the authors.

right side of the jog cross-slips onto (c) plane (Fig. 8.22 (b)). Cross-slipped dislocation $\boldsymbol{BD}$ glides on (c) plane and forms a stair-rod dislocation $\boldsymbol{B\beta}$ at the intersection of (c) and (b) planes (Fig. 8.22 (d)). When dislocation $\boldsymbol{BD}$ which cross-slips onto (c) plane reaches D, it cross-slips back to $(\bar{a})$ plane (Fig. 8.22 (e)). A Frank loop $\boldsymbol{B\beta}$ thus left behind transforms itself into a SFE by the process shown in Fig. 8.20.

Whether or not a SFT is formed depends on the kinds of stacking fault (intrinsic or extrinsic), the stacking fault energy, energy of a stair-rod dislocation and an external stress. However, all the SFTs observed so far are composed of the intrinsic stacking fault (vacancy type).

8.4 Reactions between Dissociated Dislocations

8.4.1 *Dislocation dissociated on the same slip plane*

When two dislocations of different Burgers vectors $(\boldsymbol{b}_1, \boldsymbol{b}_2)$ react, a dislocation network is formed. The Burgers vector of the product dislocation $\boldsymbol{b}_3$ follows Kirchhoff's law.

Consider a reaction between $\boldsymbol{b}_1 = CA$ and $\boldsymbol{b}_2 = CB$ on (d) plane. As a result, dislocation of $\boldsymbol{b}_3 = AB$ is formed as shown in Fig. 8.23(a).

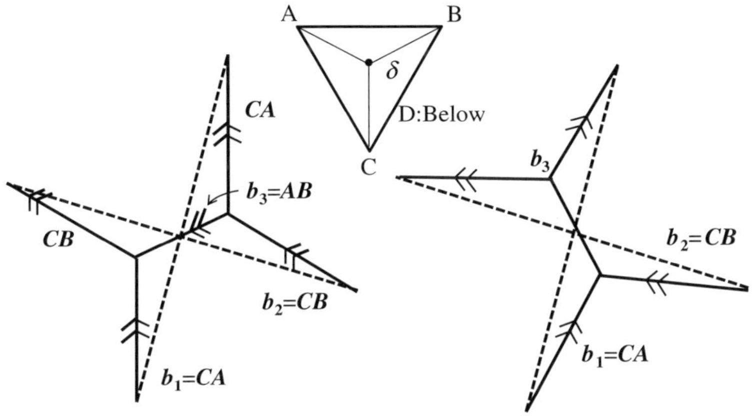

(a) (b)

Fig. 8.23 Reactions between two dislocations.

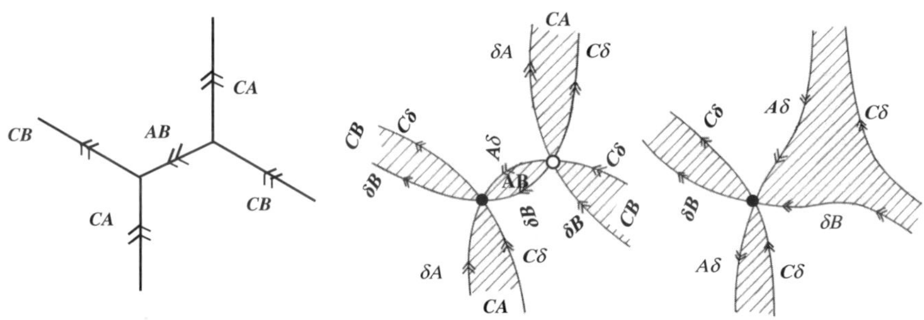

(a) (b) (c)

Fig. 8.24 Reactions between two dissociated dislocations.

Exercise 8.15 Show that the reaction shown in Fig. 8.23(b) does not take place.

Now, suppose that a dislocation network composed of three perfect dislocations is formed (Fig. 8.24(a)). When these three perfect dislocations are dissociated on (d) plane, two kinds of three-fold nodes (● and ○) are formed as shown in Fig. 8.24(b). At the node ○, Shockley partials are continuously connected, while at the node ● the respective Shockley partials cross each other, so that they meet at one point. As can be seen from Fig. 8.24(c), the nodes ● and ○ are formed alternately. Figure 8.25

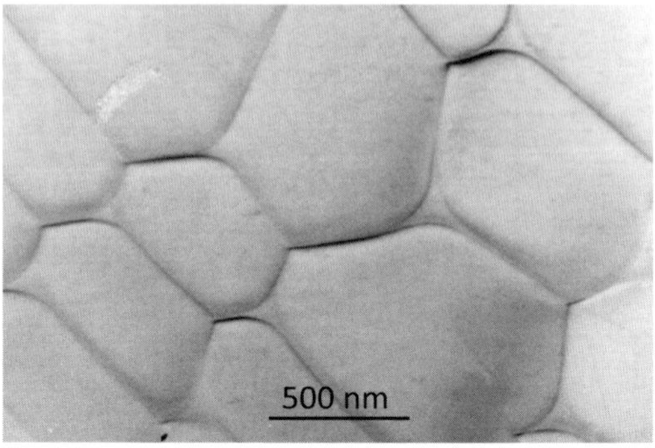

Fig. 8.25 Extended SF nodes in a Cu–Al alloy (Reproduced with permission from [7]).

shows a typical TEM of such networks [7]. The threefold node of ◯ is widely extended and covered completely with the stacking fault. With decreasing the SFE, the dissociation width of dissociated dislocations increases and hence the extended area at the node ◯ increases.

Indeed, before the dark-field weak-beam (DFWB) method [8] of electron microscopy was developed, the resolving power of a TEM had been too poor for each of the partials of a dissociated dislocation to be resolved. At that time many attempts were carried out to estimate the SFE by observing extended nodes. Figure 8.26 shows two methods to estimate the SFE from the radius (W) of the inscribed circle and the radius of curvature (R) of the extended node.

The greatest drawback of these methods is that they contain the cut-off parameter r_o of the dislocation core introduced to avoid the difficulty of divergence. As shown in Eq. (4.5′), there is ambiguity about the value of r_o, so that the value of SFE evaluated using Eq. (4.5′) is also ambiguous. However, once it becomes possible to measure the SFE precisely by means of the DFWB method, r_o can be estimated precisely. It has been shown that $r_o \risingdotseq 3b$ (Fig. 8.26(c)) [9]. The value of r_o is larger for edge than screw, in good agreement with results of computer simulation [10].

8.4.2 *Reactions between dissociated dislocations on different slip planes: Stair-rod dislocations*

Suppose that dislocation $\boldsymbol{b} = \boldsymbol{CB}$ gliding on (d) plane meets another dislocation gliding on a plane other than (d) plane, say, ($\bar{c}$) plane at the intersection ($AD = [\bar{1}10]$) of (d) and ($\bar{c}$) (Figs. 8.27(a) and 8.27(b)). Now we are

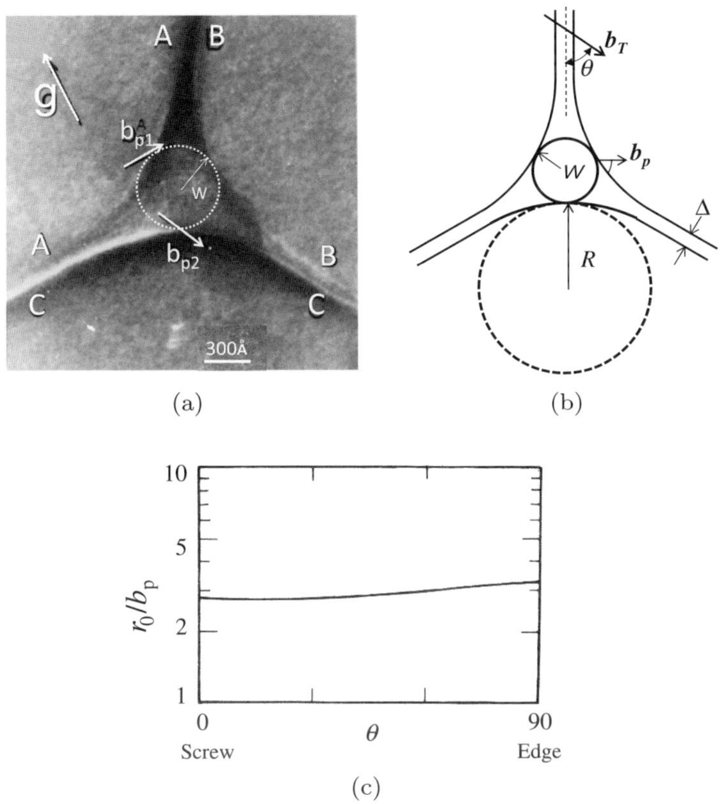

Fig. 8.26 (a) An extended SF node (Reproduced with permission from [9]), (b) Estimation of SFE from an extended node, (c) Dislocation core cut-off radius as a function of dislocation orientation (Reproduced with permission from [9]).

viewing (d) plane from the outside of the Thompson tetrahedron and (c) plane is viewed from the inside, hence it is $(\bar{c})$ plane.

First let us consider reactions between $\boldsymbol{CB}(d)$ and dislocations $\boldsymbol{BD}$, $\boldsymbol{DA}$ on $(\bar{c})$ plane.[6]

CB*, *BD and ***DA*** are dissociated as follows:

$$\begin{aligned}\boldsymbol{CB}(d) &= \boldsymbol{C\delta}(d) + \boldsymbol{\delta B}(d),\\ \boldsymbol{BD}(\bar{c}) &= \boldsymbol{B\gamma}(\bar{c}) + \boldsymbol{\gamma D}(\bar{c}),\\ \boldsymbol{DA}(\bar{c}) &= \boldsymbol{D\gamma}(\bar{c}) + \boldsymbol{\gamma D}(\bar{c}),\end{aligned}$$

[6]Another dislocation on $(\bar{c})$ plane $\boldsymbol{AB}$ need not be considered because this leads to coplanar reaction described already in Sec. 8.4.1.

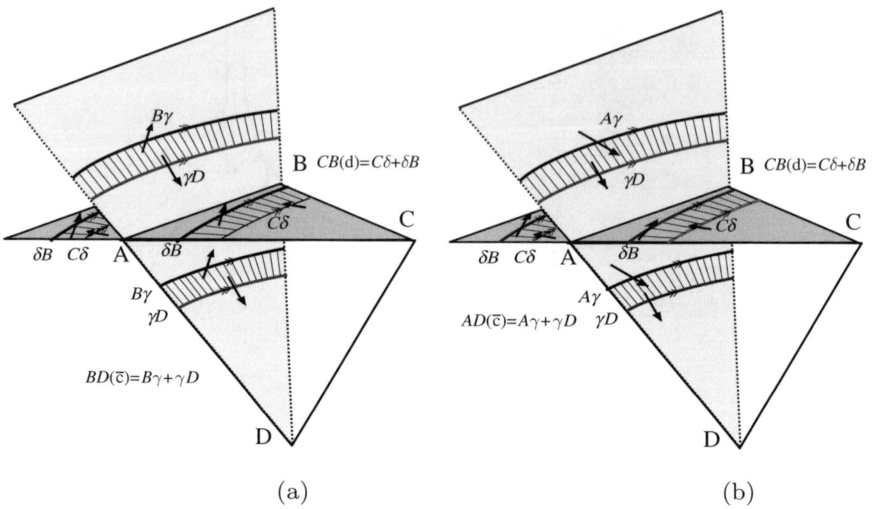

Fig. 8.27 Stair-rod dislocations.

Table 8.1 Formation of stair-rod dislocations.

		$\delta B = 1/6[\bar{2}11]$			$C\delta = 1/6[\bar{1}\bar{1}2]$		
$BD = B\gamma + \gamma D$	$B\gamma = 1/6[1\bar{2}\bar{1}]$	$1/6[\bar{1}\bar{1}0]$	$b^2 = 1/18$	***LC***	$1/6[0\bar{3}1]$	$b^2 = 5/18$	
	$\gamma D = 1/6[\bar{1}\bar{1}\bar{2}]$	$1/6[\bar{3}0\bar{1}]$	$b^2 = 5/18$		$1/6[\bar{2}\bar{2}0]$	$b^2 = 4/18$	***B4***
$DA = D\gamma + \gamma A$	$D\gamma = 1/6[112]$	$1/6[\bar{1}23]$	$b^2 = 7/18$		$1/6[004]$	$b^2 = 8/18$	
	$\gamma A = 1/6[2\bar{1}1]$	$1/6[002]$	$b^2 = 2/18$	***HL***	$1/6[1\bar{2}3]$	$b^2 = 7/18$	
$DB = D\gamma + \gamma B$	$\gamma B = 1/6[\bar{1}21]$	$1/6[\bar{3}32]$	$b^2 = 11/18$		$1/6[\bar{2}13]$	$b^2 = 7/18$	
	$D\gamma = 1/6[112]$	$1/6[\bar{1}23]$	$b^2 = 7/18$		$1/6[004]$	$b^2 = 8/18$	
$AD = A\gamma + \gamma D$	$\gamma D = 1/6[\bar{1}\bar{1}\bar{2}]$	$1/6[\bar{3}0\bar{1}]$	$b^2 = 5/18$		$1/6[\bar{2}\bar{2}0]$	$b^2 = 4/18$	***B2***
	$A\gamma = 1/6[\bar{2}1\bar{1}]$	$1/6[\bar{4}20]$	$b^2 = 10/18$		$1/6[\bar{3}01]$	$b^2 = 5/18$	

where (d) and $(\bar{c})$ are slip planes on which the relevant dislocations lie. Reactions among these partial dislocations are summarized in Table 8.1. Here, instead of Thompson vectors, conventional Burgers vectors are used to describe stair-rod dislocations.

Summation of energies (b^2) of the reactant two Shockley partials before reactions is given by

$$1/36(4+1+1) \times 2 = 6/18.$$

There are five reactions after which the energy of the product dislocation is smaller than this value (i.e., $b^2 < 6/18$). For $\boldsymbol{b} = 1/6\langle 310\rangle$ dislocation,

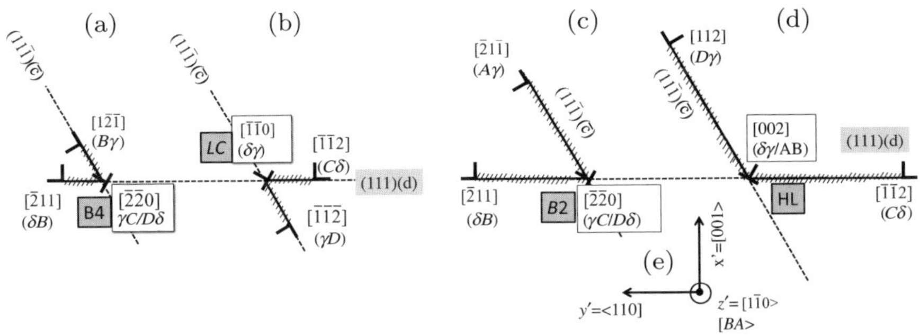

Fig. 8.28 (a)–(e) Formation of stair-rod dislocations.

however, the amount of decrease in energy is small and also it is an asymmetric barrier, so that it is usually ignored. The remaining four reactions lead to stair-rods dislocations along $[\bar{1}10]$: They are referred to as Lomer-Cottrell (LC), B4, Hirth Lock[7] (HL) and B2 [11]. Figures 8.28(a)–8.28(d) shows these stair-rod dislocations.

Exercise 8.16 Discuss on $\boldsymbol{BC}$ instead of $\boldsymbol{CB}$ and show that similar result is obtained.

Exercise 8.17 Show that $\boldsymbol{b}$ of these sessile dislocations is perpendicular to the direction of dislocation line $[1\bar{1}0]$.

These stair-rod dislocations are edge dislocations and as can be seen from Figs. 8.28(a)–8.28(d), their slip planes (their normals being $\boldsymbol{e} = (\boldsymbol{b} \times \xi)/|\boldsymbol{b} \times \xi|$) are different from those before reactions. Therefore, the stair-rod dislocations block the reactant dislocations. They are referred to as sessile dislocations.[8] In Fig. 8.28 stair-rod dislocations are illustrated according to their relative dimension. For details the reader is referred to Jossang *et al.* [11].

When Shockley partials on (d) and $(\bar{c})$ planes react over a long distance along the intersection of (d) and $(\bar{c})$, as shown in Fig. 8.28, one kind of stair-rod dislocation is formed over a long distance. By contrast, when a dislocation lying, say, on $(\bar{c})$ plane is inclined by a large angle with respect to the other plane ((d) plane) (this type of inclined dislocation is referred to as a forest dislocation), the geometry of the reaction is rather complex.

[7]This reaction was first pointed out by M. J. Whelan [12].
[8]Those dislocations which can glide are referred to as glissile dislocations.

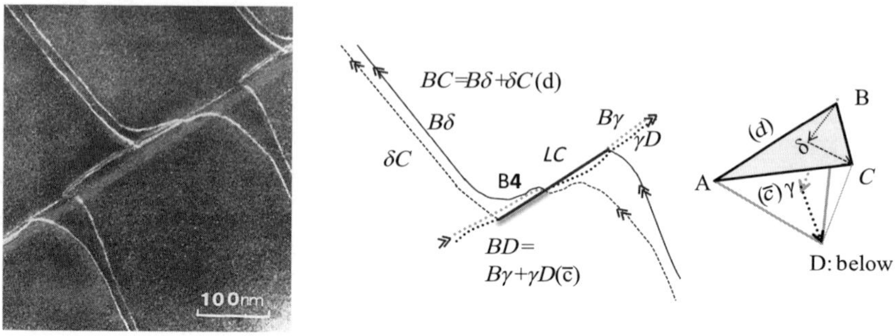

(a) (b)

Fig. 8.29 (a), (b) Formation of stair-rod dislocations by the reaction with forest dislocations (Reproduced with permission from [13]).

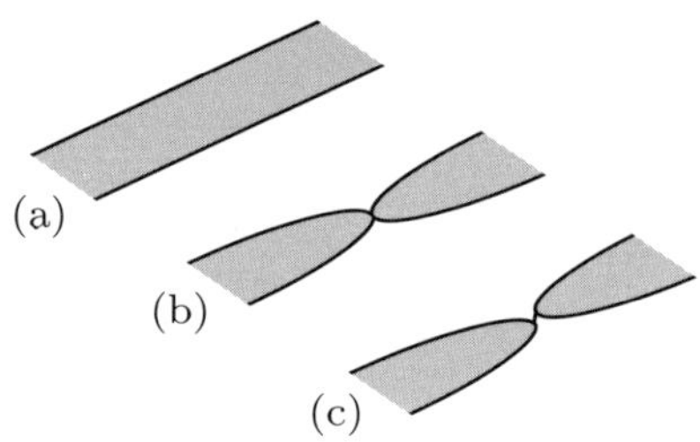

Fig. 8.30 Climb of a narrowly dissociated dislocation.

Figures 8.29(a) and 8.29(b) shows an example. In any case, once dislocations belonging to the different slip systems react to form sessile barriers, all the dislocations involved are blocked. This is the mechanism of work-hardening (see Chapter 12).

8.5 Climb Motion of Dissociated Dislocations

When a dislocation reacts with supersaturated point defects, it climbs non-conservatively. When the dissociation distance is narrow, the dislocation is constricted and then climbs (Fig. 8.33).

However, when the dislocation is widely dissociated, the energy required to constrict is so large that the dislocation climbs while dissociated. The mechanisms by which widely dissociated dislocations climb are studied by Cherns, Hirsch and Saka [14]. According to [14], the individual Shockley partial reacts with point defects independently and climbs. Figure 8.31

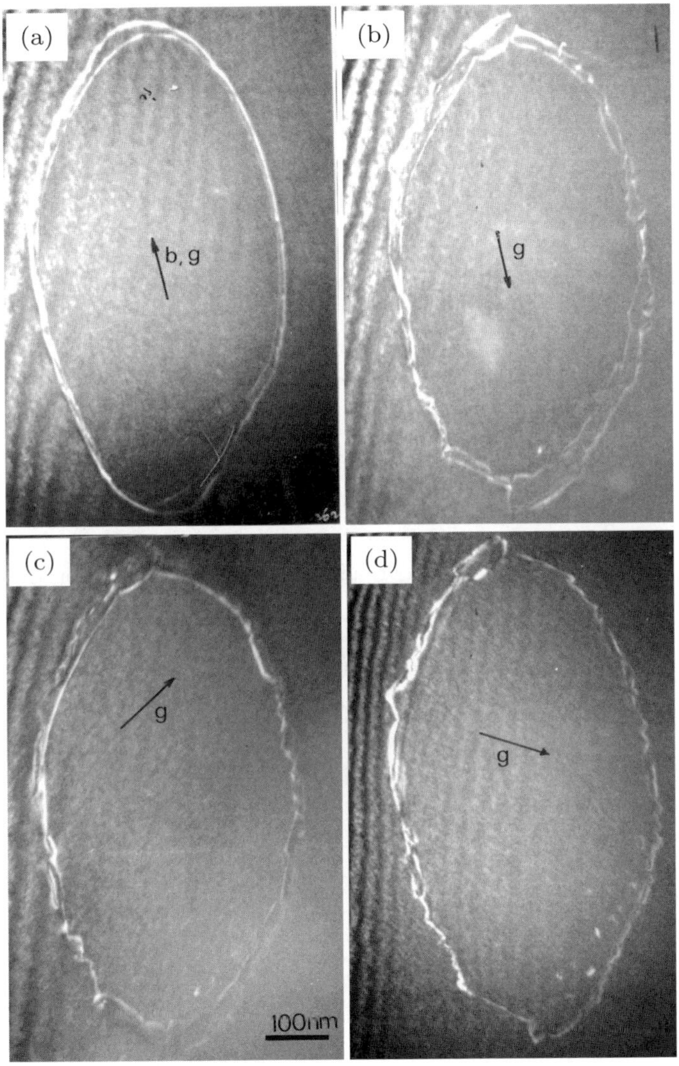

Fig. 8.31 (a)–(d) (a) is before electron radiation, (b)–(d) are after radiation. ***g*** stands for reflection vectors used in imaging (Reproduced with permission from [15]).

shows a series of TEM micrographs of climb motion of a dislocation loop on (111) plane by absorbing point defects, which are interstitials (I) introduced by electron radiation in a high-voltage electron microscope (HVEM). The dislocation is dissociated into two Shockley partials and initially curved smoothly before electron radiation (Fig. 8.31(a)). On introducing point defects (I), the Shockley partials are decorated with small dislocation loops and transform into complex shapes (Figs. 8.31(b)–8.31(d)).

Prismatic loops react with the individual Shockley partials independently.

(1) Dislocation loops formed as a result of precipitation of I are prismatic loops with $\boldsymbol{b}(= 1/2\langle 110\rangle)$, i.e., corresponding to $\boldsymbol{AB}$, $\boldsymbol{AC}$, $\boldsymbol{AD}$, $\boldsymbol{BC}$, $\boldsymbol{BD}$ and $\boldsymbol{CD}$ of the Thompson tetrahedron.
(2) The Burgers vectors of the nucleated loops (the candidates are $\boldsymbol{AB}$, $\boldsymbol{AC}$, $\boldsymbol{AD}$, $\boldsymbol{BC}$, $\boldsymbol{BD}$, $\boldsymbol{CD}$) are function of the dislocation character and are such as to minimize the total energy of the configuration (partial plus loop).
(3) The prismatic loops rotate so as to maximize their edge component.

Now consider the case where prismatic loops of I nucleate on a 60° dislocation of $\boldsymbol{b} = \boldsymbol{AB}$ (Fig. 8.32). First let us consider the nucleation of prismatic loops on $\boldsymbol{A\delta}$. Among the candidates $\boldsymbol{AB}$, $\boldsymbol{BC}$, $\boldsymbol{CA}$, $\boldsymbol{AD}$, $\boldsymbol{BD}$ and $\boldsymbol{CD}$, only the following two reactions result in formation of Shockley partials:

$$\boldsymbol{\delta A} + \boldsymbol{AB} = \boldsymbol{\delta B},$$

$$\boldsymbol{\delta A} + \boldsymbol{AC} = \boldsymbol{\delta C}.$$

Therefore, according to rule (2), only $\boldsymbol{AB}$ and $\boldsymbol{AC}$ are the candidates.

Next, consider rule (3). $\boldsymbol{A\delta}$ is a pure edge dislocation, so that the edge components of prismatic loops $\boldsymbol{AB}$ and $\boldsymbol{AC}$ nucleating on $\boldsymbol{A\delta}$ are same. Thus, both of $\boldsymbol{AB}$ and $\boldsymbol{AC}$ nucleate. On the other hand, on $\boldsymbol{\delta B}$ only prismatic loop $\boldsymbol{AB}$ nucleates (Fig. 8.32(b)).

Exercise 8.18 Discuss the reason for this difference.

These prismatic loops rotate as shown in Fig. 8.32(c) so as to maximize their edge components. Then, the upper parts of the prismatic loops are dissociated as shown in Fig. 8.32(d). Its plan view is shown in Fig. 8.32(e). As prismatic loops $\boldsymbol{AC}$ and $\boldsymbol{AB}$ expand further, pole dislocations $\boldsymbol{AC}$ and $\boldsymbol{AB}$ react to form $\boldsymbol{CB}$, resulting in formation of a dislocation network which looks like a screen (Fig. 8.32 (f)). At this point $\boldsymbol{A\delta}$ is relieved from the pole dislocation $\boldsymbol{CB}$ and straightens itself due to the line tension, assuming eventually a shape as shown in Fig. 8.32(g). $\boldsymbol{A\delta}$ has climbed to the screen-like network. Figure 8.33 shows an example of TEM. X and W correspond

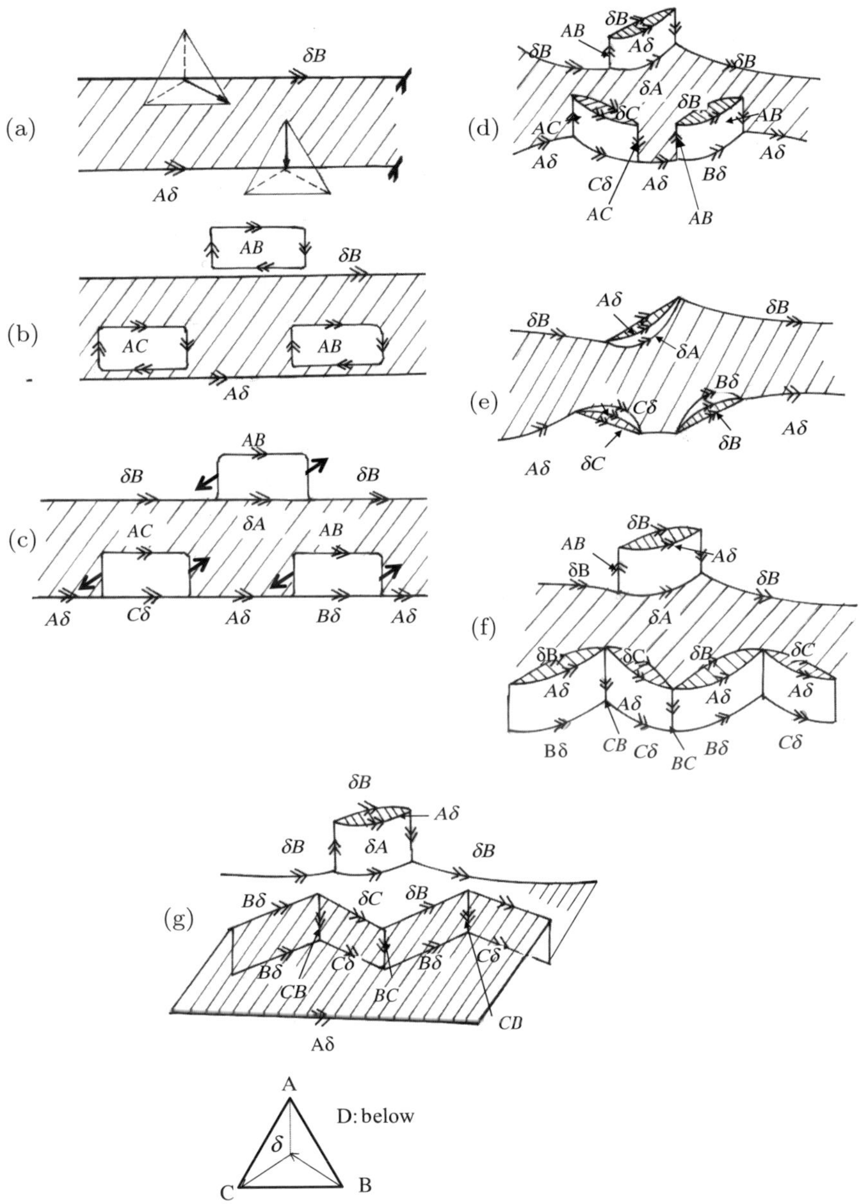

Fig. 8.32 (a)–(g) Mechanism of climb of a dissociated dislocation (I) at room temperature (Reproduced with permission from [14]).

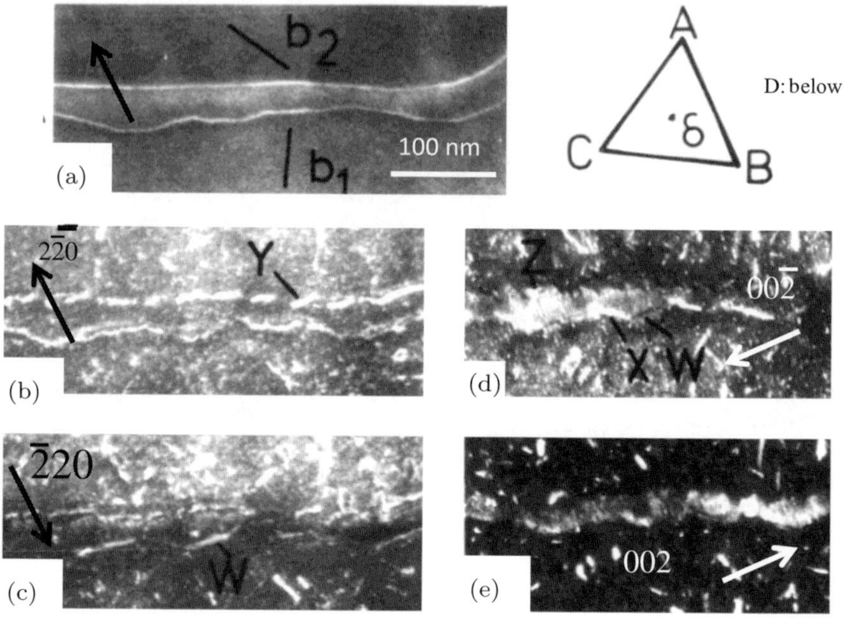

Fig. 8.33 (Reproduced with permission from [14]).

to the screen and Y corresponds to isolated island of $\boldsymbol{\delta}\mathbf{B} - \mathbf{A}\boldsymbol{\delta}/\boldsymbol{\delta}\mathbf{A}$ in Fig. 8.32(g).

Exercise 8.19 Assuming that V is precipitated instead of I, draw what corresponds to Fig. 8.32.[9,10]

At high temperatures, the density of the prismatic loops decreases. In the case of a 60° total dislocation, only on $\boldsymbol{A\delta}$ which is pure edge dislocation, the prismatic loops nucleate. Suppose that prismatic loops $\boldsymbol{AB}$ and $\boldsymbol{AC}$ nucleate alternately as shown in Fig. 8.34(a). The screen-like dislocation network moves from $\boldsymbol{A\delta}$ side and reaches the pre-existing $\boldsymbol{\delta B}$ (in front). They react and annihilate themselves (Fig. 8.34(d)). Then, the pole

[9]When vacancies are precipitated, not only prismatic loops but also Frank type loops are involved [16]. However, here, we assume that only prismatic loops are nucleated for the sake of simplicity.

[10]It has been shown that a dissociated dislocation musters vacancies while it is gliding at room temperature and prismatic loops of vacancy type precipitate on the Shockley partial [17].

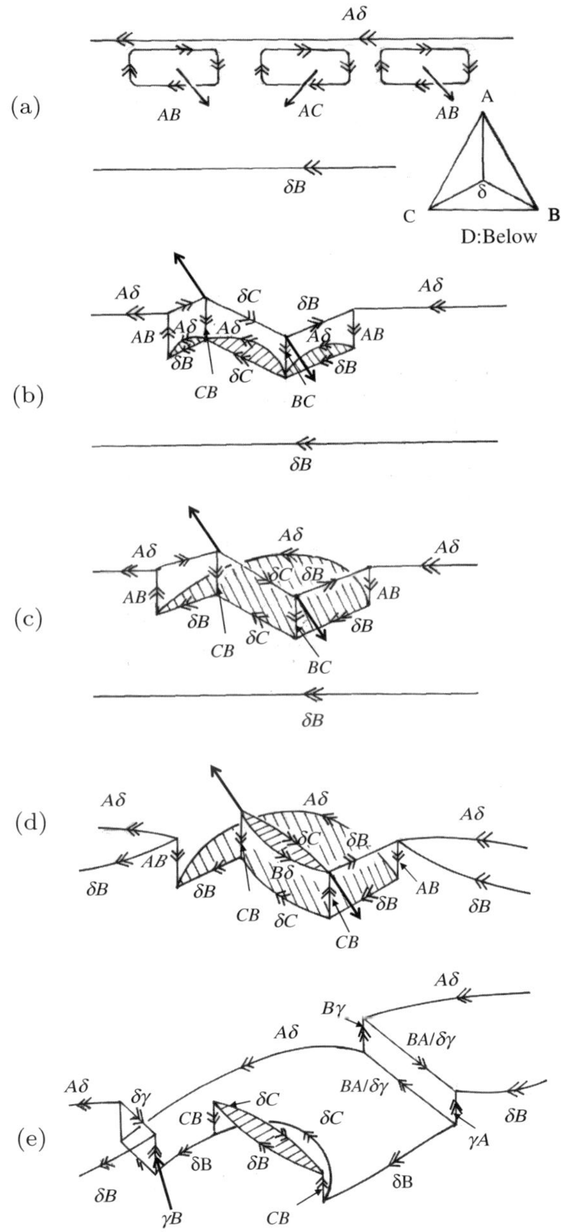

Fig. 8.34 (a)–(e) Mechanism of climb of a dissociated dislocation (I) at high temperature (Reproduced with permission from [14]).

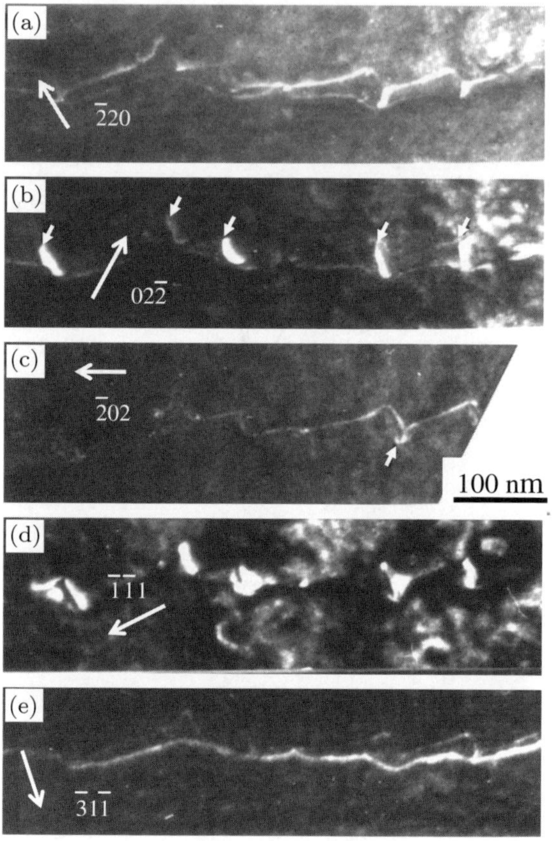

Fig. 8.35 (Reproduced with permission from [14]).

dislocation $\boldsymbol{AB}$ is dissociated and eventually assumes the shape as shown in Fig. 8.34(e). Figure 8.35 shows an example (TEM micrographs).

In summary, prismatic loops nucleating on the individual Shockley partials are shown in Fig. 8.36.

Exercise 8.20 Confirm this. Discuss whether the prismatic loops of I precipitate above or below Shockley partials.

8.6 Composition and Temperature Dependence of the Dissociation Distance

The dissociation distance of a dissociated dislocation affects behaviour of dislocations in a variety of ways. The climb becomes more complex. Perfect

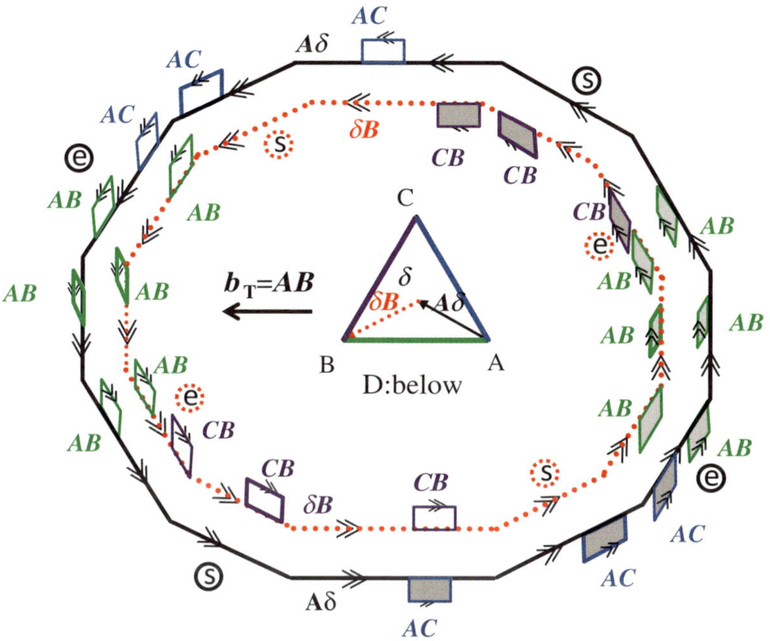

Fig. 8.36 Cherns–Hirsch–Saka model for climb motion of a dissociated dislocation (Reproduced with permission from [15]).

screw dislocations can cross-slip onto any plane belonging to the zone axis of $\boldsymbol{b}$, however, a dissociated dislocation is confined onto the plane on which it is dissociated. This makes cross-slip difficult and the slip plane planar, resulting in a more enhanced piling-up of dislocations (see Sec. 7.1).

The dissociation distance of a dissociated dislocation is inversely proportional to the SFE. Thus, it is of academic as well as practical importance to have a detailed knowledge on how the SFE varies with temperature and composition of alloys.

8.6.1 *Composition dependence of SFE*[11]

Thermodynamic significance of SFE

The significance of a material crystallizing in the FCC structure is that the FCC structure has the lowest free energy as compared with other structures such as the HCP structure. Figure 8.37 shows free energy-composition

[11] Secs. 8.6 and 8.7 are based on [18].

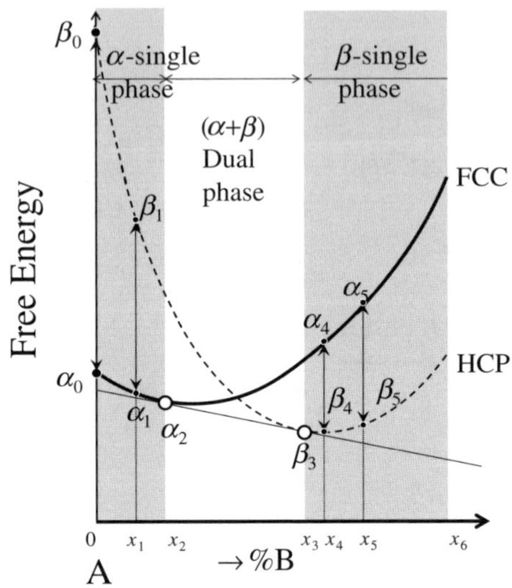

Fig. 8.37 Free energy versus composition curves for A–B binary alloy.

curves of an alloy. The solid line and broken lines show the free energy-composition curves of α phase with the FCC structure and β phase with the HCP structure, respectively.

The stacking faults in FCC crystals have the HCP structure. In that sense, SFE can be defined as a difference in the free energy between the FCC and the HCP structures. If the difference is very small, the stacking fault energy is also very small. In other words, the matrix FCC accommodates the stacking fault (HCP) and a wide stacking fault can be inserted. On the other hand, if the difference is large, the matrix FCC is not happy to accommodate the HCP structure and the dissociation is quite narrow or in the extreme the matrix rejects the stacking fault. The examples of this are Ni and Al. Thus, the SFE of a pure metal A (γ_o) is given by

$$\gamma_o = \beta_o - \alpha_o.$$

Similarly the SFE of an alloy with $x_1(\gamma_1)$ is given by $\gamma_1 = \beta_1 - \alpha_1$. It is obvious that γ decreases with increasing concentration of x.

The SFE of the HCP structure at a composition of x_4 is given by $\gamma_4 = \alpha_4 - \beta_4$. In this case, γ decreases with decreasing x. In other words, as the composition of an alloy approaches the solubility limits (α_2 in FCC, β_3 in HCP) γ decreases.

Stability of these primary solid solution is governed by Hume-Rothery's law: If solute atoms at the right side of noble metals (such as Cu and Ag) in

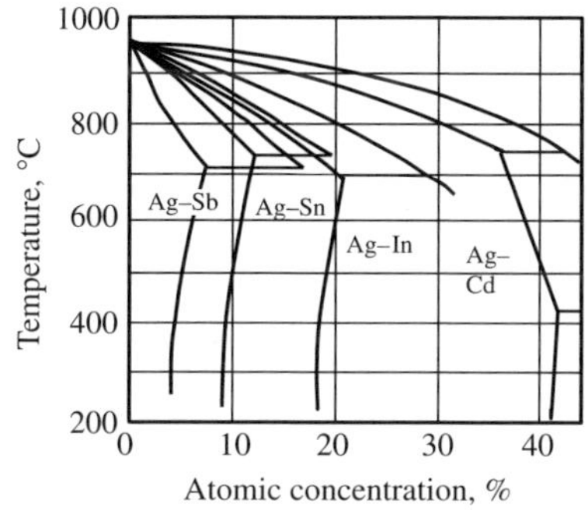

(a)

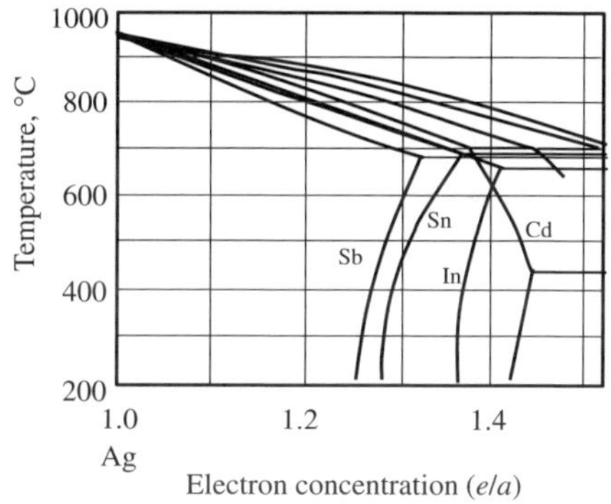

(b)

Fig. 8.38 Solubility limits of Ag-alloys expressed in terms of atomic concentration (a) and electron concentration (b) (Reproduced with permission from [19]).

the periodic table are added to the noble metal, the solubility is determined by its electron concentration (e/a).

Figure 8.38(a) shows solubility limits of primary solid-solutions of a variety of Ag base alloys. On increasing the concentration of alloying element, the solubility limit decreases but in a different manner from one alloy to another. However, plotting as functions of e/a, as shown in Fig. 8.38(b), their solubility limits fall in a similar range.

On increasing the content of an alloying element, e/a increases and the solubility limit of the primary solid solution of the FCC structure occurs at $e/a \fallingdotseq 1.4$. That is, beyond $e/a \fallingdotseq 1.4$, the FCC structure is stable no longer. Thermodynamic significance of this is that the free energy of the FCC structure increases and becomes similar to the free energy in the adjacent phase (for instance HCP). That is, the SFE decreases. Indeed, the SFE of Cu and Ag base solid-solution alloys decreases rapidly with increasing e/a (alloying content), as shown in Fig. 8.39(a).

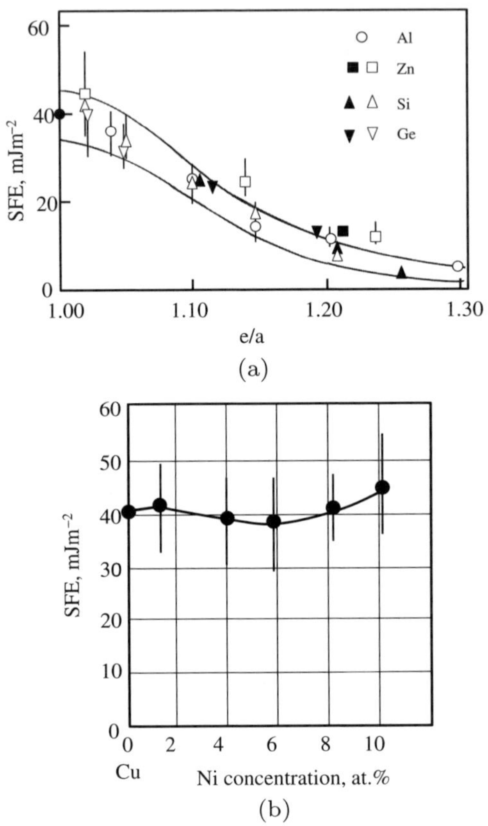

Fig. 8.39 (a) Dependence of γ on e/a in Cu alloys (Reproduced with permission from [20]. See also [21]). (b) Dependence of γ on Ni content in Cu–Ni alloys (after C.B.Carter [22]).

By contrast when alloying elements at the left side in the periodic table such as Ni is added, e/a does not change, and hence the SFE does not change virtually (Fig. 8.39(b)).

Austenitic stainless steel

Of practical importance are austenitic stainless steels. There are two types of stainless steel. One is Fe–Cr binary alloys and the other is Fe–Cr–Ni ternary alloys. In the Fe–Cr phase diagram, γ (austenite phase with FCC structure[11]) forms a loop (γ-loop), that is, on increasing Cr content,

[11]Here γ stands for crystal structure of Fe, not SFE.

γ (FCC) phase disappears. On the other hand, the addition of Ni stabilizes γ phase. Here, we treat Fe–Cr–Ni ternary austenitic stainless steels with the FCC structure.

A typical austenitic stainless steel, 18–8 steel (Fe–18wt%Cr–8wt%Ni) transforms from γ(FCC) to α(BCC) by sub-zero treatment (cooled below 0°C) or plastic deformation at room temperature. This is the reason why 18–8 steel is called a metastable austenitic stainless steel. By contrast, in 19–9 steel the $\gamma \rightarrow \alpha$ transformation does not take place and 19–9 steel is called a stabilized stainless steel. Addition of Ni and Cr stabilizes γ phase.

Stabilizing γ phase (FCC) results in an increase in SFE, as shown in Fig. 8.40.

Yonezawa *et al.* [24] proposed following equations.

For water-quenched

$\gamma = -7.1 + 2.8\%\text{Ni} + 0.49\text{Cr} + 2.0\text{Mo} - 2.0\text{Si} + 0.75\text{Mn} - 5.7\text{C} - 24\text{N}$

For furnace cooled

$\gamma = -4.8 + 2.8\%\text{Ni} + 0.44\text{Cr} + 2.0\text{Mo} - 2.0\text{Si} + 0.75\text{Mn} - 2.17\text{C} - 17\text{N}$

For aged

$\gamma = -4.0 + 2.8\%\text{Ni} + 0.39\text{Cr} + 2.2\text{Mo} - 2.0\text{Si} + 0.75\text{Mn} - 4.7\text{C} - 12\text{N}$

Here, the right hand gives chemical equivalent with weighed contents of each alloying element. The metallurgical significance is as follows: Fe-X binary phase diagrams are classified into four groups as shown in Fig. 8.40(b). On alloying, γ-region increases in A-I; in B-I the γ-region becomes smaller and forms a loop (γ-loop). Obviously A-I stabilizes γ, and B-I destabilizes γ. Similar is true for A-II and B-II, the former stabilizing γ and the latter destabilizing γ. Alloying elements forming A-I and A-II types are referred to as γ stabilizers and those for B-I and B-II are referred to as γ destabilizers.

Cr, Ni and N belong to B-I, A-I and A-II types, respectively. As is described in Sec. 8.6.1, the SFE is considered as a difference in free energy between the FCC matrix and the HCP structure inserted as the stacking fault. That is, SFE = $G_{HCP} - G_{FCC}$, where G_{HCP} and G_{FCC} are free energies of HCP and FCC phases. Stabilizing γ phase means lowering G_{FCC}. This results in increasing SFE.

8.6.2 *Temperature dependence of SFE*

Temperature dependence of SFE of pure materials

The only pure materials whose temperature dependence of the SFE has been measured so far are silver [25] and Si [26]. The temperature

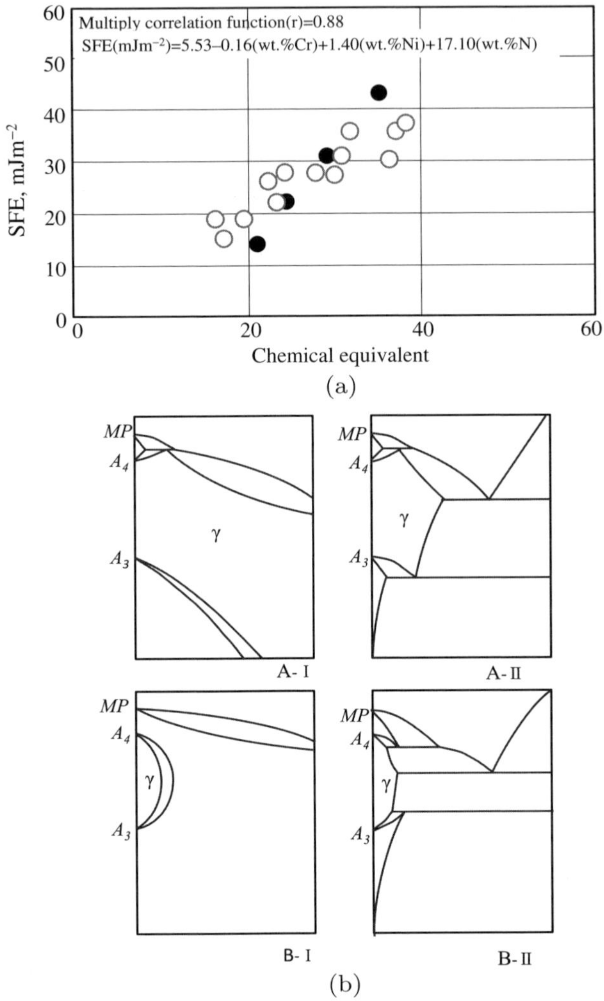

Fig. 8.40 (a) Composition dependence of SFE in stainless steel (Reproduced from [23]), (b) Fe-X phase diagrams (Reproduced with permission from from [31]).

dependence of SFE of these materials is negligibly small. The SFE in compound semiconductors such as GaAs decreases with decreasing temperature (Fig. 8.41) [27, 28].

Temperature dependence of SFE of alloys

First let us consider cases where there is no interaction between solute atoms and the stacking fault. The SFE is inversely proportional to the dissociation

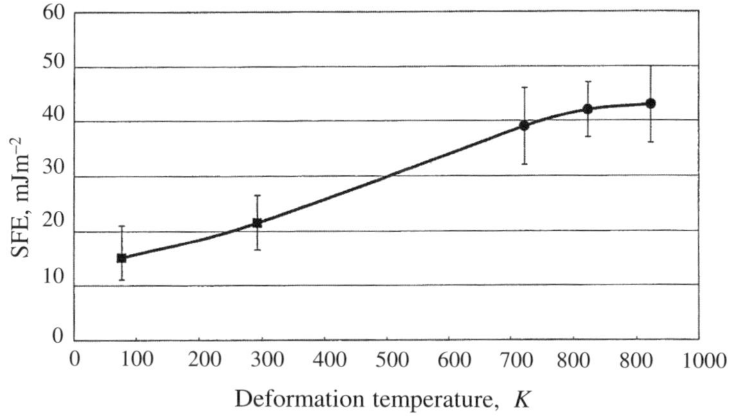

Fig. 8.41 Temperature dependence of γ in GaAs (Reproduced with permission from [28]). γ has been measured at as low as liquid nitrogen temperature.

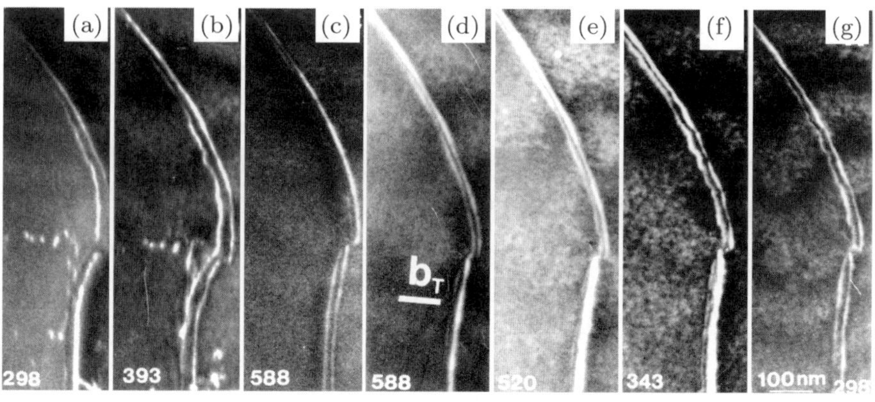

Fig. 8.42 An example of *in situ* heating experiment in which the dissociation distance of a dissociated dislocation was measured at various temperatures (indicated in Kelvin). a is in the as-deformed state (at room temperature). The specimen was heated from room temperature to 588 K (a→b→c), and held at 588 K (c→d), followed by cooling (d→e→f→g) (Reproduced with permission from [29]).

distance. Therefore, the temperature dependence of SFE is most correctly measured by measuring the dissociation distance in a specimen heated *in situ* inside a TEM as a function of temperature. Figure 8.42 shows an example obtained on a Ag–Al alloy. With increasing temperature the dissociation distance decreases, that is, the SFE increases.

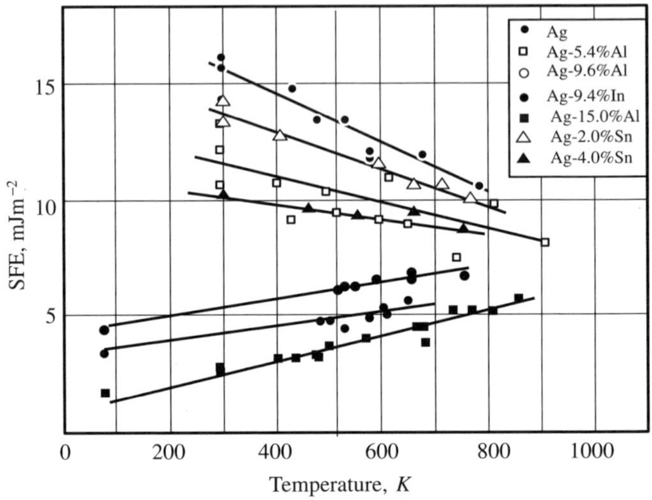

(a)

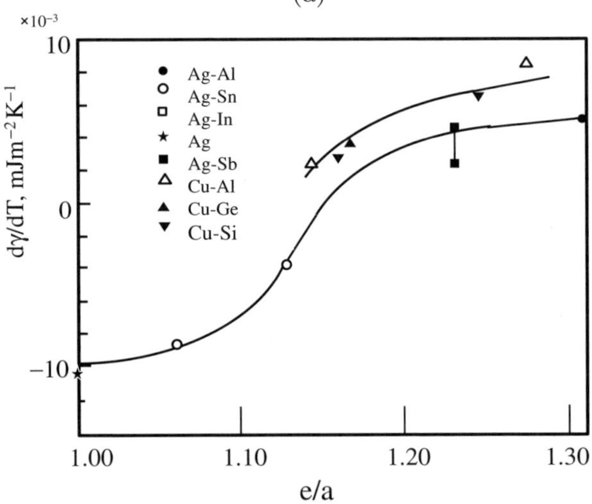

(b)

Fig. 8.43 (a) Temperature dependence of SFE (γ) in Ag alloys. (Reproduced with permission from [29]). (b) $d\gamma/dT$ versus electron concentration (e/a) in Ag and Cu base alloys (Reproduced from [30]).

The temperature dependence of the SFE in noble metal base alloys is summarized in Fig. 8.43(a). The temperature dependence of the SFE depends on the content of alloying element in such a way that $d\gamma/dT$ depends on the electron concentration (e/a) as shown in Fig. 8.43(b).

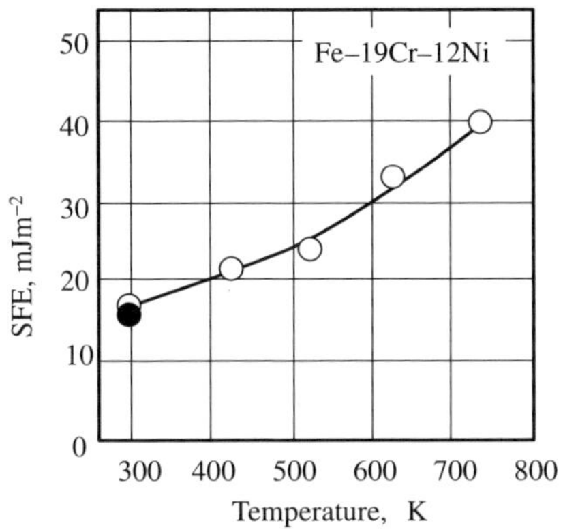

(a)

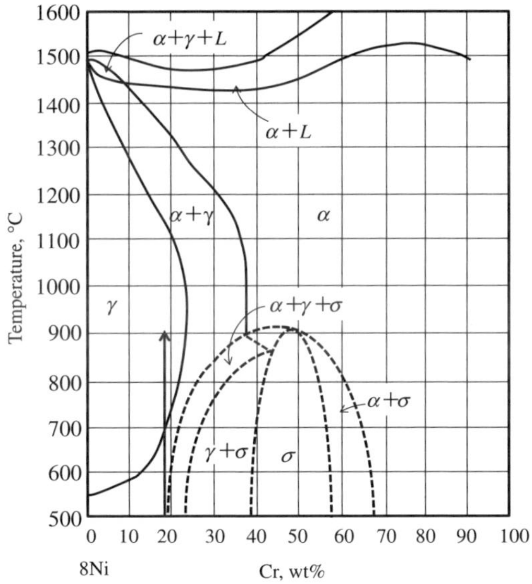

(b)

Fig. 8.44 (a) Temperature dependence of SFE in an austenitic stainless steel [31], (b) The sectional phase diagram passing through 8 wt.%Ni in the Fe–Cr–Ni ternary system(b) (Reproduced with permission from [32]).

An experimental result obtained on an austenitic stainless steel is shown in Fig. 8.44(a) [31]. The SFE increases with increasing temperature. Figure 8.44(b) shows the vertical sectional diagram passing through 8 wt.%Ni of the Fe–Cr–Ni ternary phase diagram [32]. The composition of 18%Cr–8%Ni is indicated by an arrow. As is obvious from this diagram, the single-phase region of γ-phase expands with increasing temperature from room temperature up to ~900°C. That is, with increasing temperature, γ-phase is stabilized and therefore the SFE increases. Austenitic stainless steels are often used as heat-resisting steels. It is necessary to take into account the temperature dependence of the SFE for the proper evaluation.

8.7 Interaction between the Stacking Fault and Solute Atoms

8.7.1 *Suzuki effect*

Theory

The stacking fault in FCC crystals can be regarded as a very thin layer of HCP phase. In the case of a solid-solution alloy the equilibrium concentration of solute atoms in the stacking fault may be different from the equilibrium concentration in the FCC matrix. This was pointed out by Hideji Suzuki [33], and referred to as chemical interaction or Suzuki effect.

Let us consider the case where the SFE decreases with increasing the content of solute atoms, as shown in Fig. 8.39(a). The dissociation distance is determined by a balance between two forces, that is, the elastic repulsive force between the two Shockley partials and the attractive force due to the SFE (Sec. 8.2). Consideration from the viewpoint of energy is shown in Fig. 8.45. Interaction energy between the two Shockley partials ($E_{\rm int}$) decreases rapidly with increasing the dissociation distance. An additional

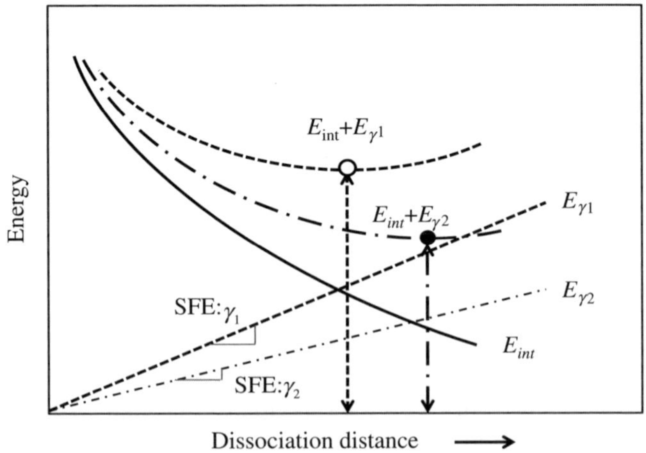

Fig. 8.45 Mechanism of Suzuki effect.

energy due to the stacking fault (E_γ) increases linearly with the dissociation distance. The equilibrium dissociation distance corresponds to the point where the total energy, i.e., (E_{int} plus E_γ) is the minimum. Figure 8.45 illustrates two cases for different SFE (γ_1 and γ_2). The total energy of the alloy is summation of the two energies, so that the equilibrium dissociation distance for γ_1 and γ_2 are indicated by ○ and ●, respectively.

Let us compare the cases of a pure metal and an alloy with reference to Fig. 8.45. Alloying results in a decrease in the SFE. Thus, γ_1 corresponds to the pure metal and γ_2 to the alloy in Fig. 8.45. The slope of E_γ is higher in the pure metal than in the alloy. E_{int} does not change significantly on alloying. Therefore, the dissociation distance is smaller in the pure metal than in the alloy. This is exactly what Eq. 8.9′ means. The total energy (E_{total}) is also lower in the alloy than in the pure metal.

In such an alloy, if the concentration of solute atoms increases locally (or solute atoms segregate) near the stacking fault, the SFE decreases apparently and also the total energy of the system (alloy) decreases. That is, in Fig. 8.45 the free energy decreases from ○ to ●. Figure 8.46 illustrates

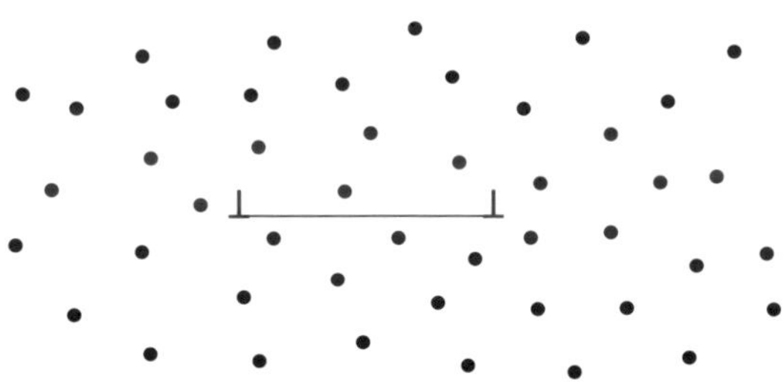

(a)

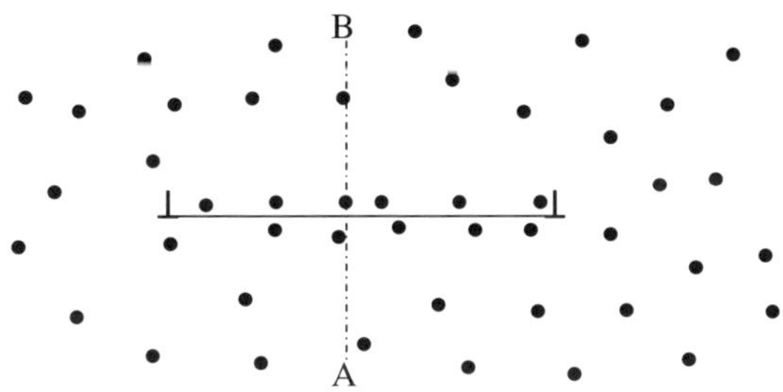

(b)

Fig. 8.46 Suzuki segregation. (a) before segregation, (b) after segregation.

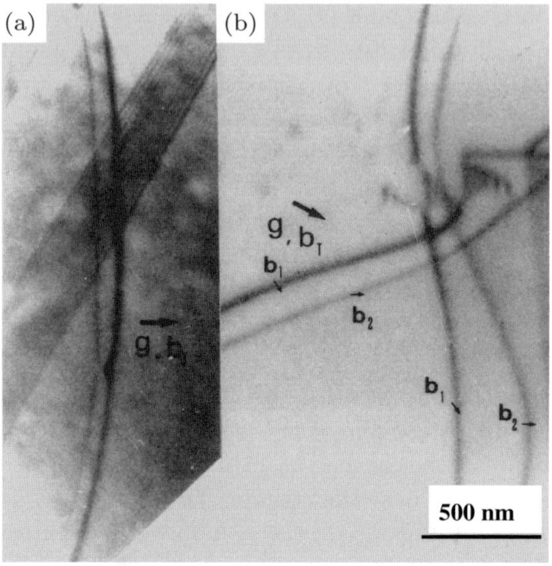

Fig. 8.47 Indirect evidence for Suzuki segregation, (a) was taken just after deformation at room temperature. (b) was taken after annealing at 653 K followed by a very slow cooling (10 K/day), the dissociation distance being much larger than (a) (Reproduced with permission from [33]).

schematically this phenomenon; the phenomenon is referred to as chemical interaction or Suzuki effect and the segregation is referred to as Suzuki atmosphere.

Experimental confirmation of Suzuki effect

Suzuki effect has been studied experimentally by TEM.

① The first way is to confirm that the dissociation distance increases by segregation. Figure 8.47 is an example to show, albeit indirectly, Suzuki effect in a Cu–Si alloy [34]. Figure 8.50(a) shows dissociated dislocations in the as-deformed state of a Cu–Si alloy where solute atoms are distributed randomly. At room temperature, the diffusion is too sluggish for segregation to take place actually even if the total energy is to decrease by segregation. In order for the segregation to take place, it is necessary to heat at temperatures where diffusion rate is high enough. If segregation takes place, the dissociation distance should be larger than that in the as-deformed state (corresponding change from ○ to ● in Fig. 8.48). Figure 8.50(b) shows the dissociated dislocations after the heat treatment. Obviously the

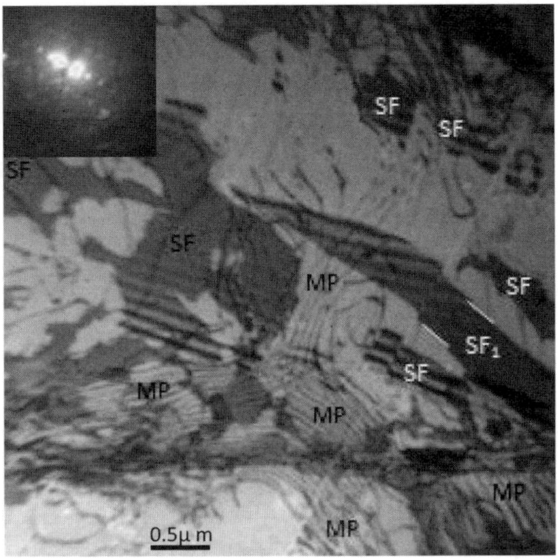

Fig. 8.48 Dissociated dislocations aged at room temperature for 38 years (Reproduced with permission from [35]). Those in the as-deformed state are shown in Fig. 8.11.

dissociation distance after the heat treatment is much larger than that in the as-deformed state (Fig. 8.50(a)).[12]

② More direct way is to detect segregation by an EDX analysis in TEM.[13] Figure 8.49(a) shows an example [36]. SF denotes the stacking fault viewed end on, and Fig. 8.49(b) shows a result of a line analysis of the concentration of the solute (Si) along AB. Si segregates on the stacking fault with denuded zones around.

Suzuki effect in austenitic stainless steels

It has been confirmed that Suzuki effect takes place in an austenitic stainless steel [31].

---

[12]Usually experimental confirmation of Suzuki effect is performed by annealing specimens above room temperature. However, recently (2014), it was confirmed that Suzuki effect takes place when held at room temperature for a very long time. Figure 8.51 [35] shows dissociated dislocations in a Cu–13.43at.%Al alloy which was aged at room temperature for 38 years; the dislocations are extremely widely dissociated. Compare with the as-deformed dislocations shown in Fig. 8.11.

[13]EDX analysis: A very small area is illuminated by an electron beam in an electron microscope and the characteristic X-ray emitted from there is used to analyze the chemical composition.

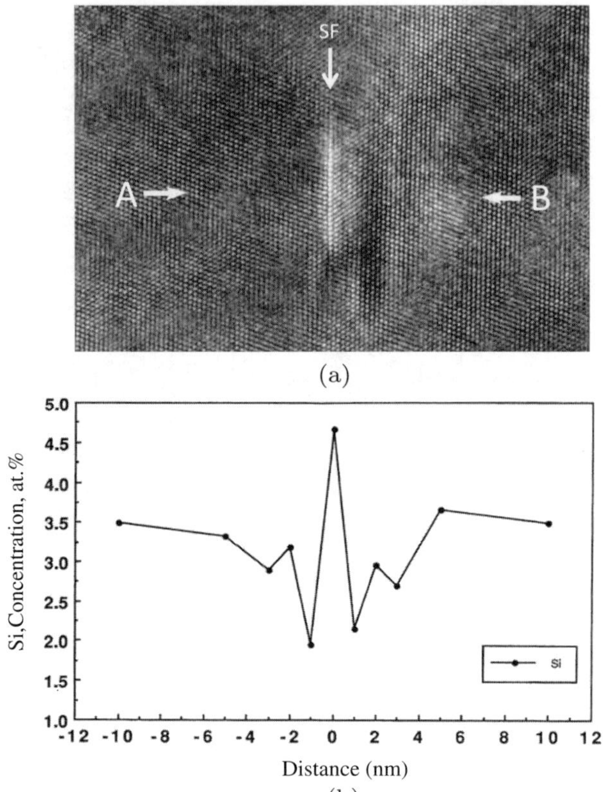

Fig. 8.49 Direct evidence for Suzuki effect by an EDX analysis. (a) is a high resolution micrograph where a stacking fault (denoted by SF) is viewed end on. (b) is the line analysis along AB in (a) (Reproduced with permission from [36]).

8.7.2 *Radiation Induced Segregation*

As mentioned above, when an alloy is deformed at room temperature, segregation does not take place practically. In order for segregation to take place, solute atoms must move (diffuse). For this to occur it is necessary to increase temperature or to enhance diffusion under a special condition such as electron radiation. Under electron irradiation, not only radiation-enhanced diffusion (RED) but also radiation-induced segregation (RIS) take place. RED is an equilibrium event but RIS is a non-equilibrium event. Figure 8.50 shows an example of widening of a dissociated dislocation under electron radiation. RIS results in both widening due to segregation (Cu–Si, Cu–Ge and Ag–Al alloys) and narrowing due to desegregation (Ag–In alloy) [37].

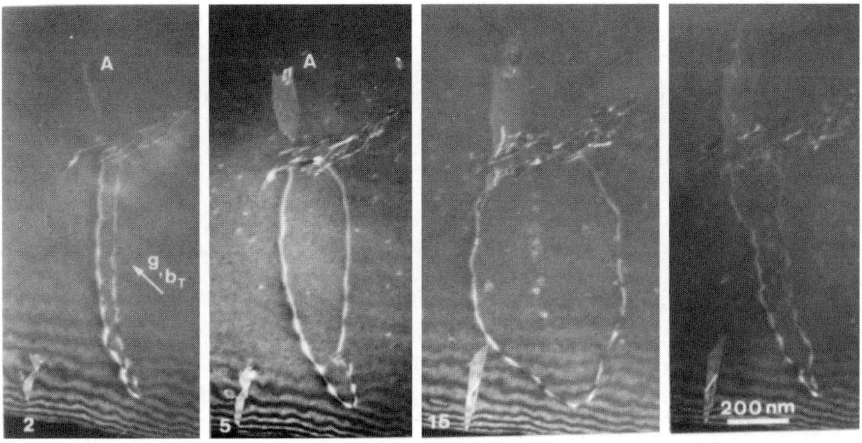

(a) (b) (c) (d)

Fig. 8.50 (a) was taken at room temperature before radiation, (b) and (c) were taken after radiation by 1 MV electrons at 573 K, (d) was taken after interruption of electron radiation: Si atoms diffused away from the dislocation and the dislocation was less segregated by Si (Reproduced with permission from [37]).

Many mechanisms have been proposed to explain RIS but of importance is that the dissociation distance can change significantly as a result of RIS (see Fig. 8.50(b), 8.50(c)).

Further Reading

Hirth, J. P., Lothe, J. *Theory of Dislocations*, 1st ed., McGraw-Hill, (1967); 2nd ed., Wiley, (1982).

References

1. Saka, H. *Electron Microscopy of Crystals* (in Japanese), Uchidarokakuho, 1997, p. 210.
2. Heidenreich, R. D. and Shockley, W. in *Report of a Conference on Strength of Solids*, Physical Society, London, 1948, p. 57.
3. Saka, H., Kawamura, K. and Hashimoto, M. *Phil. Mag.*, A **59** (1989), 687.
4. Saka, H., Sueki, Y. and Imura, T. *Phil. Mag.*, A **37** (1978), 273.
5. Oshima, R., Hori, F., and Mori, H. *Jpn. J. Appl. Phys.* **37** (1998), Part 2, No. 12A, L 1430.
6. Loretto, M. H., Clarebrough, L. M., and Segall, R. L. *Phil. Mag.* **11** (1965), 459.
7. Imura, T. and Enokido, Y. *J. Phys. Soc. Japan* **26** (1969), 869.

8. Cockayne, D. J. H., Ray, I. L. F., and Whelan, M. J. *Phil. Mag.* **20** (1969), 1265.
9. Saka, H. *Phil. Mag.* A **42** (1980), 185.
10. Cotterill, D. M. J. and Doyama, M. *Phy. Rev.* **145** (1966), 465; Doyama, M. and Cotterill, R. M. *J. Phy. Rev.* **150** (1966), 448.
11. Jossang, T., Hirth, J. P. and Hartley, C. S. *J. Appl. Phys.* **36** (1965), 2400.
12. Whelan, M. J. *Proc. Roy. Soc., London, Ser.* A **249** (1959), 114.
13. Karnthaler, H. P. and Winter, E. *Acta Met.* **23** (1975), 1501,
14. Cherns, D., Hirsch, P. and Saka, H. *Proc. Roy. Soc. London, Ser. A* **371** (1980), 213.
15. Saka, H., Kondo, T., and Imura, T. *Toyota Research Reports* **35** (1982), 47 (in Japanese).
16. Decamps, B. and Cherns, D. *Phil. Mag.* A **48** (1983), 123.
17. Saka, H. and Sugawara, S. *Phil. Mag. Lett.* **95** (2015), 433.
18. Saka, H. *J. Mater. Sci.* **51** (2016), 405–424.
19. Fujita, E. F. in Physical Properties of Metals (in Japanese), ed. Physical Society of Japan, (Shokabo, Tokyo 1968), 327.
20. Carter, C. B. and Ray, I. L. F. *Phil. Mag.* **35** (1977), 189.
21. Gallagher, P. C. J. *Metall. Trans.* **1** (1970), 1.
22. Carter, C. B. *Dissertation*, University of Oxford, 1975.
23. Ojima, M. *et al. Steel Research Inter.* **80** (2009), 477.
24. Yonezawa, T., Suzuki, K., Ooki, S., and Hashimoto, A. *Met. Materi. Trans,* A **44** (2013).
25. Saka, H., Iwata, T., and Imura, T. *Phil. Mag.* A **37** (1978), 291.
26. Iwata, H., Kutsuna, M., Okuno, T., Saka, H. *Phil. Mag. Lett.* **96** (2016) http://dx.doi.org/10.1080/09500839.2016.1198052.
27. Nakada, Y. and Imura, T. *Phys. Stat. Solidi.* A **102** (1987), 625.
28. Moon, W.-J, Umeda, T., and Saka, H. *Phil. Mag. Lett.* **83** (2003), 233.
29. Saka, H., Kondo, T., and Imura, T. *Phil. Mag.* A **47** (1983), 859.
30. Saka, H., Kondo, T., and Imura, T. *Dislocation in Solids*, 1985, University of Tokyo Press, 255.
31. Kaneko, Y., Kaneko, K., Nohara, A., and Saka, H. *Phil. Mag.* A **71** (1995), 399.
32. Tsukamoto, F. and Nishi, Y. in *Kinzoku Kogaku Koza* (in Japanese), Vol. 1, Iron and Steels, ed. Imai,Y., (Asakura, Tokyo, 1960), 211.
33. Suzuki, H. *Sci. Rep. Inst.* Tohoku Univ. A **4** (1952), 452.
34. Saka, H. *Phil. Mag.* A **47** (1983), 131.
35. Saka, H. *Phil. Mag. Lett.* **94** (2014), 455.
36. Kamino, T., Ueki, Y. *et al. Phil. Mg. Lett.* **66** (1992), 27.
37. Saka, H. *Phil. Mag.* A **48** (1983), 239.

Chapter 9

Dissociated Dislocations in HCP

9.1 Stacking Faults in Hexagonal Close Packed Structure

In the hexagonal indices, x- and y-(or equivalently $\boldsymbol{a}$ and $\boldsymbol{b}$) axes make an angle of $120°$, so it is somewhat confusing as compared with cubic systems such as the face-centered cubic (FCC) and body-centered cubic (BCC) structures. Before starting this chapter, it would be worth reviewing the crystal structure of hexagonal closed packing (HCP) as shown in Fig. 9.1(a) [1]. The plane in the HCP structure which corresponds to the close packed $\{111\}$ planes in the FCC structure is the basal plane (0001). In the FCC structure, $\{111\}$ planes are stacked in the order of ***abcabc***, while in the HCP structure, the (0001) basal planes are stacked in the order of ***ababab*** (or equivalently ***bcbcbc*** or ***cacaca***). In Frank symbols the FCC structure is denoted by $\triangle\ \triangle\ \triangle\ \triangle\ \triangle\ \triangle\ \triangle\triangle$, while the HCP structure is denoted by $\triangle\triangledown\triangle\triangledown\triangle\triangledown\triangle\triangledown\triangle\triangledown\triangle\triangledown\triangle\triangledown$. Here, $\triangle$ indicates the stacking order of ***ab***, ***bc*** and ***ca***, while $\triangledown$ indicates ***ba***, ***cb*** and ***ac***.

As is shown in Fig. 9.1(a), the close packed basal planes (A, B, C, D) and (A', B', C', D') are equivalent, while those of (A_0, B_0, C_0, D_0) and (A'_0, B'_0, C'_0, D'_0) are displaced by $C\alpha$. Throughout this book the coordinates of HCP are defined as shown in Fig. 9.1(a). Miller indices in this definition are shown in Fig. 9.1(b). It is to be noted that symbols such as A, B, C and D in Fig. 9.1(a) are similar to the symbols indicating atomic planes; the reader is requested to be careful and not to be confused.

Atomic arrangements depicted in Fig. 9.1(a) are redrawn in Figs. 9.1(c) and 9.1(d). Here, tetrahedra $\underline{\boldsymbol{b}}\boldsymbol{c}$ and $\boldsymbol{cb}$ have a twin relationship.

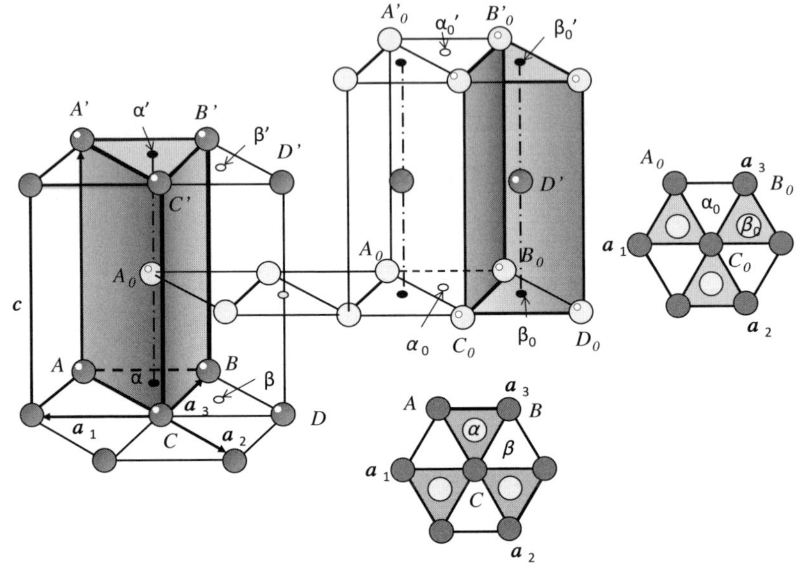

(a)

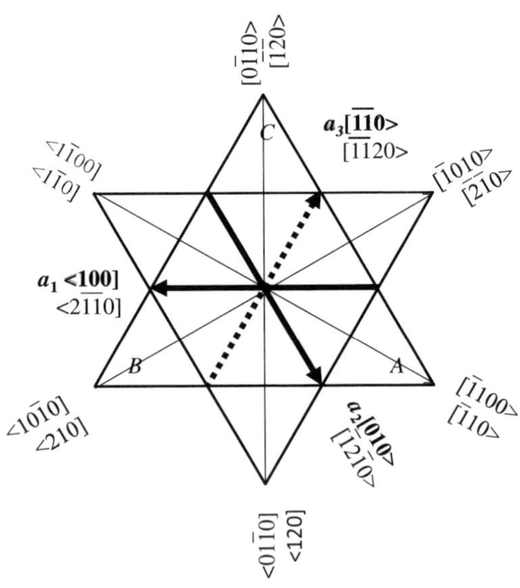

(b)

Fig. 9.1 (a) Atomic arrangements in HCP, (Redrawn based on [1]), (b) Directions (same as Fig. 1.5(b)), (c) and (d) Two equivalent ways to show the twin relationship of tetrahedra.

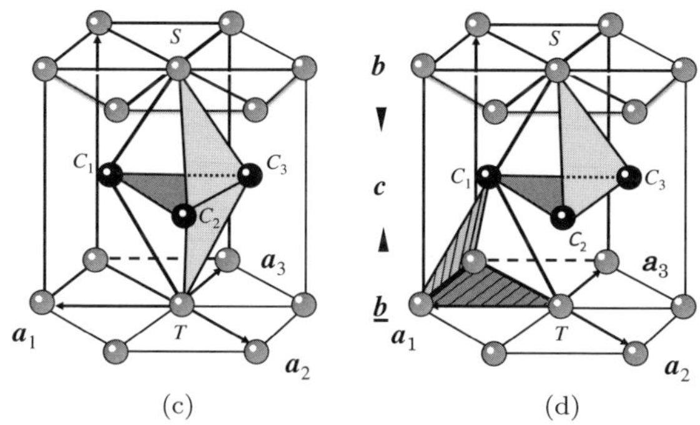

Fig. 9.1 (*Continued*)

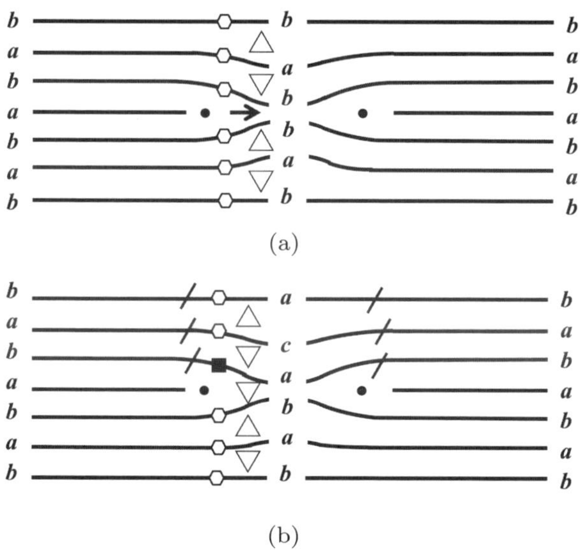

Fig. 9.2 (a), (b) I_1 type.

9.1.1 *Frank type stacking fault in HCP*

Frank type SF in the HCP structure is formed by removing or inserting one (0001) plane, the former being of vacancy type and the latter of interstitial type.

Vacancy type

Suppose that one ***a*** plane is removed, then the SF shown in Fig. 9.2(a) is generated. However, the stacking of ***bb*** indicated by → is very unstable and

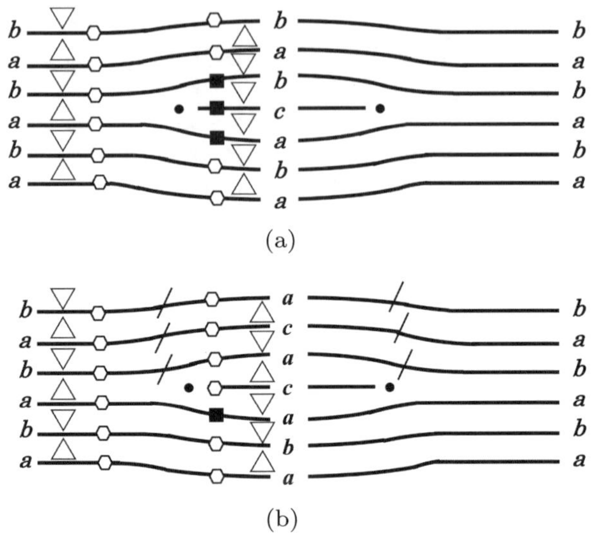

Fig. 9.3 (a) E type, (b) I_1 type.

has quite a high energy. As a result, a slip takes place between upper and lower parts of the stacking of ***bb***, resulting in displacement of the upper ***b*** plane to ***a*** (***b***→***a***) and the SF shown in Fig. 9.2(b) is generated. Here let us introduce a new symbol for HCP. Let △▽ and ▽△ be denoted by ⬡, and ▽▽ and △△ by ■.[1] In this notation, FCC is denoted by ■■. . . , and HCP by ⬡⬡. . . . As is clear from Fig. 9.2(b), in the SF of vacancy type, one plane of ■ is inserted. This type of SF is referred to as I_1, since a vacancy-type (intrinsic) SF with one wrong plane (■) is inserted.

Interstitial type

If one (0001) plane is to be inserted between, say ***a*** and ***b*** planes, the plane is restricted to ***c*** plane, as shown in Fig. 9.3(a). This type of SF is referred to as the E type (E means the extrinsic type). The E type SF contains three wrong planes ■, so that its energy is quite high.

If ***b*** plane on top of ***a*** plane slips like ***b***→***a***, the number of the wrong planes (■) is reduced to one as is shown in Fig. 9.3(b) and becomes I_1 type.

[1]It is noted that Frank symbol indicates the way the two neighbouring (111) planes are stacked. In other words, it indicates the environment of the two neighbouring (111) planes. By contrast, this new symbol indicates whether a (111) plane in consideration actually assumes the FCC structure or the HCP structure.

Thus, in the HCP structure both vacancy and interstitial types of Frank SF end up low-energy I_1 type.

Exercise 9.1 Apply this symbol to Eqs. (8.5) and (8.7) describing the stacking faults in the FCC structure.

9.1.2 *Burgers vectors of Shockley partials*

The slip direction on the basal slip in the HCP structure is $1/3 <\bar{1}2\bar{1}0>$; this is equivalent to $1/2 <110>$ slip in the FCC structure. In Shockley type SF, an atom in ***a*** plane, for instance, is displaced to the position of ***c*** and two wrong planes (■) are inserted as shown in Fig. 9.4. This type of SF is referred to as I_2.

The Thompson tetrahedron is used to describe Shockley partials in the FCC structure. In the HCP structure, the close packed plane is restricted to the basal plane (0001), so that there is no equivalent to Thompson tetrahedron. However, from the viewpoint of stacking of atoms, the Thompson tetrahedron has a close relationship with a slip on (0001) plane in the HCP structure.

Now let us review the Thompson tetrahedron. There are three (111) planes ①,②, ③[2] in the unit cell of the FCC structure (see Fig. 9.5(a)).

Plane ① is located at the origin (0,0,0) or equivalently at the coordinate (1,1,1), and indicated by ●.

Plane ② passes through the coordinates $(1/2, 1/2, 0)$, $(1/2, 0, 1/2)$ and $(0, 1/2, 1/2)$, and indicated by ○.

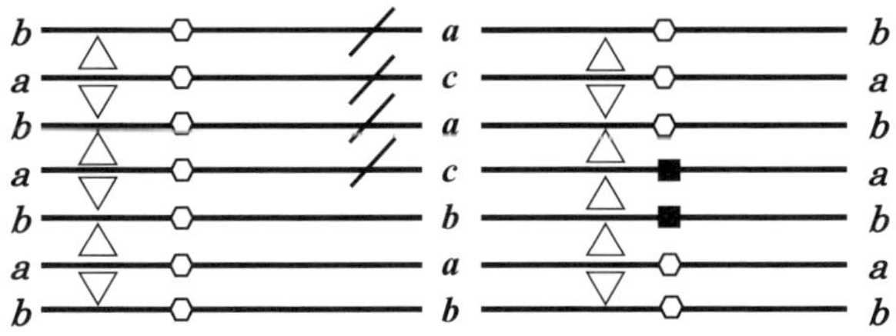

Fig. 9.4 I_2 type.

[2]So far, close packed planes are indicated by ***a***, ***b*** and ***c***. In order to distinguish them from (*a*), (*b*) and (*c*) planes in Thompson tetrahedron and Burgers vector ***b***, close packed planes will be indicated by ①, ② and ③ here.

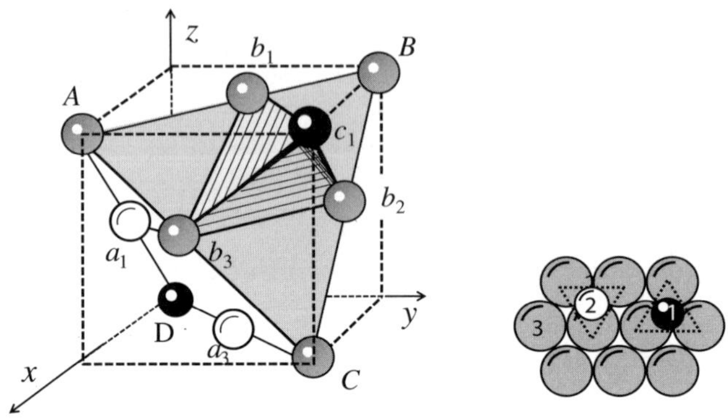

(a) (b)

Fig. 9.5 (a), (b) Stacking of close packed planes in FCC (same as Fig. 8.2).

Plane ③ passes through the coordinates (1,1,0), (1,0,1) and (0,1,1), and indicated by ●.

In the Thompson tetrahedron, atom c_1 in plane ① (indicated by ●) slides on top of plane ③ (△ABC, indicated by ●): This corresponds to viewing the Thompson tetrahedron from the outside. The arrangements of individual atoms and the Thompson tetrahedron are shown in Fig. 9.5(a). It is noted that the apex of the Thompson tetrahedron is downwards, while that of tetrahedron composed of actual atoms is upwards. In other words, the order of stacking of close packed planes of (111) in the FCC structure, i.e., ①②③①②③①②③.... is viewed from the origin along the positive normal of (111) plane. This order ①②③①②③①②③.... is written in Frank symbol by △ △ △ △ △ △ △ △ △...; the sense of △ is coincident with the sense of the apex of the tetrahedron composed of actual atoms.

Projection of these close packed planes is shown in Fig. 9.5(b). Let AB of the Thompson tetrahedron be located horizontally with C below AB (Fig. 9.5(a)). The atom belonging to ① (●) is projected on the midpoint of the triangle composed of atoms belonging to ③ (●). It is noted that this triangle composed of atoms ③ (●) has its horizontal lateral below its midpoint (let us call this triangle an upward triangle).

Exercise 9.2 Discuss the stacking of ② on top of ① and also ③ on top of ②.

Exercise 9.3 Consider stacking of ③ on top of ①, ① on top of ② and ② on top of ③. In these cases, atoms above are projected on the midpoints of the downward triangles (where the horizontal laterals are located above).

Let us apply this concept to HCP. Here, only $\triangle ABC$ is considered because $\triangle ADB$, $\triangle ADC$ and $\triangle ACD$ are not present in the HCP structure. Let only plane ③(◍) and plane ①(●) be present, that is, plane ②(○) is missing. In other words, let us consider only stacking of the order ①③①③①③...... (◍●◍●◍●◍●). In the Thompson tetrahedron the midpoint of △ABC is indicated by δ, but in the HCP structure it will be indicated by σ to distinguish between the FCC and the HCP structures [2].

Let us consider that plane ①(●) slips over plane ③(◍) (Fig. 9.6(a)). Atom (●) on plane ① is located just above the midpoint of the upward triangle composed of atoms (◍) on plane ③. This atom (●) moves on plane ① from A to B ($A \rightarrow B$) rightward. If the dislocation is to be dissociated, the only possible direction of motion of this atom (●) is northeast by east as indicated by a short arrow p_1.[3] This is shown by σB on the *Thompson* tetrahedron.[4] That is, for the slip in stage 1, the Burgers vector of the leading partial ($\boldsymbol{b}_{p1}$) is σB. Taking the direction of the dislocation line upward, the slipped area is in the right side. That is, the left-positioned partial is the leading partial. It is obvious that the Greek (σ) is outside.

Exercise 9.4 When plane ③ glides above plane ①, the Greeks are inside. Prove this (Fig. 9.6(b)).

Consider that a basal dislocation is dissociated into partials bounding the intrinsic SFI_2,

(1) If the slip takes place between (0001) planes indicated by ▲ of Frank symbol, Burgers vector of the Shockley partial is such that σ is outside.
(2) If the slip takes place between (0001) planes indicated by ▼ of Frank symbol, Burgers vector of the Shockley partial is such that σ is inside. (Figs. 9.7(a) and 9.7(b))

[3]Displacement to southeast by east brings ● just on top of ◍, resulting in quite a high energy, so that this displacement is unlikely.

[4]When applied to HCP, the *Thompson* tetrahedron is printed in *italic* to distinguish from the genuine Thompson tetrahedron applied to FCC.

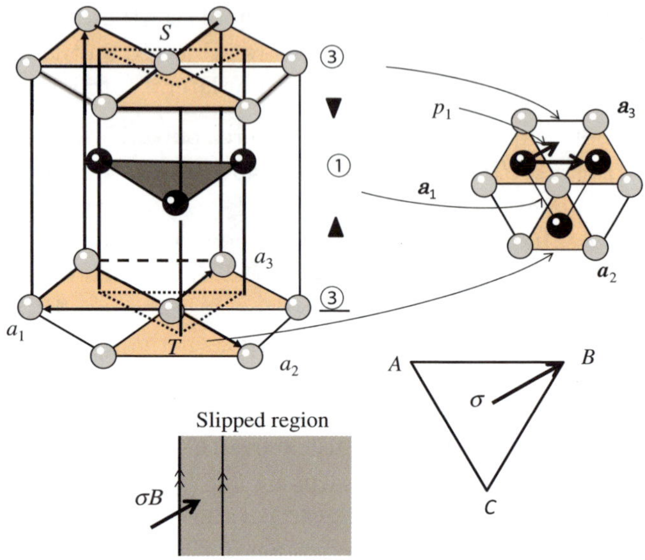

(a)

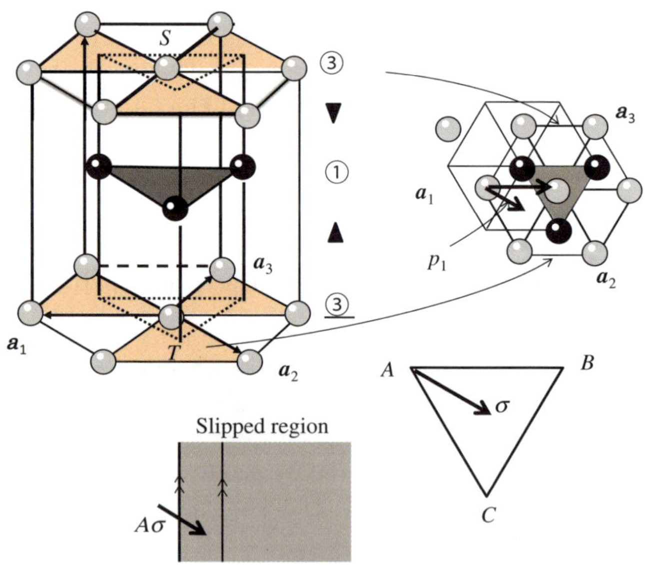

(b)

Fig. 9.6 (a), (b) Shockley partials in the HCP structure.

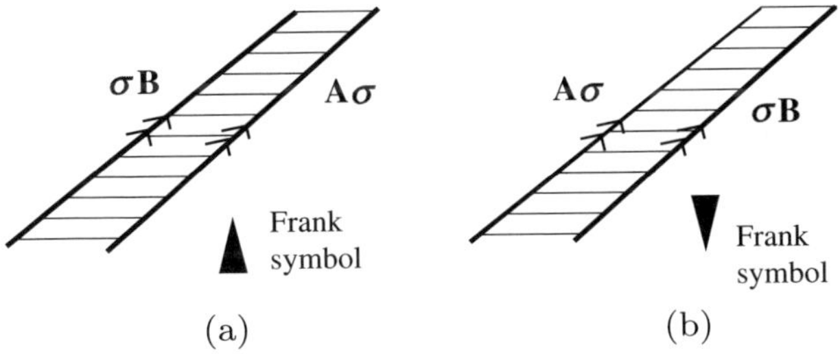

Fig. 9.7 (a), (b).

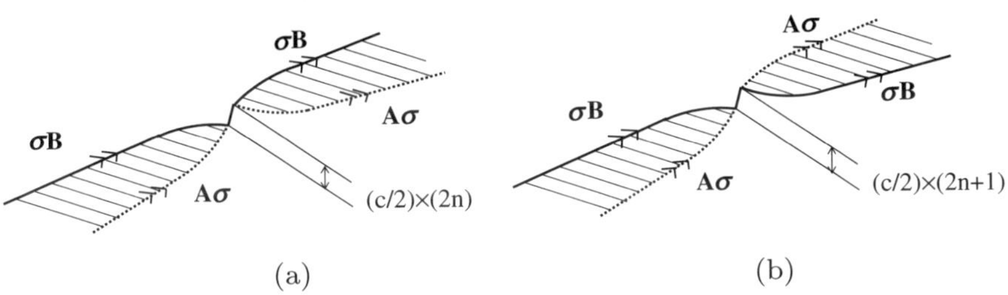

Fig. 9.8 (a), (b).

When a dissociated dislocation contains a jog, the dissociation mode is different between the case where the jog height is even of (c/2) and that where the jog height is odd (Figs. 9.8(a) and 9.8(b)).

Vectors connecting A, B, C and σ and their indices are shown in Figs. 9.9(a) and 9.9(b), where Figs. 9.9(a) and 9.9(b) corresponds to Figs. 9.6(a) and 9.6(b), respectively.

9.2 Composition Dependence of SFE in HCP Alloys

HCP structure is closely related with FCC structure. This is true for the composition dependence of the SFE. The SFE decreases as the composition approaches the solubility limits [3].

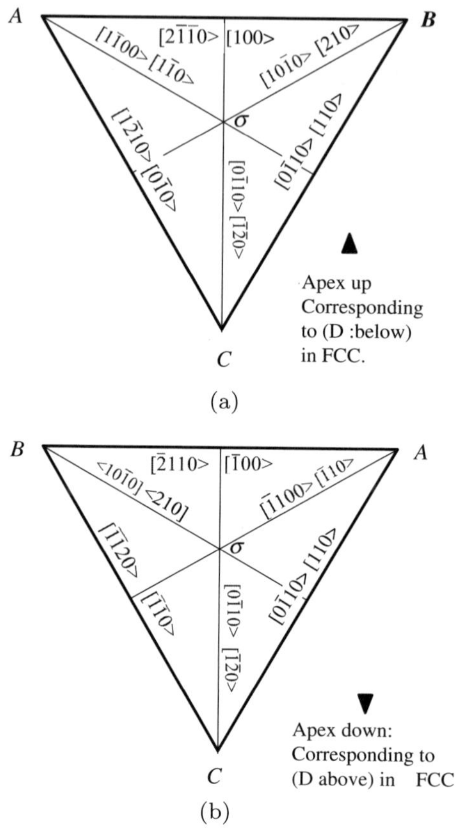

Fig. 9.9 (a), (b) Burgers vectors of perfect dislocations and Shockley partials in the HCP structure.

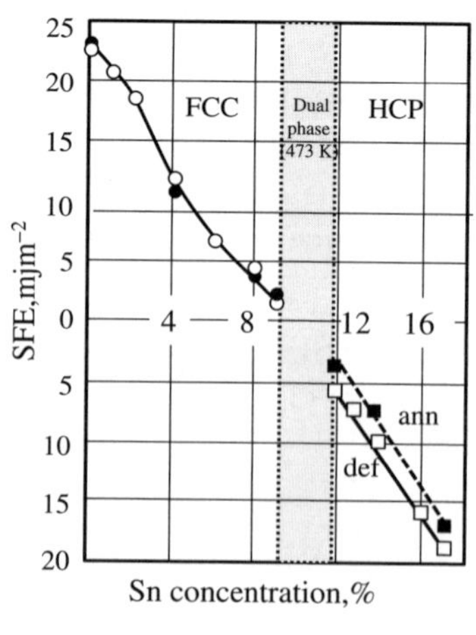

Fig. 9.10 Composition dependence of SFE in FCC and HCP structures in the Ag–Sn system. "ann" stands for annealed state and "def" stands for deformed state [3].

References

1. Hirth, J. P. and Lothe, J. *Theory of dislocations*, 1st ed., MaGraw-Hill (1968), p. 336.
2. Berghezan, A., Fourdeux, A. and Amelinckx, S. *Acta Met.*, **9** (1961), 464.
3. Ruff, A. W. and Ives, I. K. *Phys. Stat. Solidi*(a), **16** (1973), 133.

Chapter 10

Dislocations in Ordered Alloys and Intermetallic Compounds and the Inverse Temperature Dependence of Strength

10.1 General Theory

10.1.1 *Superdislocations and superpartial dislocations*

An ordered alloy (or superlattice) is an alloy where the constituent atoms are arranged regularly. Let us consider a binary solid solution alloy of A-50 at.% B for the sake of simplicity. An alloy in which A and B atoms are arranged in a complete random manner is referred to as a disordered alloy (Fig. 10.1(a)). On the other hand, an alloy where A and B atoms are arranged regularly is referred to as an ordered alloy (Fig. 10.1(b)). Figures 10.1(a) and 10.1(b) shows the disordered body-centered cubic (BCC) and a BCC-based ordered alloy, which is referred to as B2, respectively.

At high temperatures because of the effects of entropy, the ordered state changes into the disordered state, as shown in Fig. 10.2. This is referred to as an order-disorder transformation and the temperature at which the degree of order becomes zero is referred to as the Curie temperature or the Curie point, denoted by T_c.

On the other hand, an alloy where the ordered state is maintained up to the melting point is referred to as an intermetallic compound. In this case, the Curie point (T_c) is higher than the melting point (T_m), i.e., $T_c > T_m$. Therefore, there is no essential difference between an ordered alloy and an intermetallic compound from the viewpoint of atomic arrangement.

Let us introduce a dislocation into an ordered alloy. Let A and B atoms be indicated by ● and ○, respectively (Fig. 10.3(a)) and atoms in the

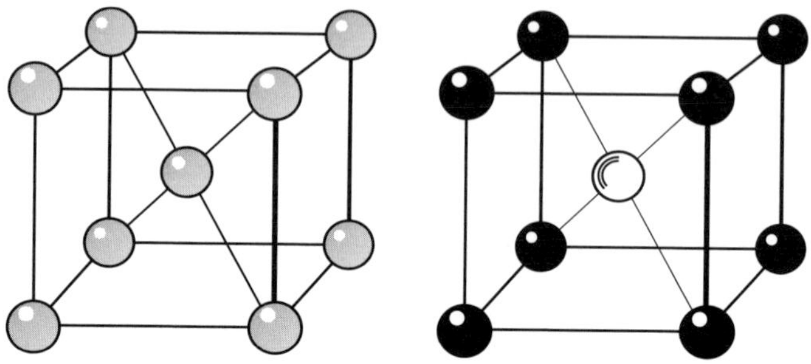

Fig. 10.1 (a) BCC disordered alloy, (b) B2 ordered alloy.

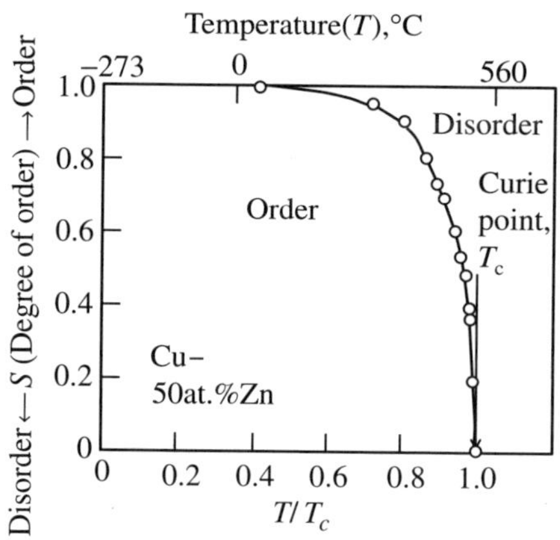

Fig. 10.2 Definition of the Curie point.

disordered state be indicated by ● (Fig. 10.3(b)). In the disordered state, A and B atoms occupy lattice sites statistically at the same probability, therefore, A and B atoms cannot be distinguished from each other.

Now let a perfect dislocation be introduced into a disordered alloy (Fig. 10.3(c)). However, when this dislocation (with the same length of Burgers vector $\boldsymbol{b}$) is introduced into an ordered alloy, the dislocation leaves behind wrong bonds such as A–A and B–B (Fig. 10.3(d)). This is referred to as the anti-phase boundary (APB). The wrong bonds have a higher free energy than the right ones (this is the reason why an ordered alloy is formed); the APB has an additional interfacial energy, γ_{APB}. Therefore, if an external force is to move the dislocation further, the area of APB increases, resulting in an increase in energy, hindering the motion of the dislocation.

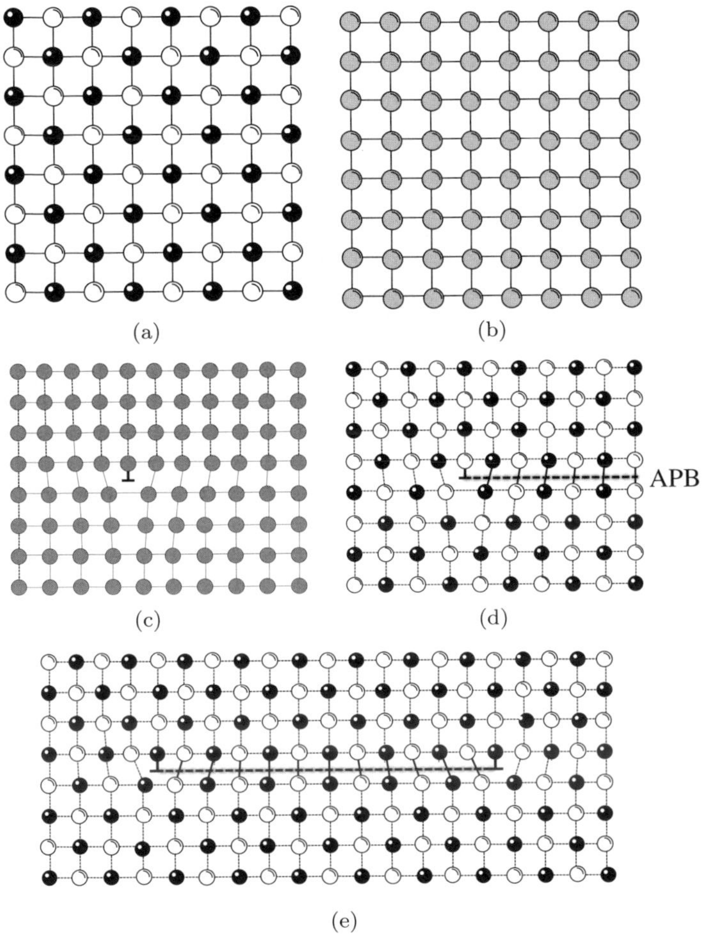

Fig. 10.3 Dislocations in ordered and disordered states. (a) Ordered state, (b) Disordered state, (c) A perfect dislocation in disordered state, (d) A superpartial with APB left behind, (e) A superdislocation consisting of a pair of superpartials with APB in between.

However, what happens if another dislocation with the same nature follows the first dislocation exactly on the same slip plane? Then, after a pair of these dislocations glides, the wrong bonds return back to the right bonds (Fig. 10.3(e)). That is, the APB disappears and the energy decreases again.

When the spacing between a pair of dislocations (referred to as superpartial dislocations or, in contracted form, superpartials) decreases, the

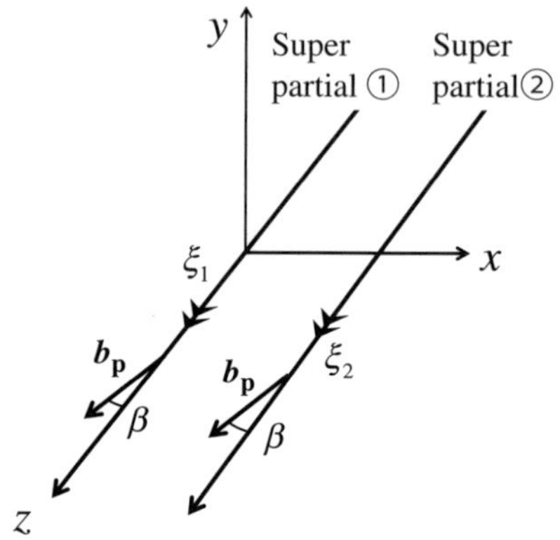

Fig. 10.4 Equilibrium separation between a pair of superpartials.

additional energy due to APB decreases. In other words, APB attracts the two partials, however, on the other hand a repulsive force between the two partials increases. The balance between the attractive and repulsive forces determines the equilibrium separation between the superpartials. This is exactly the same as the discussion on the dissociation distance of dissociated dislocations in the face-centered cubic (FCC) and the hexagonal close-packed (HCP) structures.

Exercise 10.1 Equilibrium separation of superpartials Δ is given by Eq. (10.1). Show this (Fig. 10.4).

$$\Delta = \frac{\mu b_p^2}{2\pi\gamma_{\mathrm{APB}}}\left(\cos^2\beta + \frac{\sin^2\beta}{1-\nu}\right), \tag{10.1}$$

where γ_{APB} is the APB energy (per unit area) and given by

$$\gamma_{\mathrm{APB}} = E_{AB} - \frac{E_{AA}+E_{BB}}{2}. \tag{10.2}$$

Here, E_{AB}, E_{AA} and E_{BB} are the bonding energies of A–B, A–A and B–B bonds, respectively.

Figure 10.5 shows a superdislocation in an ordered alloy of β'-CuZn with $B2$ structure [1]. In Fig. 10.6 is plotted the superpartial separation against the dislocation direction. The superpartial separation in the edge orientation is larger than that in the screw orientation, as is the case for Shockley partials in the FCC and the HCP structures [2].

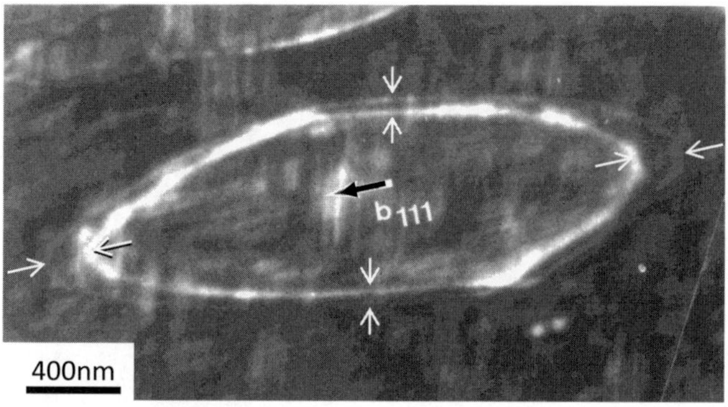

Fig. 10.5 A superdislocation consisting of a pair of superpartials in β'-CuZn (Reproduced with permission from [1]).

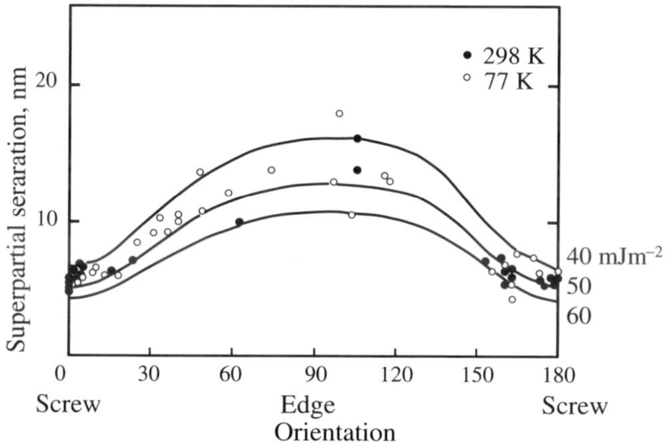

Fig. 10.6 The variation of superpartial separation versus dislocation orientation (Reproduced with permission from [2]).

10.1.2 *Reaction of superdislocations with APB*

Figure 10.7(a) shows an APB which is not associated with a dislocation. Suppose that, during a disorder-order transformation, ordering takes place at two different places (domains) independently. Suppose that these two domains grow and impinge. When the ordering of these two domains happens to be out of phase, an APB is formed (Fig. 10.7(b)). This is referred to as the grown-in APB and shown in Fig. 10.7(c)[3].

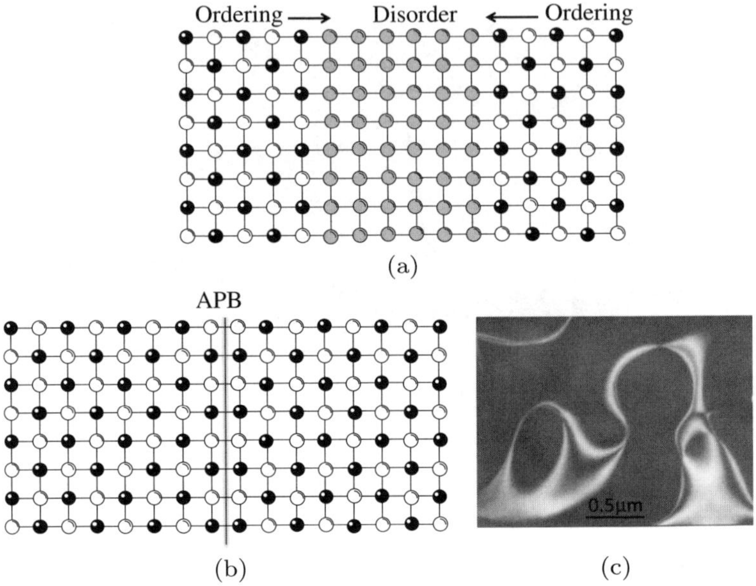

Fig. 10.7 Grown-in APB. (c) TEM of grown-in APB (after Kuroda).

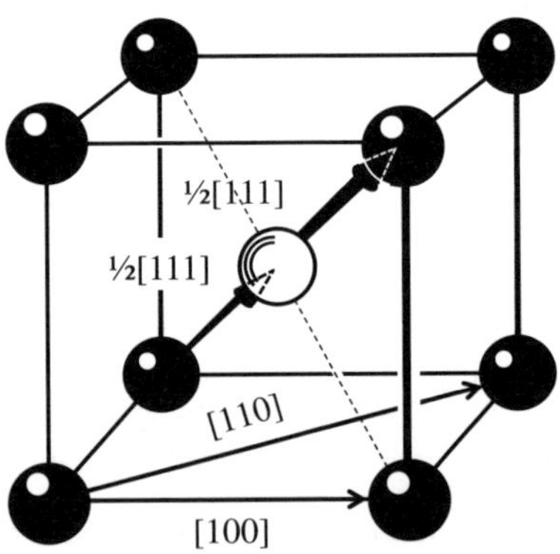

Fig. 10.8 Possible Burgers vectors of dislocation in *B*2.

10.2 Dislocations in Ordered Alloys

10.2.1 *B2 structure*

The structure of B2 is shown in Fig. 10.8. This structure is based on BCC, and the translational vectors are $\langle 111\rangle$, $\langle 001\rangle$ and $\langle 110\rangle$.

Among them, $\langle 111 \rangle$ can be dissociated as follows:

$$\langle 111 \rangle = 1/2\langle 111 \rangle + \mathrm{APB} + 1/2\langle 111 \rangle$$

Dislocations of $\boldsymbol{b} = \langle 001 \rangle$ and $\langle 110 \rangle$ are not dissociated.

Relative energies of these three dislocations are given by

$\langle 111 \rangle$ perfect dislocation: 3

$\langle 111 \rangle$superdislocation: $1/4 \times 3 + 1/4 \times 3 + \gamma_{\mathrm{APB}} = 3/2 + \gamma_{\mathrm{APB}}$

$\langle 001 \rangle$ perfect dislocation: 1

$\langle 110 \rangle$ perfect dislocation: 2.

Thus, possibility of $\langle 111 \rangle$ perfect dislocation and $\langle 110 \rangle$ perfect dislocation is low. The energy of $\langle 111 \rangle$ superdislocation is higher than that of $\langle 001 \rangle$ perfect dislocation. However, in reality both of them occur.

In an intermetallic compound the Curie point of the order-disorder transformation (T_{c}) is higher than its melting point (T_{m}). That is, energy of A–B bonding is much higher than those of A–A and B–B bonding. Therefore, as Eq. (10.2) shows, γ_{APB} is high. As a result, $\langle 001 \rangle$ perfect dislocations are more likely to occur than $\langle 111 \rangle$ superdislocations.

10.2.2 *DO_3 structure*

Figure 10.9 shows the DO_3 structure. Eight unit cells of BCC (or more exactly B2) are stacked. As in the case of B2, the fundamental translational vectors are $\langle 111 \rangle$, $\langle 001 \rangle$ and $\langle 110 \rangle$.

All of them can be dissociated as follows:

$$\begin{aligned}\langle 111 \rangle &= 1/4\langle 111 \rangle + \mathrm{APB} + 1/4\langle 111 \rangle + \mathrm{APB} + 1/4\langle 111) + \mathrm{APB} + 1/4\langle 111 \rangle,\\ \langle 001 \rangle &= 1/2\langle 001 \rangle + \mathrm{APB} + 1/2\langle 001 \rangle,\\ \langle 110 \rangle &= 1/2\langle 110 \rangle + \mathrm{APB} + 1/2\langle 110 \rangle.\end{aligned}$$

Figure 10.10 shows a four-fold dissociated dislocation in an Fe_3Al.

10.2.3 *$L1_2$ structure*

Figure 10.11 shows $L1_2$ structure. $L1_2$ is based on FCC, and its slip is $\langle 110 \rangle$. Using the Thompson vectors it can be written by

$$\langle 110 \rangle = 2\mathbf{AB} = \mathbf{AB} + \mathrm{APB} + \mathbf{AB}$$

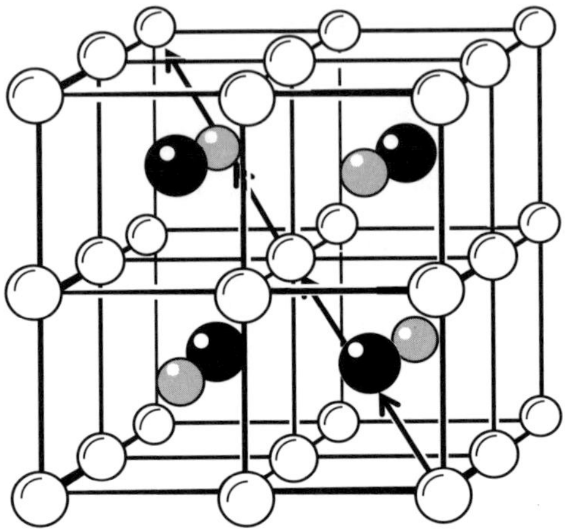

Fig. 10.9 Burgers vector of a superdislocation in DO_3.

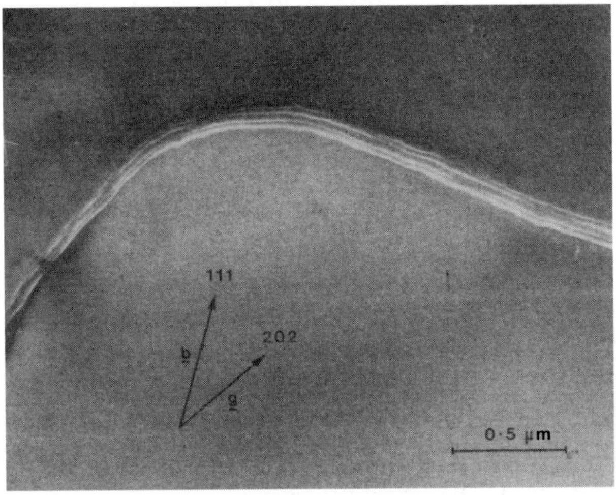

Fig. 10.10 A superdislocation composed of four superpartials in DO_3 (Reproduced with permission from [4]).

However, as described in Chapter 8, the superpartials of $\mathbf{AB} = 1/2\langle 110\rangle$ themselves are dissociated into two Shockley partials.

That is,

$$2\mathbf{AB} = \mathbf{AB} + \text{APB} + \mathbf{AB} = (\mathbf{A}\delta + \text{SF} + \delta\mathbf{B}) + \text{APB} + (\mathbf{A}\delta + \text{SF} + \delta\mathbf{B})$$

(Fig. 10.12)

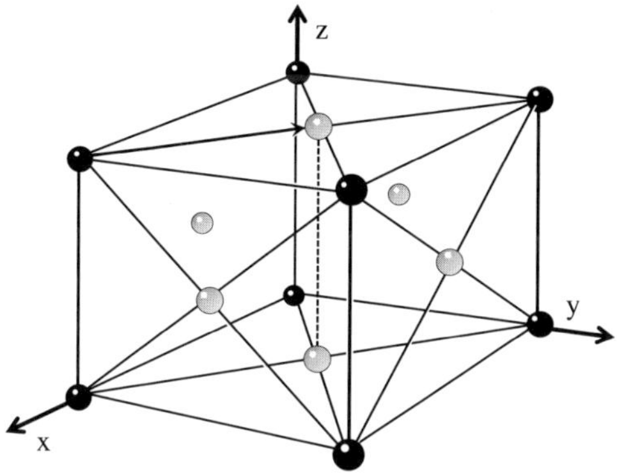

Fig. 10.11 $L1_2$ structure.

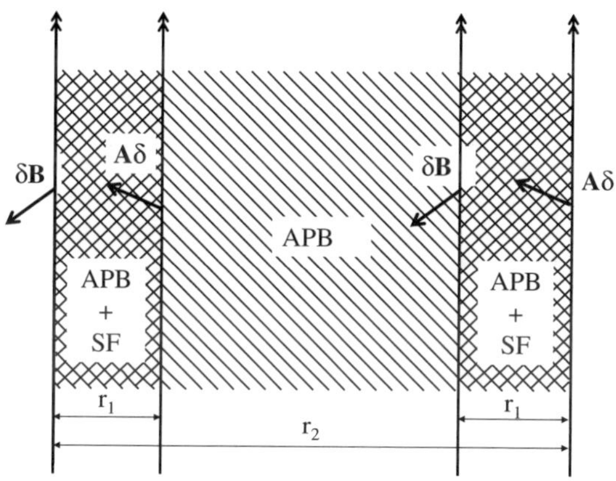

Fig. 10.12 A superdislocation in $L1_2$ structure (Reproduced with permission from [5]).

Figure 10.13 shows an example of such a dissociated dislocation of $\boldsymbol{b} = \langle 110 \rangle$ in a Ni_3Fe alloy of $L1_2$ structure.[1]

10.3 Strengthening by Ordering

10.3.1 *Reactions with grown-in APB*

As is discussed in Sec. 10.1.2, domains, and hence APB's (grown-in APB) are present before deformation. Let us suppose that a superdislocation consisting of two superpartials crosses a grown-in APB (Fig. 10.14).

[1]This particular sample was radiated by electrons in a HVEM, so that the degree of order is reduced, resulting in an increase in r_2.

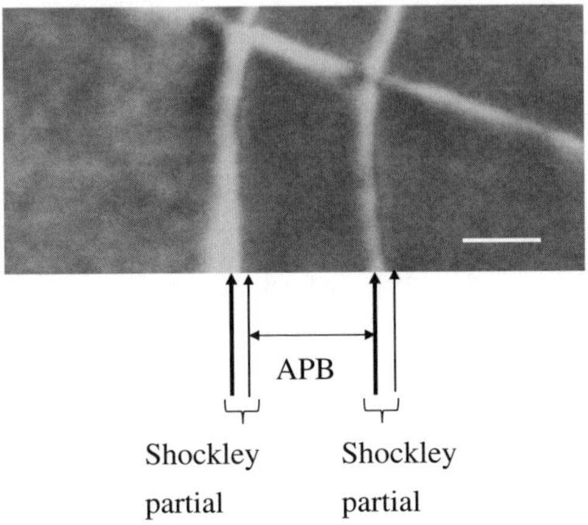

Fig. 10.13 Superdislocations in Ni_3Fe ($L1_2$) [6].

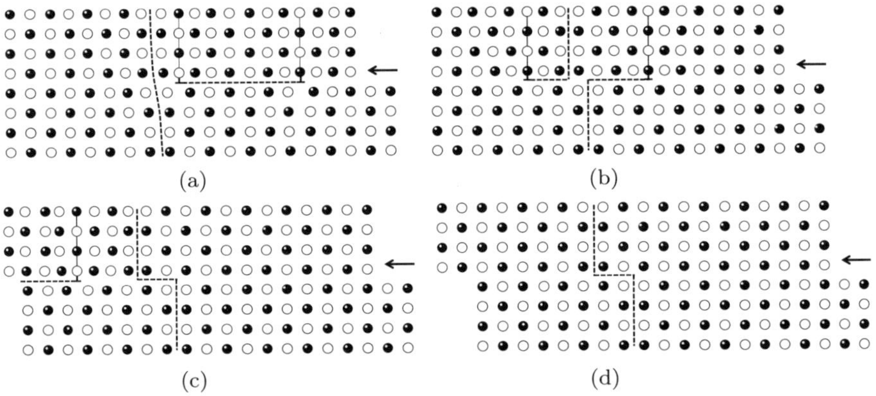

Fig. 10.14 Reaction between a superdislocation and APB.

At a stage shown in Fig. 10.14(a) a superdislocation is approaching the grown-in APB from the right. In (b) the superdislocation crosses the APB. In (c) the superdislocation cuts the grown-in APB, leaving leftward (d). As a result, the length (i.e., area) of the grown-in APB increases. Total energy of the alloy increases and an additional force is required for the dislocation to move further. Obviously, the smaller domain results in more strengthening of the alloy.

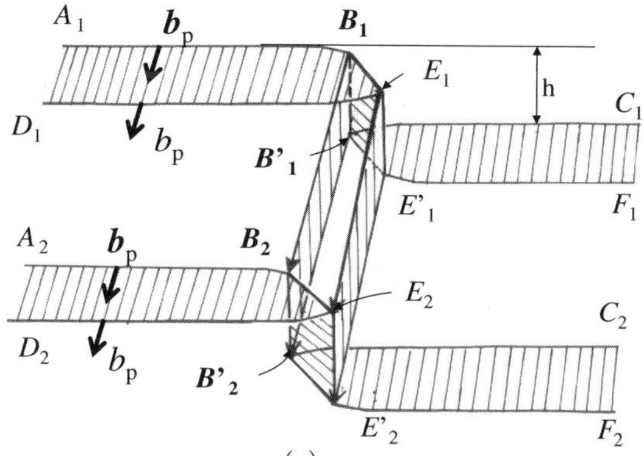

(a)

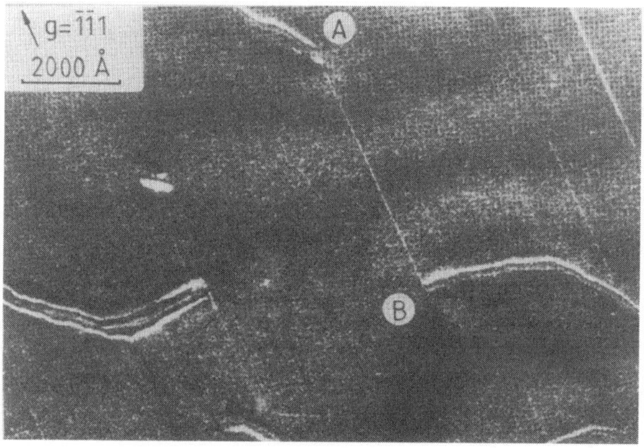

(b)

Fig. 10.15 (a) APB tube formed by the cutting of forest superdislocations [7], (a) was redrawn based on [7], (b) TEM of APB tubes (Reproduced with permission from [8]).

10.3.2 *Strengthening by APB tube formed as a result of cutting superdislocations*

Suppose that a superdislocation consisting of two superpartials cuts a forest dislocation to form a jog (Fig. 10.15). The jog formed by the cutting is also dissociated; the areas bounded by $A_1{-}B_1{-}B_1'{-}C_1$ and $D_1{-}E_1{-}E_1'{-}F_1$ are APB's. The area bounded by $B_1{-}E_1{-}E_1'{-}B_1'$ is an APB associated with an extended jog. $A_1{-}B_1$ and $D_1{-}E_1$ move abreast with $B_1'{-}C_1$ and $E_1'{-}F_1$, therefore the area of the APB remains unchanged. B_1E_1 (or $B_1'E_1'$) of the extended $B_1{-}E_1{-}E_1'{-}B_1'$ is coincident with $\boldsymbol{b}_p$, then the APB newly formed as a result of movement of E_1E_1' is cancelled by B_1B_1', thus the width of B_1E_1 (or $B_1'E_1'$) remains unchanged. However, when B_1E_1 (or $B_1'E_1'$) is not coincident with $\boldsymbol{b}_p$, the newly formed APB by E_1E_1' is not cancelled by B_1B_1'. Therefore, when the dislocation moves from $A_1{-}B_1{-}B_1'{-}C_1(D_1{-}E_1{-}E_1'{-}F_1)$ to $A_2{-}B_2{-}B_2'{-}C_2(D_2{-}E_2{-}E_2'{-}F_2)$, a tunnel connecting $B_1{-}E_1{-}E_1'{-}B_1'$ with $B_2{-}E_2{-}E_2'{-}B_2'$ is formed, the walls being covered with APB. This tunnel is referred to as an APB tube. Figure 10.15(b) shows an example of APB tube. Along the direction $[\bar{1}\bar{1}1]$ indicated by an arrow, white contrasts such as Ⓐ−Ⓑ can be seen. These are the APB tunnels.

10.4 Inverse Temperature Dependence of the Strength

Almost all substances weaken with increasing temperature. However, there are a few exceptions. Typical examples are β'-CuZn (brass) with $B2$ structure and $L1_2$ structured alloys, a typical of which is Ni_3Al. In particular, the latter is the base alloy of modern Ni-base heat-resisting super alloys. This exceptional phenomenon, the inverse temperature dependence of strength, is attributed to peculiar structures of dislocations in ordered structures.

10.4.1 *L*1₂ *structure*

Ni_3Al is of most practical importance among $L1_2$ structure metallic materials. Ni_3Al supports the modern civilization as the most important strengthener (γ': gamma prime) of heat-resisting superalloys. Figure 10.16 shows its temperature dependence of strength.

Figure 10.17 shows the structure of Ni_3Al ($L1_2$). It is based on the FCC structure. Two types of (001) planes, one consisting entirely of Ni and the other consisting of a mixture of Ni and Al with a ratio 1:1, are stacked alternately. Since it is based on the FCC structure, $\boldsymbol{b}$ is $\langle 110\rangle$ and dissociated as follows:

$$\langle 110\rangle = 1/2\langle 110\rangle + \text{APB} + 1/2\langle 110\rangle$$

There are two possibilities for its slip plane, that is, $\{111\}$ and $\{100\}$ planes.

APB energy

$\langle 110\rangle/\{001\}$ *slip*

Figure 10.17 shows a slip direction $[\bar{1}10\rangle$ and a slip plane (001), viewed along $[\bar{1}0\bar{1}]$, (101) plane being hatched. Burgers vector of the leading superpartial

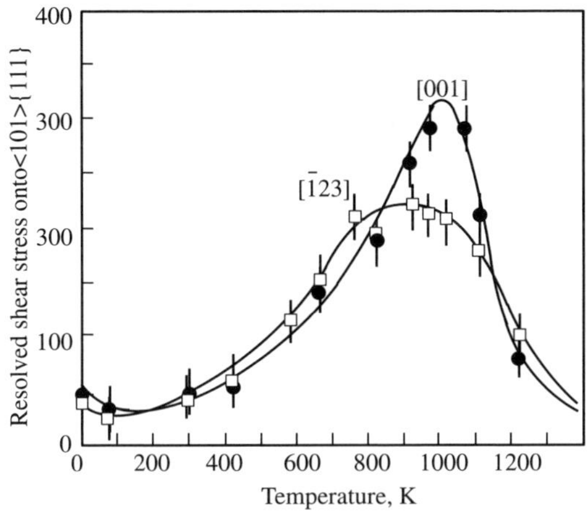

Fig. 10.16 Inverse temperature dependence of the strength of Ni_3Al single crystals (After Bontemps [9]).

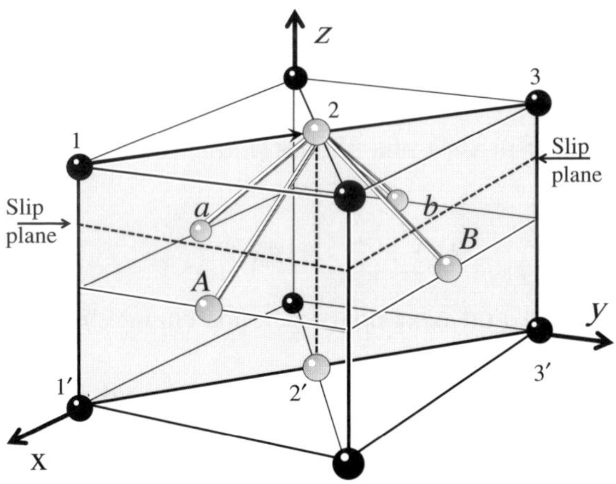

Fig. 10.17 $[\bar{1}10\rangle$ slip in $L1_2$.

$\boldsymbol{b}_{p1} = 1/2[\bar{1}10\rangle$ is indicated by $1 \to 2$. Let the actual slip plane be located as indicated by an arrow. Now let us focus on atom 2, and count the number of bonds across the slip plane.

The bonds of nearest neighbours (indicated by columns) change as follows:

$$2\bigcirc\text{-}A\bigcirc \to 1\bullet\text{-}A\bigcirc,$$

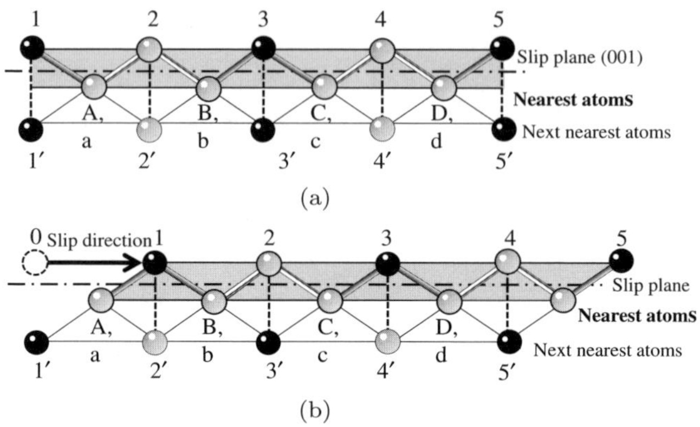

Fig. 10.18 Variation in bonds resulting from ⟨110⟩/{001} slip in $L1_2$. (a) before slip, (b) after slip.

$$2\,○\text{-}B○ \rightarrow 1\,●\text{-}B○,$$
$$2\,○\text{-}a○ \rightarrow 1\,●\text{-}a○,$$
$$2\,○\text{-}b○ \rightarrow 1\,●\text{-}b○$$

(Fig. 10.18(a)).

The bond of next nearest neighbours (indicated by broken line) changes as follows:

$$2\,○\text{-}2'○ \rightarrow 1\,●\text{-}2'\,○\,.$$

That is, both nearest and next nearest bonds change from ○-○ to ●-○ (Fig. 10.18 (b)).

However, if one counts the total number of bonds over the whole area of the slip plane, the number of the nearest bonds does not change as can be seen from Fig. 10.18(a). This is because the atoms just beneath the slip plane are all ○, so that if atoms just above the slip plane move, say, from ○ to ●, the total number of each bond remains unchanged. On the other hand, next nearest bonds change from ○-○ plus ●-● to ●-○ (Fig. 10.18(b)).

⟨110⟩/{111} *slip*

Now let us consider a slip $[\bar{1}10\rangle$ on (111) plane (Fig. 10.19). Nearest bonds prior to motion of a superdislocation are all ○-● type (Fig. 10.19(a)). After the motion they all change either into ○-○ or ●-● (Fig. 10.19(b)).

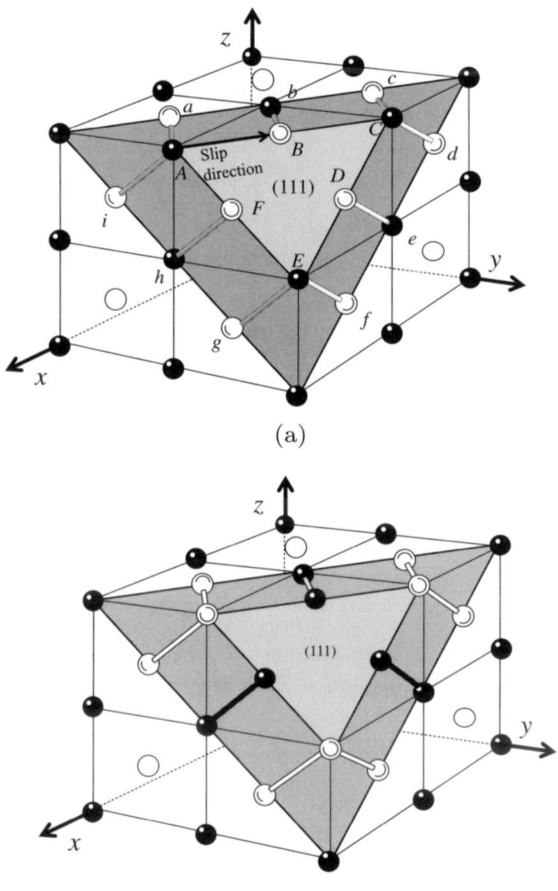

Fig. 10.19 Variation in bonds resulting from $\langle 110\rangle/\{111\}$ slip in $L1_2$. (a) Before slip, (b) After slip.

Mechanism of inverse temperature dependence of strength

Aforementioned consideration leads to the conclusion that the APB introduced by a superpartial of $\boldsymbol{b} = 1/2\langle 110\rangle\{111\}$ has wrong bonds of nearest neighbour, while the APB introduced by a superpartial of $\boldsymbol{b} = 1/2\langle 110\rangle\{001\}$ has wrong bonds of next nearest neighbours alone, that is, the nearest bonds remain to be right bonds. In other words, a $\langle 110\rangle$ superdislocation dissociated on $\{001\}$ has a lower energy than that dissociated on $\{111\}$. On the other hand, from the viewpoint of slip, it is much

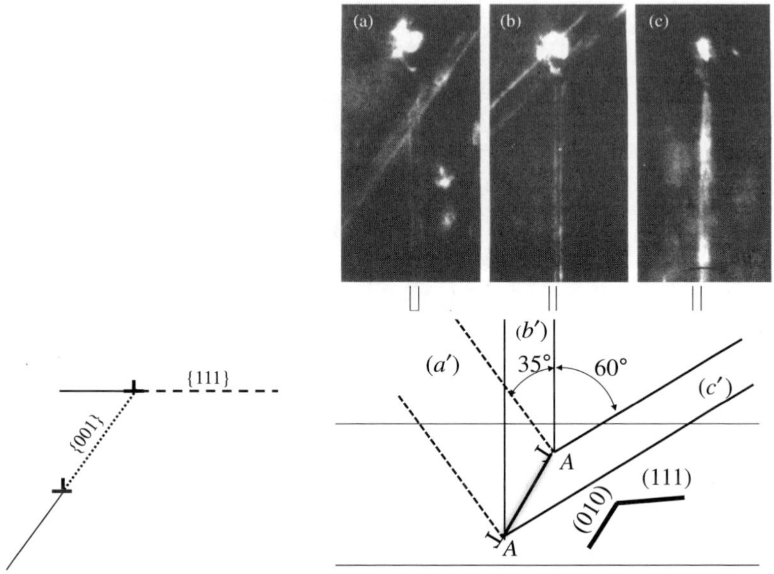

Fig. 10.20 A superdislocation dissociated on (001) plane according to K-W mechanism. The above three micrographs (a), (b) and (c) were viewed along three directions (a′), (b′) and (c′) depicted below. The variation in the superpartial separation (indicated by ||) shows that the superdislocation is dissociated on (001) plane (Reproduced with permission from [11]).

easier for a slip to take place on {111} planes than on {001} planes. As a result, a superdislocation moves on {111} plane, but when it comes to halt, it is likely dissociated onto {001} plane (Fig. 10.20(a)). At low temperature this conversion is unlikely, but on increasing temperature it becomes easier and easier thanks to thermal energy. Once a dislocation is dissociated on {001} plane, it must return from a lower-energy configuration to a higher-energy one in order to move further, and an additional force is required, that is, the alloy becomes harder. This mechanism is referred to as Kear-Wilsdorf's mechanism [10]. Figure 10.20 shows transmission electron microscopy (TEM) micrographs of a dislocation dissociated on {010} plane.

Above the peak temperature, slip on {001} planes becomes easier and the strength decreases monotonously with increasing temperature. Takeuchi and Kuramoto [12] and Paider, Pope and Vitek [13] have proposed revised versions of the K-W mechanism.

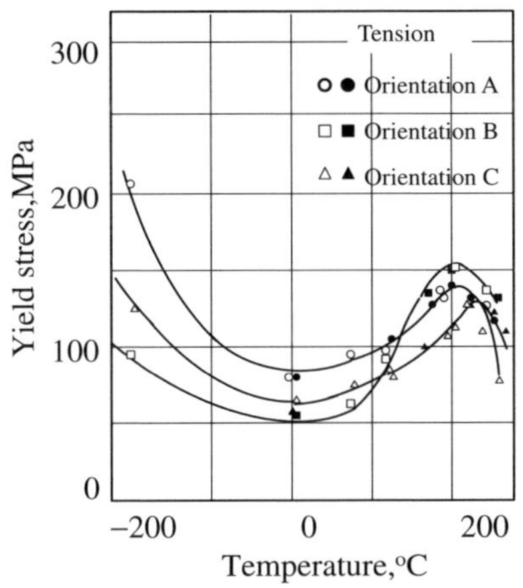

Fig. 10.21 Orientation dependence of yield stress of β'-CuZn. For orientations A, B and C refer to Fig. 10.28 (Reproduced with permission from [14]).

10.4.2 *Inverse temperature dependence of strength in β'-brass*

β-brass is an alloy of Cu–Zn. In the Cu–Zn binary system there are two phases of practical importance. One is α-brass of the primary solid solution of FCC containing maximum ~30% Zn and the other is β-brass containing 40–60% Zn. β-brass crystallizes in the (disordered) BCC structure at high temperatures but below ~460°C it transforms into an ordered alloy of B2 (β'-brass). It is β'-brass that shows the inverse temperature dependence of strength.

Figure 10.21 shows yield stress of β'-brass against temperature. There is a peak at around ~250°C in the yield stress, showing the inverse temperature dependence of strength.

Stage I (Below room temperature)

Dislocations operating in this temperature range are superdislocations of $\boldsymbol{b} = \langle 111 \rangle$ dissociated on the slip plane. Figure 10.5 shows an example. Figure 10.22(a) shows a low-magnification TEM micrograph; superdislocations are elongated along $\langle 111 \rangle$, i.e., $\boldsymbol{b}$. The density of screw dislocations is much higher than that of edge; this implies that the mobility of edge dislocations is much larger than that of screws. This is the same as disordered BCC crystals excepting that the dislocations are dissociated.

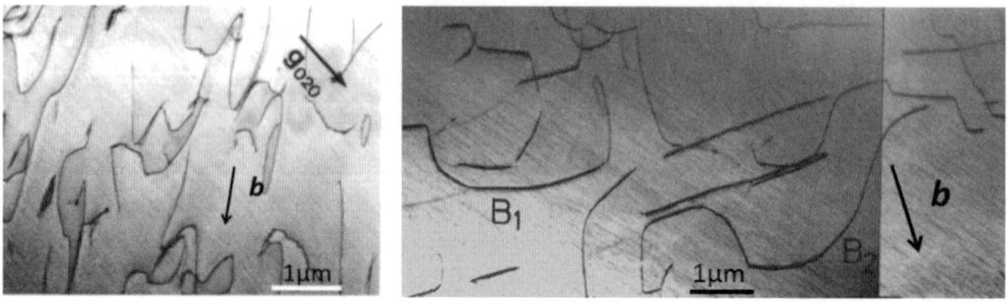

(a) (b)

Fig. 10.22 (a) Dislocation in stage I (Reproduced with permission from [15]) and (b) stage II (Reproduced with permission from [16]).

Stage II (Between room temperature and peak temperature (∼250°C))

In this temperature range, superdislocations of $\boldsymbol{b} = \langle 111 \rangle$ are activated, but the two superpartials are not on the same slip plane (Fig. 2.23(a)). They are aligned on a plane perpendicular to the slip plane (Fig. 10.23(c)). In order to distinguish between these two configurations, the former is referred to as the glide dissociation and the latter to the climb dissociation (Fig. 10.23(b)).

As is obvious from Fig. 2.23(c) the configuration of the climb-dissociated edge superdislocation is the same as that of the configuration of two edge dislocations of the same sign (Sec. 4.4.2).

$$\begin{aligned} E_c &= E_1 + E_2 + E_{12}^c + \gamma_c \times y, \\ E_g &= E_1 + E_2 + E_{12}^g + \gamma_g \times x, \\ E_{12}^c &= \frac{\mu b^2}{2\pi}\left(\ln \frac{R}{y} - \sin^2\theta\right), \\ E_{12}^g &= \frac{\mu b^2}{2\pi} \ln \frac{R}{x}. \end{aligned}$$

Here, γ_g, and γ_c are APB's formed by glide and climb dissociation, respectively, and E_{12}^c and E_{12}^g are elastic interaction energies between the two partials in the climb and glide configurations, respectively.

Substituting into

$$\Delta E = E_c - E_g = \frac{\mu b^2}{2\pi}\left(\ln \frac{x}{y} - \sin^2\theta\right) + \gamma_c \times y - \gamma_g \times x,$$

the equilibrium values of x and y are:

$$x = \frac{\mu b^2}{2\pi\gamma_g}$$

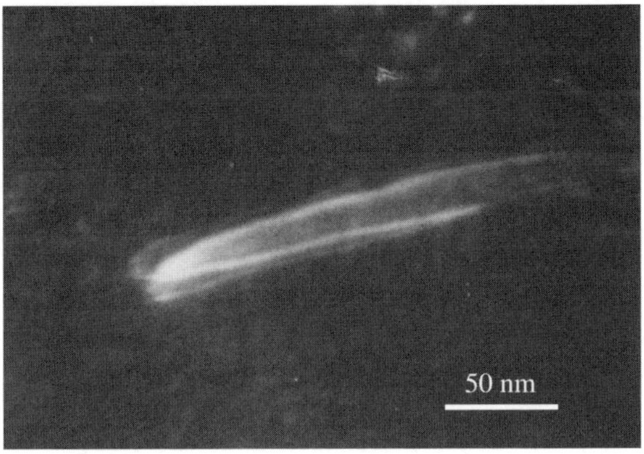

(a)

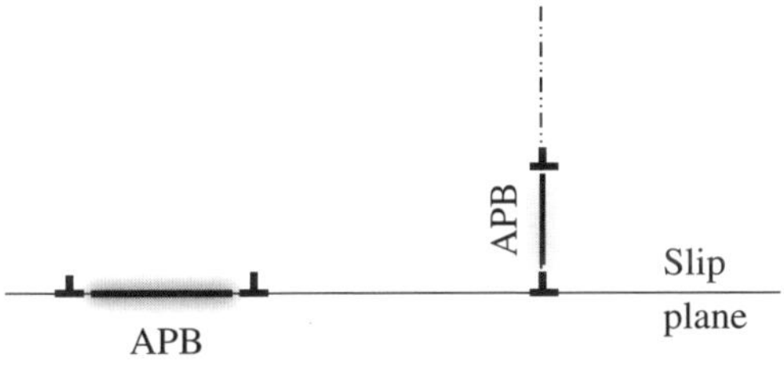

(b) (c)

Fig. 10.23 (a) TEM of a climb dissociated superdislocation [17], (b) Glide dissociation, (c) Climb dissociation.

$$y = \frac{\mu b^2}{2\pi\gamma_c},$$

$$\Delta E = \frac{\mu b^2}{2\pi}\left(-\ln\Gamma + \sin^2\theta\right),$$

is obtained, where $\Gamma = \gamma_c/\gamma_g$.

When $\Delta E < 0$, a $\langle 111\rangle$ superdislocation has lower energy in the climb dissociation than in the glide dissociation. Figure 10.24 shows stability of climb versus glide dissociation as functions of θ and Γ.

Except for the case $\Gamma \gg 1$, an edge dislocation has lower energy in the climb dissociation than in the glide dissociation. Figure 10.23(a) shows a TEM micrograph of a $\langle 111\rangle$superdislocation viewed slantwise. What looks like a forked tongue of a snake is the two superpartials in the climb dissociation. For screw dislocation, climb dissociation is not necessarily favoured from the viewpoint of energy (see Fig. 10.24), however, they are dissociated

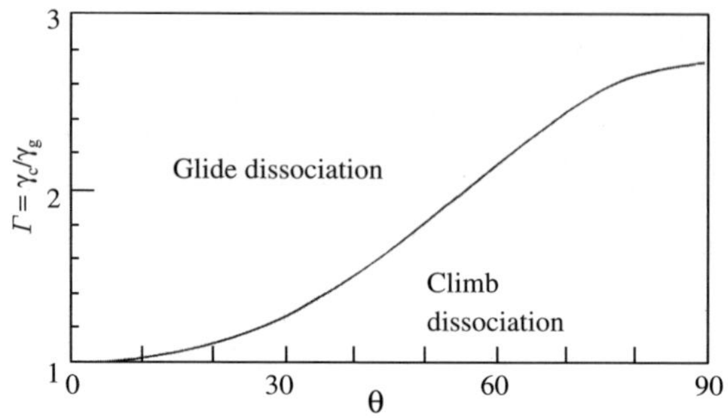

Fig. 10.24 Glide dissociation versus climb dissociation (Reproduced with permission from [18]).

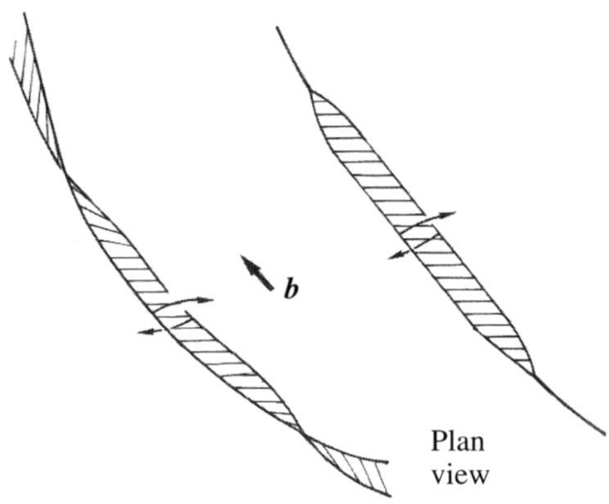

Fig. 10.25 Mechanism by which a screw superdislocation climbs (Reproduced with permission from [1]).

off the original slip plane. This is rationalized by considering that pure screws are rotated by the line tension of the adjacent non-screw dislocations which have already climb-dissociated as shown in Fig. 10.25.

Now let us consider a process by which a climb-dissociated superdislocation starts to glide under the action of an external stress. The superdislocation must experience a process shown in Figs. 10.26(a)–10.26(c). Here, in stage (b) the superpartial denoted by ○ is required to glide against the shear stress to the direction of ●. In other words, a climb-dissociated edge dislocation has difficulty in gliding. On the other hand, a screw dislocation can continue to move by cross-slip. As a result, in contrast to stage I, dislocations observed are mostly of edge type, as shown in Fig. 10.22(b).

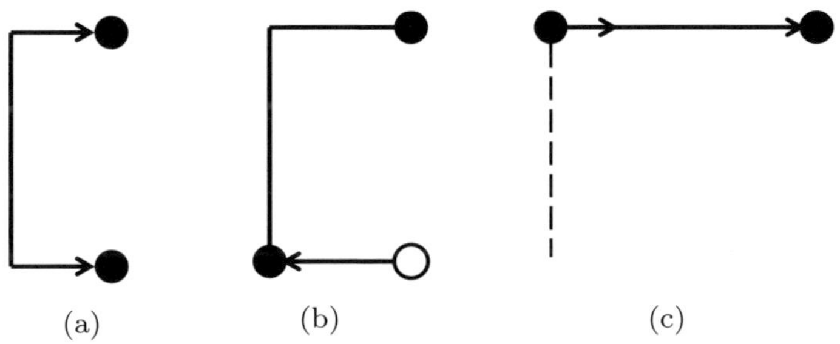

Fig. 10.26 Motion of a climb-dissociated dislocation (Reproduced with permission from [1]).

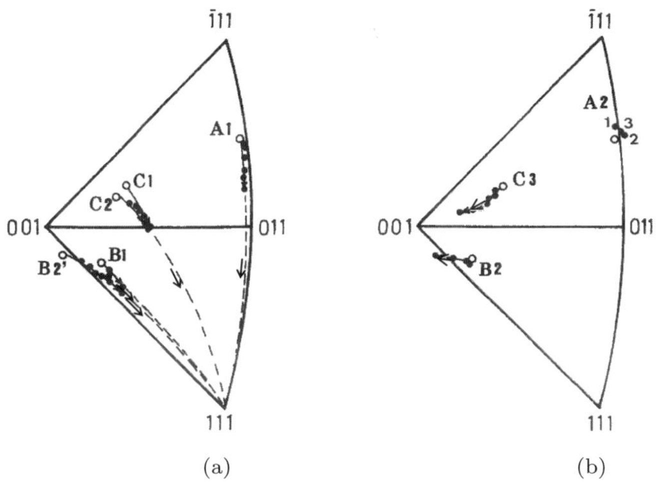

Fig. 10.27 (a) Variation in orientations during tensile deformation below T_p(a) and above T_p(b) (Reproduced with permission from [19]).

Stage III (above T_p)

In stage III, i.e., above the peak temperature (T_p), slip directions transit to non-⟨111⟩ directions. As will be described in detail in Sec. 12.1.2, the slip direction can be determined from the rotation of the longitude axis of a single crystal deformed in tension (see Fig. 12.4). Figure 10.27 shows rotations of the tensile axes of single crystals of β'-CuZn. Below T_p (Fig. 10.27(a)) the axes rotate towards ⟨111⟩; this indicates that the slip direction is ⟨111⟩. However, above T_p (Fig. 10.27(b)), the slip directions have apparently changed. The slip direction of C3 is regarded as ⟨001⟩. B2 slips towards ⟨001⟩ but deviates considerably. In particular, A2 remains

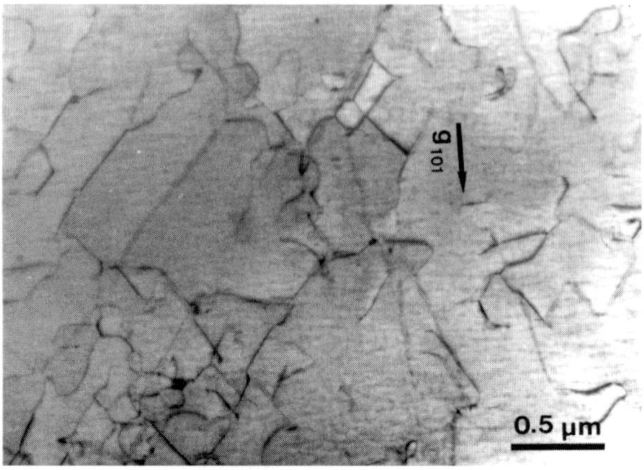

Fig. 10.28 Dislocations in stage III (Reproduced with permission from [20]).

virtually unchanged; this suggests that many slip systems are activated simultaneously in A2.

Dislocation structures revealed by TEM are in keeping with this. Figure 10.28 shows dislocation structure of a specimen deformed above T_p. It is quite different from those deformed below or at T_p (Figs. 10.22(a) and 10.22(b)). Comprehensive analysis has shown that below T_p almost all the dislocations are $\langle 111 \rangle$ dislocations, while above T_p non-$\langle 111 \rangle$ dislocations, that is, $\langle 100 \rangle$ and $\langle 110 \rangle$ dislocations are abundant (Figs. 10.29(a) and 10.29(b)).

Figure 2.18 shows an example of dislocation reactions near T_p. $\langle 111 \rangle$ dislocation is dissociated as follows:

$$\langle 111 \rangle = \langle 001 \rangle + \langle 110 \rangle [18].$$

This reaction transforms a $\langle 111 \rangle$ dislocation into a set of $\langle 001 \rangle$ and $\langle 110 \rangle$ dislocations. Anyway, non-$\langle 111 \rangle$ dislocations, i.e., $\langle 001 \rangle$ and $\langle 110 \rangle$ dislocations are perfect dislocations; these superdislocations are not dissociated and hence do not show peculiar behaviour such as the inverse temperature dependence of strength: The strength decreases monotonously until it drops precipitously at T_c due to the order-disorder transformation.

Another interesting experiment to show the importance of the role played by edge dislocations in the inverse temperature dependence of strength of β'-CuZn is mechanical tests using thin crystals [2]. Two types of thin crystals (E, S), 100-μm in thickness, which have a common tensile axis but have their surface orientations corresponding to the side face and

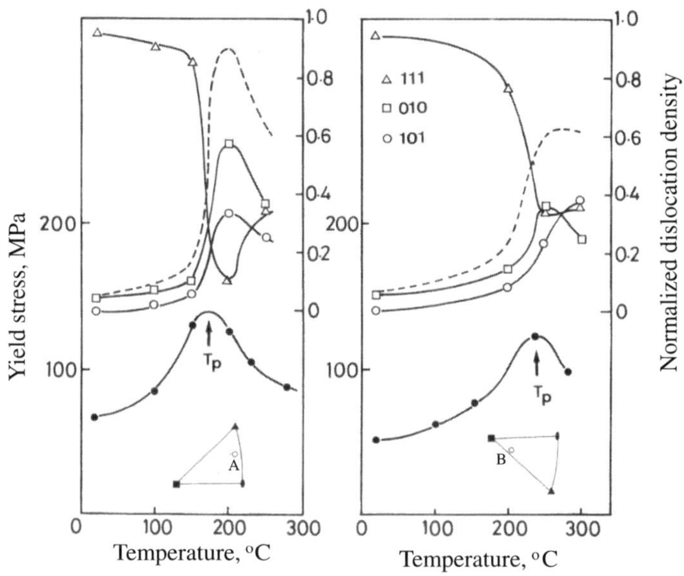

Fig. 10.29 Temperature dependence of characters of active dislocations. Broken line shows total density of non-⟨111⟩ dislocations. (Reproduced with permission from [21]).

top faces, (see Appendix A.5.2) (Fig. 10.30) respectively, were deformed at various temperatures to measure the temperature dependence of the yield stress (Fig. 10.31). Geometry of these crystals suggests that, in crystal E, a strain is carried mostly by edge dislocations, while, in crystal S, it is carried mostly by screw dislocations. Figure 10.31 clearly shows that, in crystal E, the strength shows the inverse temperature dependence, while in crystal S it does not.

Since the pioneering work by Ardley and Cottrell [22] in 1953 many theories have been proposed on the mechanisms of the inverse temperature dependence of strength of β'-CuZn; they could not stand the test of time. It is to be attributed to the climb dissociation of edge superdislocations and change in slip direction. However, their orientation dependence and tension-compression asymmetry etc. are still unsolved. For detail readers are referred to literature [23].

10.4.3 *Summary*

The inverse temperature dependence of strength is a result of competition between two phenomena, i.e., mobility and equilibrium energy of

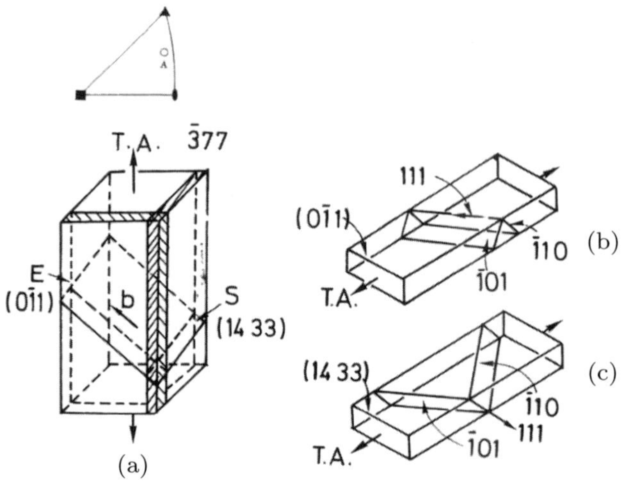

Fig. 10.30 A method to distinguish between roles played by edge and screw dislocations (Reproduced with permission from [1]).

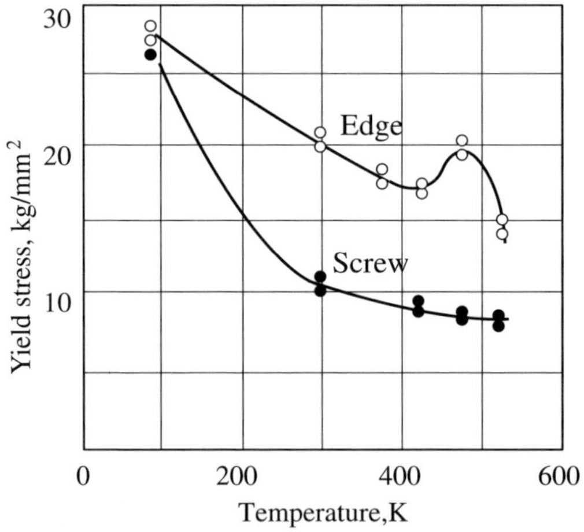

Fig. 10.31 Temperature dependence of strength in β'-CuZn. While edge dislocation shows the inverse temperature dependence of strength, screw dislocations do not (Reproduced with permission from [1]).

dislocations. When a dislocation (or the slip system) with high mobility is coincident with a dislocation with low equilibrium energy, the inverse temperature dependence of strength does not occur. However, when they are not coincident, at low temperatures slips of type 1 (with higher mobility and

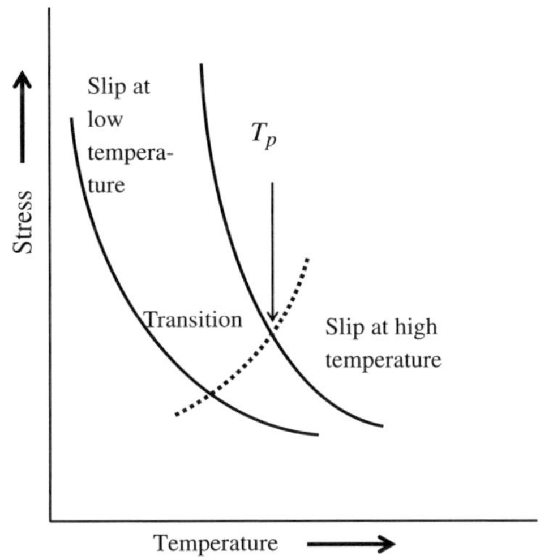

Fig. 10.32 Mechanism of the inverse temperature dependence of strength.

higher energy) occurs and at high temperature slips 2 (with lower mobility and lower energy) occurs, and at the transition the inverse temperature dependence of strength occurs, as is shown in Fig. 10.32.

References

1. Zhu, Y. M. and Saka, H. *Phil. Mag. A* **59** (1989), 661.
2. Saka, H., Kawase, M., Nohara, A., and Imura, T. *Phil. Mag. A* **50** (1984), 65.
3. Kuroda, K. private communication.
4. Crawford, R. C., Ray, I. L. F., and Cockyane, D. J. H. *Phil. Mag.* **27** (1973), 1.
5. Marchinkovski, M. J., Brown, N., and Fisher, R. M. *Acta Met.* **9** (1961), 129.
6. Korner, A. and Saka, H. Unpublished work.
7. Chou, C. T. and Hirsch, P. B. *Phil. Mag. A* **44** (1981), 1415.
8. Chou, C. T., Hazzledine, P. M., Hirsch, P. B., and Anstis, G. R. *Phil. Mag. A* **56** (1987), 799.
9. Bontemps, C. Cited in *Dislocations in Solids*, vol. 10, edited by Nabarro.
10. Kear, B. H. and Wildsdorf, H. G. *Trans. Metall. Soc. AIME* **224** (1962), 382.
11. Suzuki, K., Ichihara, M., and Takeuchi, S. *Acta Mat.* **27** (1979), 193.
12. Takeuchi, S. and Kuramoto, E. *Acta Met.* **21** (1973), 415
13. Paider, V., Pope, D. P., and Vitek, V. *Acta Met.* **32** (1984), 435.
14. Nohara, A., Izumi, M., Saka, H., and Imura, T. *Phys. Stat. Sol.* (*a*), **37** (1984), 719.
15. Zhu, Y. M. and Saka, H. *Phil. Mag. A* **69** (1994), 1155.
16. Zhu, Y. M. Dissertation, Nagoya University, 1988.
17. Zhu, Y. M. and Saka, H. Unpublished work.

18. Saka, H. and Zhu, Y. M. *Phil. Mag. A* **51** (1985), 629.
19. Saka, H., Zhu, Y. M., Kawase, M., Nohara, A., and Imura, T. *Phil. Mag. A* **51** (1985), 365.
20. Zhu, Y. M. and Saka, H. *Phil. Mag. A* **54** (1986), 783.
21. Saka, H. and Kawase, M. *Phil. Mag. A* **40** (1984), 525.
22. Ardley, G. W. and Cottrell, A. H. *Proc. Roy. Soc. Ser. A* **219** (1953), 328.
23. Saka, H. Memoirs School of Eng'g, Nagoya Univ., **52** (2000), 53.

Chapter 11

Dislocations in Diamond, Zincblende, Wurtzite Structures and SiC

11.1 Dislocations in Diamond Structure (Shuffle-Set and Glide-Set Dislocations)

11.1.1 *Theoretical consideration*

The diamond structure is shown in Fig. 1.1(d). Two face-centered cubic (FCC) lattices ① and ② are interpenetrate. Table 11.1 and Fig. 11.1(a) show coordinates of atoms in the unit cell. Lattice ② is displaced by (1/4, 1/4, 1/4) with respect to lattice ①. In what follows, explanation will be made on Si, which has been most extensively studied. In order to distinguish between atoms in the interpenetrating lattices ① and ②, they are indicated by ● and ○, respectively, in Table 11.1 and Fig. 11.1(a), but needless to say, both of them stand for Si atoms. In the zincblende structure (for instance GaAs) which will be discussed later, ● and ○ stand for different elements (Ga and As). There are so many analogies between the diamond and the zincblende structures that, throughout this book, even when describing the diamond structure (Si), ● stands for Si atoms in the lattice ①, which pass through the origin, and ○ stands for Si atoms in the lattice ②, which pass through (1/4, 1/4, 1/4).

Si atoms are bound by four covalent bonds I, II, III, IV as is shown in Fig. 11.1(b). Their directionality, length and angles between them are very strict.

Exercise 11.1 Calculate the distance of neighbouring atoms (bond length) and the angle between bonds.

Table 11.1 Coordinates of atoms in the diamond structure.

	FCC lattice ① with 000 as the origin A, B, C ●				FCC lattice ② with 1/4 1/4 1/4 as the origin a, b, c ○			
x	0	0	1/2	1/2	1/4	1/4	3/4	3/4
y	0	1/2	0	1/2	1/4	3/4	1/4	3/4
z	0	1/2	1/2	0	1/4	3/4	3/4	1/4

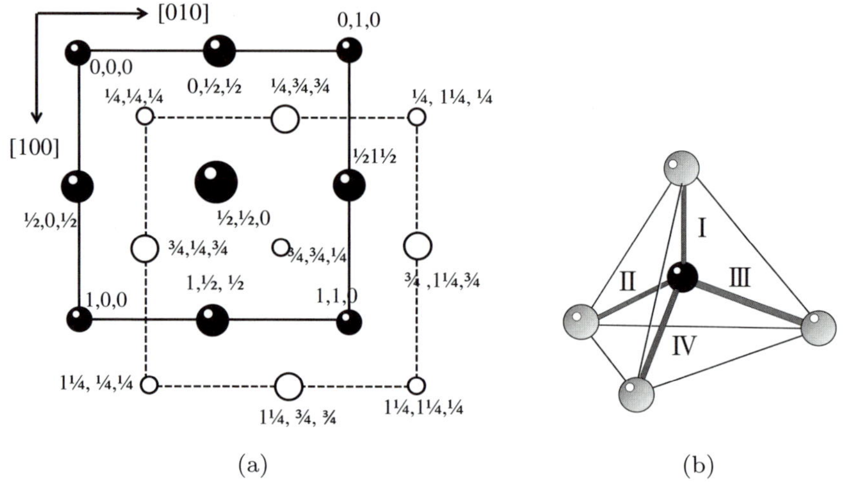

Fig. 11.1 (a) Coordinates of atoms in the diamond structure, (b) Covalent bonds in the diamond structure.

Hornstra's review [1] is the most comprehensive on dislocations in the diamond structure, on which the following explanation is based. The diamond structure is basically FCC structure, so that the slip system is $1/2\langle 110\rangle/\{111\}$. As shown in Fig. 11.2(a), there is a doubly bent tortoise-shell indicated by 1246531 along (111) plane. In Fig. 11.2(b) this shell-like (111) is placed horizontally.

Figure 11.2(c) is a projection viewed along $[1\bar{1}0]$ direction as redrawn from Fig. 11.2(b). A, B and C correspond to $\{111\}$ plane of the interpenetrating FCC lattice ①, while a, b and c correspond to the other interpenetrating FCC ②. There are two candidates for the position between $\{111\}$ planes into which a dislocation can be inserted; one is the widely separated position such as aA, bB and cC and the other is the narrowly separated position such as Ac, Ba and Cb. The dislocation in the former position

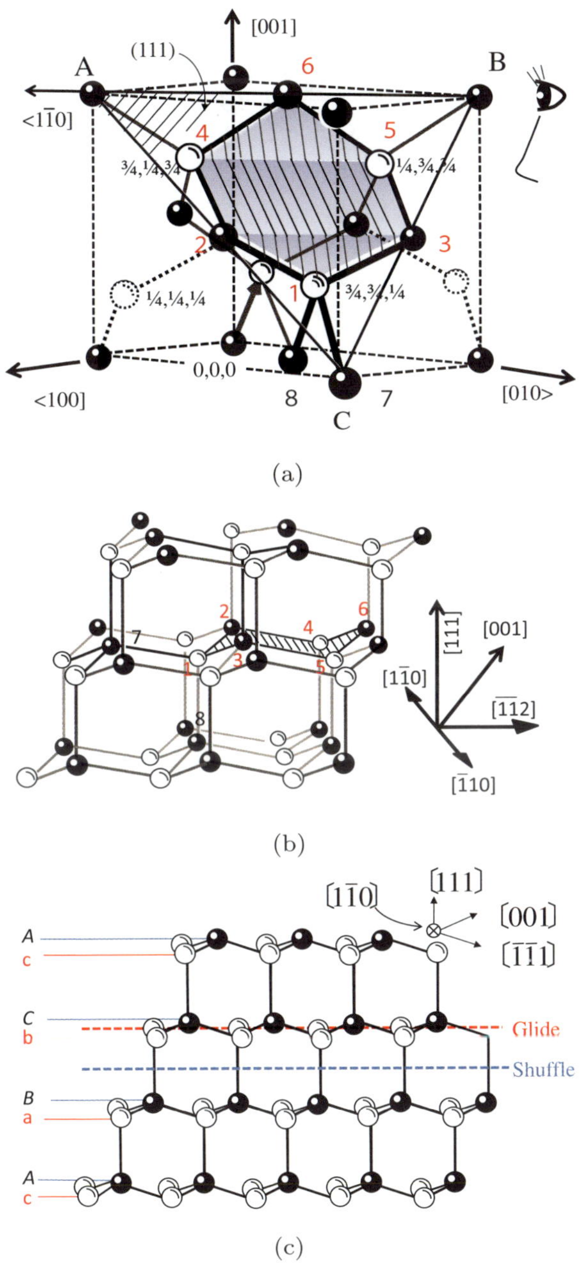

Fig. 11.2 {111} plane in the diamond structure.

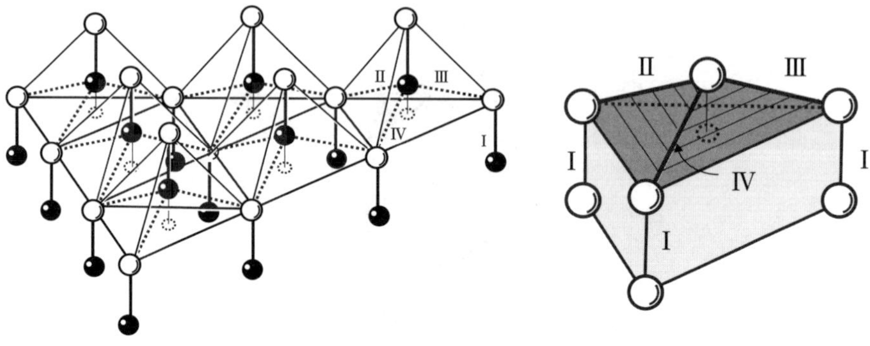

(a) (b)

Fig. 11.3 {111} plane in the diamond structure.

is referred to as the shuffle-set dislocation (SP) and the dislocation in the latter position is referred to as the glide-set dislocation (GD).

In Fig. 11.2(b), {111} planes are projected along the direction of $\langle 110 \rangle$. In Fig. 11.3(a), this is redrawn using tetrahedron shown in Fig. 11.1(b). In Fig. 11.1(b), four bonds are equivalent to each other. However, when one considers that the upper part is sheared with respect to the lower part across two adjacent {111} planes as shown in Fig. 11.3(a), the thick bond I is no longer equivalent to the other bonds II, III, IV (indicated by broken lines in Fig. 11.3(b)). In Fig. 11.3(b), bond I hangs below three bonds II, III and IV. The structure can be compared to a triangular pyramid on a triangular prism. Sliding the triangular pyramid as a whole over the prism corresponds to the SP (Fig. 11.4(a)). On the other hand, dividing the pyramid into upper and lower parts and then sliding the upper part with respect to the lower part corresponds to the GD (Fig. 11.4(b)). The problem here is to identify which is easier to slide. In other words, which P–N force is lower.

P–N force is given by Eq. (6.1).

$$\tau_p = \frac{2\mu}{1-\nu} \exp\left\{-\frac{2\pi d}{b(1-\nu)}\right\}, \qquad \text{(6-1)(Again)}$$

Lattice spacing d is much larger for the SP than for the GD. Therefore, according to Eq. (6.1) the Peierls–Nabarro (P–N) force of the shuffle-set dislocation is to be much lower than that of the GD. That is, the SP is to be much easier to move than the GD.

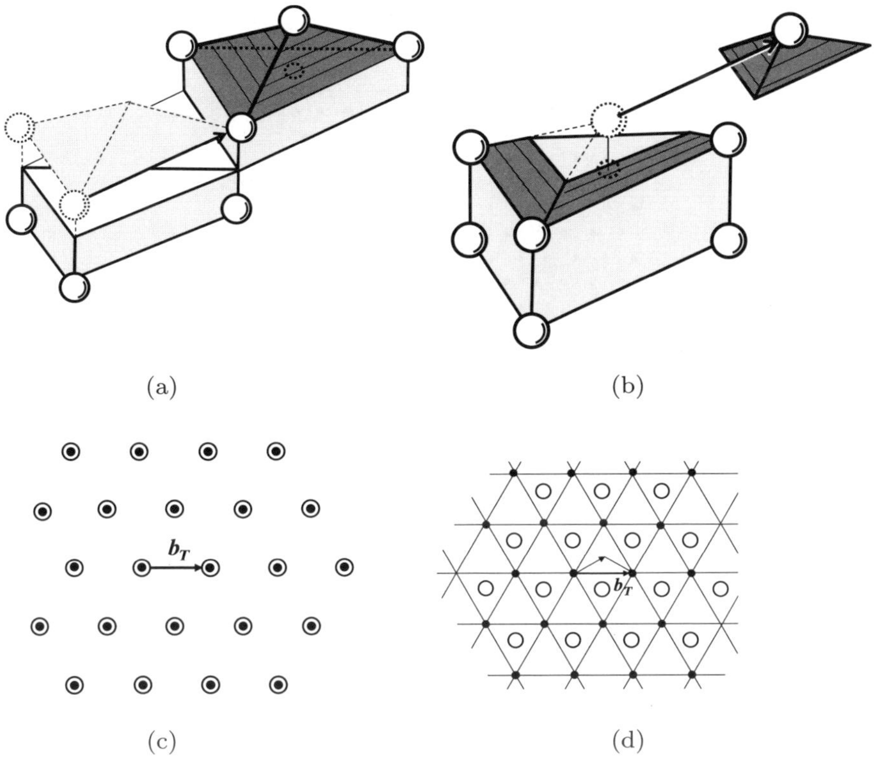

Fig. 11.4 (a)–(c) SP. (b)–(d) GD.

On the other hand, from the viewpoint of dissociation of dislocations in the FCC structure, the SP is unlikely to be dissociated because two atoms face each other across the slip plane (Fig. 11.4(c)). That is, the SP is a perfect dislocation. This is the reason why the shuffle-set dislocation is denoted by SP in a contracted form: P means a perfect dislocation. On the other hand, since atomic configuration above and below the slip plane is same as FCC, the GD is expected to be dissociated (Fig. 11.4(d)). This is the reason why the GD is denoted by GD in a contracted form: D means a dissociated dislocation.

In any case, atomic bonds in the diamond structure are strong and also their directionality is strong. Because of this directionality, or anisotropy, dislocations have the tendency to align themselves along preferential crystallographic orientations. As can be inferred from the structure of a

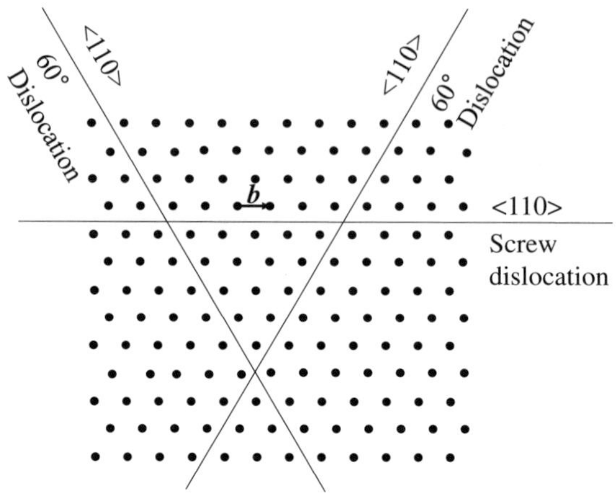

Fig. 11.5 P-N potential in the diamond structure.

tetrahedral shown in Fig. 11.1(b), the preferred orientation is ⟨110⟩.[1] There are deep valleys along ⟨110⟩, where dislocations lie. The force required for the dislocations to escape from the valleys is the P–N force. Burgers vector $\boldsymbol{b}$ of a perfect dislocation in the diamond structure is $1/2\langle 110\rangle$, so that the angle between the direction (ξ) of the dislocation line lying along the P-N valleys and $\boldsymbol{b}$ is either 0 or 60°. The former is a screw dislocation and the latter is a 60° dislocation lying along two directions corresponding to ±60° (Fig. 11.5).

11.1.2 *Experimental results*

Brittle-ductile transition

Si is quite brittle below ∼600°C,[2] above which it is ductile. The temperature at which such a transition from brittle to ductile takes place is referred to as a ductile–brittle transition temperature (DBTT). The DBTT in Si is quite sharp.

Stress-strain curves

Figures 11.6(a) and 11.6(b) shows stress–strain curves. Below DBTT, no plasticity is observed (Fig. 11.6(a)), while above DBTT, stress–strain curves are quite similar to those of FCC metals and alloys (Fig. 11.6(b)).

---

[1] As will be mentioned later, the SP have deep P–N valleys along ⟨112⟩ and ⟨123⟩ beside ⟨110⟩.

[2] DBTT is sensitive to a strain rate: It decreases with increasing a strain rate.

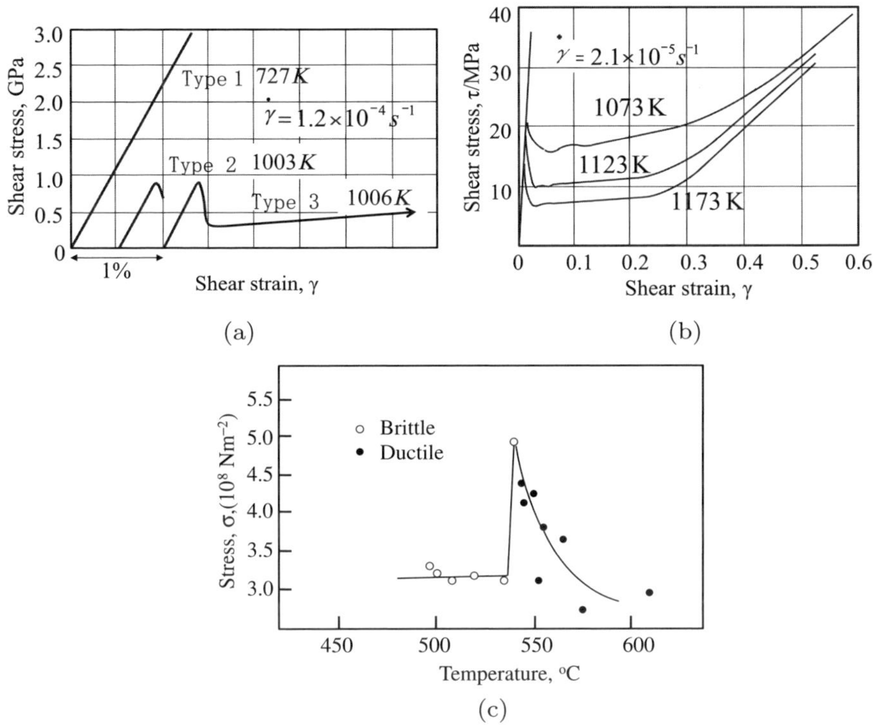

Fig. 11.6 Temperature dependence of mechanical properties in Si. (a), (b) Stress–strain curves (Reproduced with permission from [2], [3]), (c) Fracture stress below DBTT and yield stress above DBTT (Reproduced with permission from [4]).

Fracture toughness

In accordance with the transition of plasticity, fracture toughness is quite small and temperature-independent below DBTT, while above 600°C it jumps, followed by a monotonous decrease (Fig. 11.6(c)).

Dislocation structure

T > DBTT

As can be inferred from Figs. 11.6(b) and 11.6(c) many dislocations operate in Si deformed above DBTT. Figure 11.7 shows dislocations in Si deformed at 520°C. Straight dislocations are lying along ⟨110⟩ directions as shown in Fig. 11.5; this suggests strongly that the P–N force along ⟨110⟩ direction is high. Similar photographs are shown in Figs. 6.7 and 6.10. Figure 6.7 is an infrared micrograph and Fig. 6.10 is an X-ray topograph. The magnification

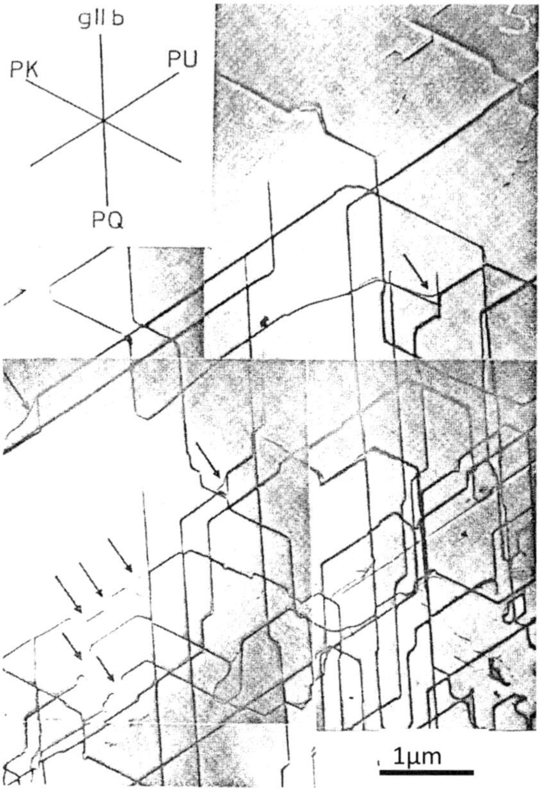

Fig. 11.7 Dislocations in Si lying along ⟨110⟩ directions (Reproduced with permission from [5]).

of these photographs is smaller than that of Fig. 11.7(a) by a factor of magnitude of 3–4.

It is to be added that dislocations lie strictly along ⟨110⟩ P–N valleys only at the initial stage of deformation. Also, when deformed at higher temperatures, the P–N force exerts less influence, allowing dislocations to curve smoothly. Figure 11.8(a) shows dislocations in Si deformed at 1173 K (900°C) and Fig. 11.8(b) shows the same dislocations annealed at a higher temperature range (1173–1273 K). In the as-deformed state (a), dislocations are bent sharply, while after being annealed at higher temperatures they are curved smoothly, indicating that they have overcome the P–N potentials.

Figure 11.9 shows a dislocation at a higher magnification. It is obvious that the dislocation is dissociated into two Shockley partials [7]. That is, above DBTT, the GD operate. On the other hand, below DBTT, Si does

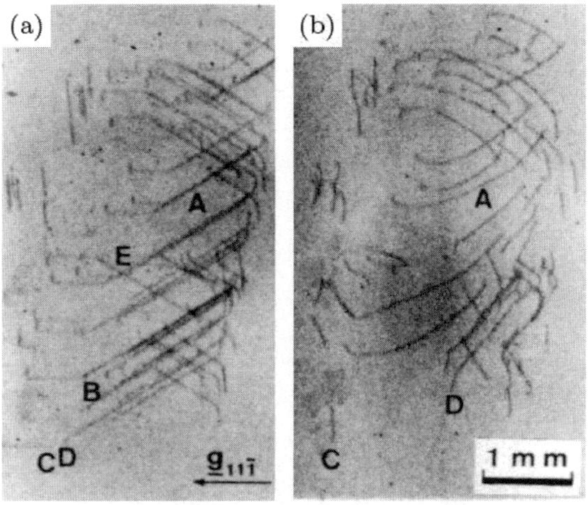

Fig. 11.8 Variations in dislocation configuration with temperature (Reproduced with permission from [6]).

not show any plasticity in a conventional tensile test, but can be deformed plastically in compression under a high hydraulic pressure. Figure 11.10 shows defect structure in Si deformed plastically at 400°C in a high-pressure cell [8]. Elongated triangles and squares with white contrasts are widely extended stacking faults, that is, the GD. 400°C is a critical temperature above which the GD can be introduced by macroscopic deformation.

Below room temperature

Below DBTT, the SP operate. Especially in a deformation mode of indentation or scratching, where a hydraulic pressure is associated with deformation, dislocations are introduced even at liquid nitrogen temperature (77 K). Figures 11.11(a) and 11.11(b) show TEM micrographs of dislocations in Si deformed locally by indentation using a diamond stylus (like in the indentation test used to meassure hardness) at 77 K. The dislocations are straight but their directions are not only $\langle 110\rangle$ but also $\langle 112\rangle$, $\langle 123\rangle$. Figure 11.12 shows a high resolution TEM micrograph and shows that the dislocation is not a dissociated but perfect one. Drawing a Burgers circuit, $\boldsymbol{b}$ is $1/2\langle 110\rangle$, that is, Burgers vector of a perfect dislocation.

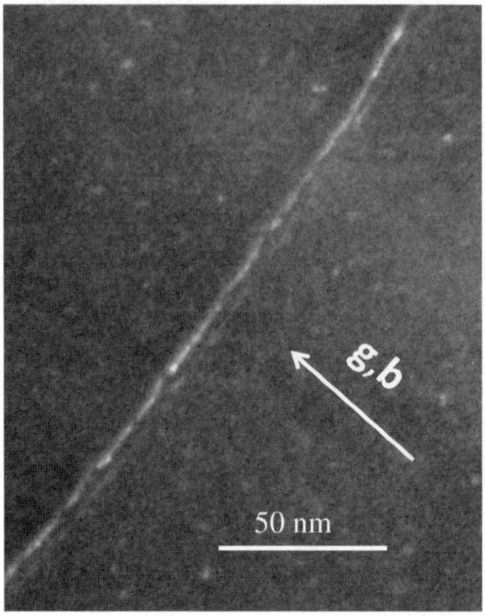

Fig. 11.9 A dissociated GD in Si (Reproduced with permission from [7]).

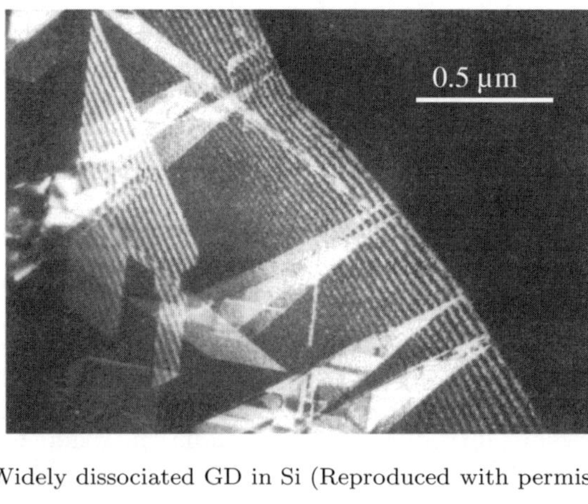

Fig. 11.10 Widely dissociated GD in Si (Reproduced with permission from [8]).

Room temperature ~ just below DBTT

Those dislocations that are introduced below DBTT are the SP but their configurations are not straight at all as shown in Fig. 11.13. Presumably, this is accounted for by assuming that part of a SP is transformed to a GD as shown Figs. 11.14(a) and 11.14(b). Two mechanisms have been proposed for transformation from the shuffle to the glide-sets (Figs. 11.15(a)

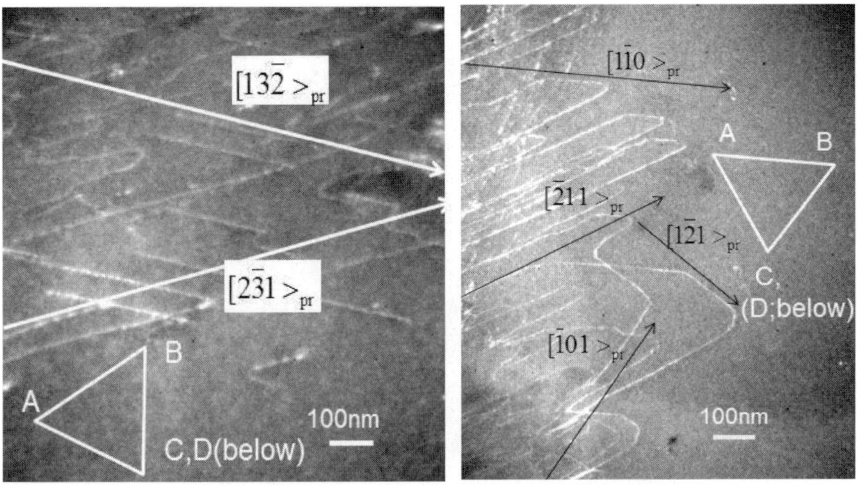

(a) (b)

Fig. 11.11 (a), (b) Shuffle-set (SP) dislocations in Si introduced at 77 K (Reproduced with permission from [9]).

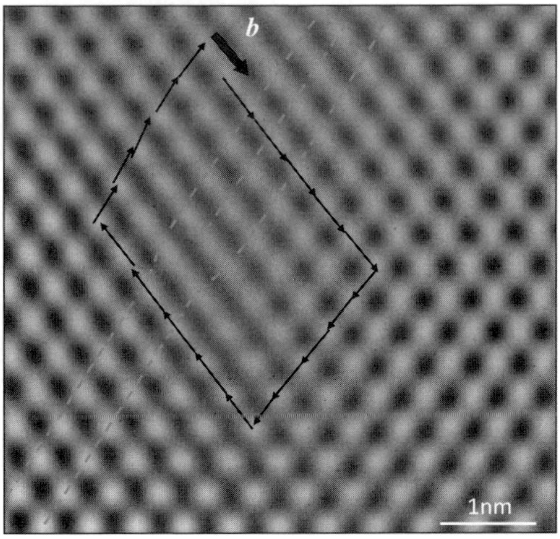

Fig. 11.12 High resolution electron micrograph of a SP dislocation and the associated Burgers circuit (Reproduced with permission from [10]).

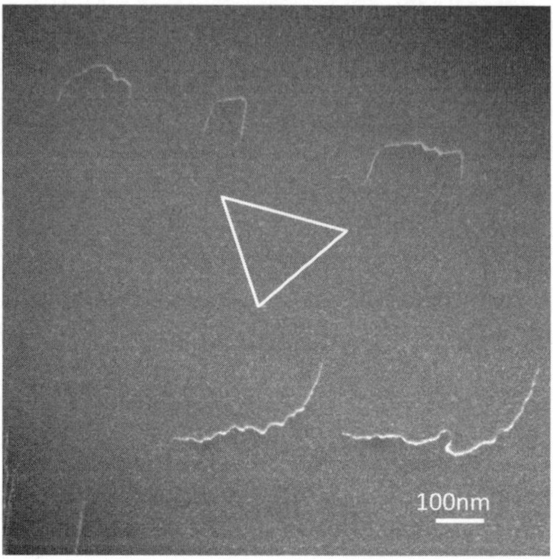

Fig. 11.13 SP dislocations introduced just below DBTT in Si (Reproduced with permission from [9]).

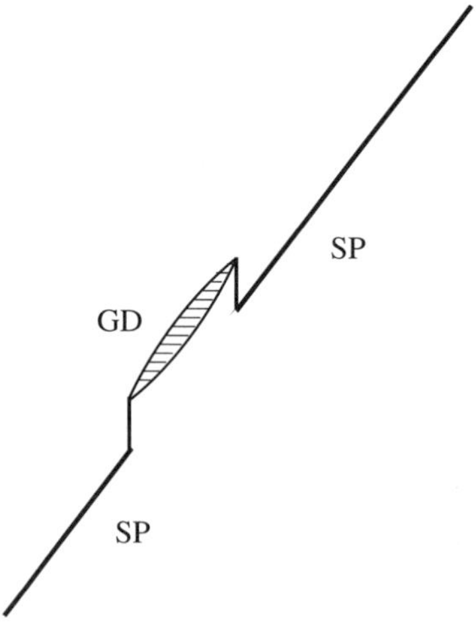

(a)

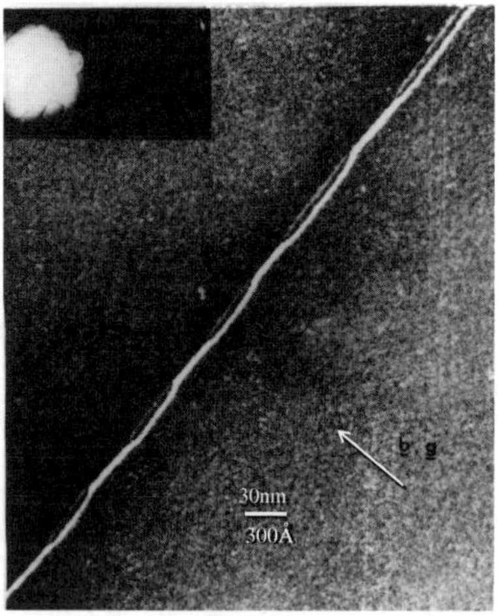

(b)

Fig. 11.14 Mixture of GD and SP dislocations. (a) Schematic illustration, (b) Dark-field weak-beam micrograph (Reproduced with permission from [11]).

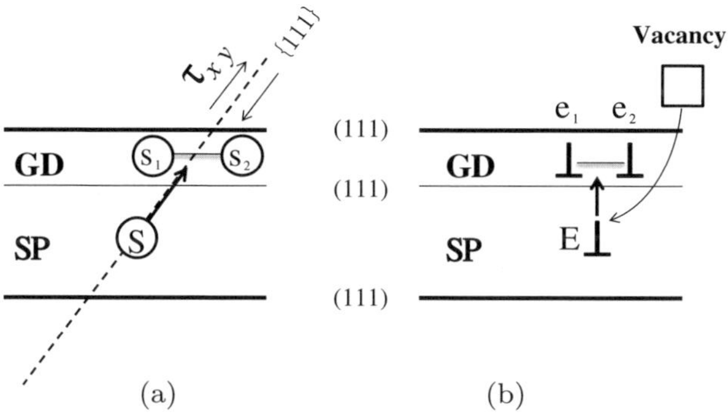

Fig. 11.15 Two mechanisms by which a SP dislocation is transformed into a GD, i.e., (a) by cross-slip and (b) by climb (Reproduced with permission from Ref. [12]).

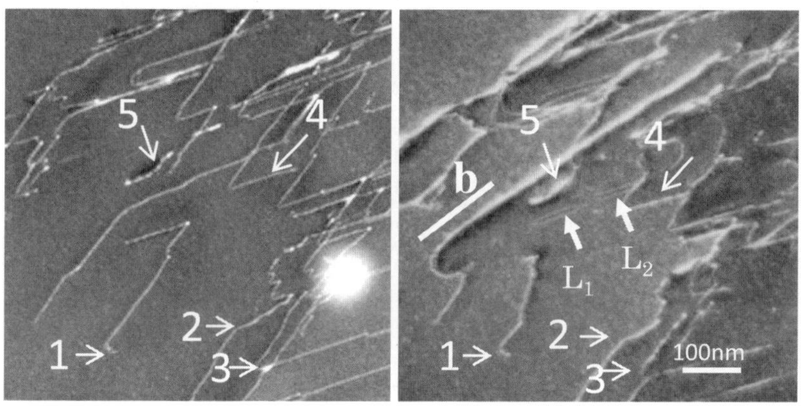

Fig. 11.16 (a), (b) Transformation of dislocations into GD during *in situ* heating in an electron microscope (Reproduced with permission from [13]). Dislocations 1, 2, 3 and 4 remained unchanged during the event.

and 11.15(b)). One is that the SP on (111) plane cross-slips onto a different {111} plane under the action of a shear stress and transforms itself into the GD at the position indicated by GD. This process requires cross-slip, so that the SP must be a screw dislocation (Fig. 11.15(a)). The other mechanism is that the SP climbs absorbing vacancies and transforms itself into the GD at the position indicated by GD.

Figures 11.16(a) and 11.16(b) show a direct evidence of the shuffle-to-glide transformation observed by an *in situ* heating experiment in a TEM. The SP introduced at a low temperature (Fig. 11.16(a)) transform into the

GD and glide away (Fig. 11.16(b)). Faint contrasts due to dislocation loops L_1 and L_2 are left behind the SP that has transformed into a GD (b).

11.2 Dislocations in the Zincblende Structure

The zincblende structure has two interpenetrating FCC lattices as in the diamond structure. While in the diamond structure, the two FCC lattices are composed of the same element (such as Si), in the zincblende structure the two FCC lattices are composed of different elements (one belonging to II or III groups and one belonging to VI or V groups) (Fig. 11.17(a)). The base consisting of two interpenetrating tetrahedra is shown in Fig. 11.17(b). The slip system is $1/2\langle 110\rangle\ \{111\}$, and the stacking of the $\{111\}$ slip planes is shown in Fig. 11.17(c).

Configurations of the SP and the GD are same as in the diamond structure. Let us take GaAs, a typical zincblende structure, as an example and follow Hornstra [1].[3]

What Fig. 11.2(c) is for Si corresponds to what Fig. 11.17(a) is for GaAs. In Si, both lattices ① and ② are occupied by Si, so that they are not distinguished. But in the zincblende structure, lattices ① and ② are occupied by different elements. In what follows, the lattice ① (origin) is occupied by elements of II or III groups (indicated by A: (in the case of GaAs by Ga (●)), the lattice ② (1/4, 1/4, 1/4) is occupied by elements of VI or V groups (indicated by B: (in the case of GaAs by As (○)).

Close inspection of Fig. 11.17(a) reveals that there are two types of tetrahedra comprising GaAs. These tetrahedra correspond to I–II–III–IV and I′–II′–III′–IV′, respectively, in Fig. 11.17(b). The tetrahedron I–II–III–IV has As(○) at the four corners and Ga(●) in the centroid. Let us refer to this tetrahedron as the As tetrahedron. The reverse is true for the tetrahedron I′–II′–III′–IV′, which will be referred to as the Ga tetrahedron.

GaAs can be depicted using either only As tetrahedra or only Ga tetrahedra as shown in Figs. 11.17(c) and 11.17(d). Needless to say, these are equivalent, but the As tetrahedron whose apex directs upwards will be used throughout this book.

11.2.1 *Polarity*

Consider a thin crystal (wafer) whose wide faces are $\{111\}$. The $\{111\}$ surfaces are terminated by either Ga or As. In the narrowly separated space

[3]Hornstra did not discuss on the zincblende structure, however, by distinguishing lattices ① and ②, one can apply his model (Fig. 11.17).

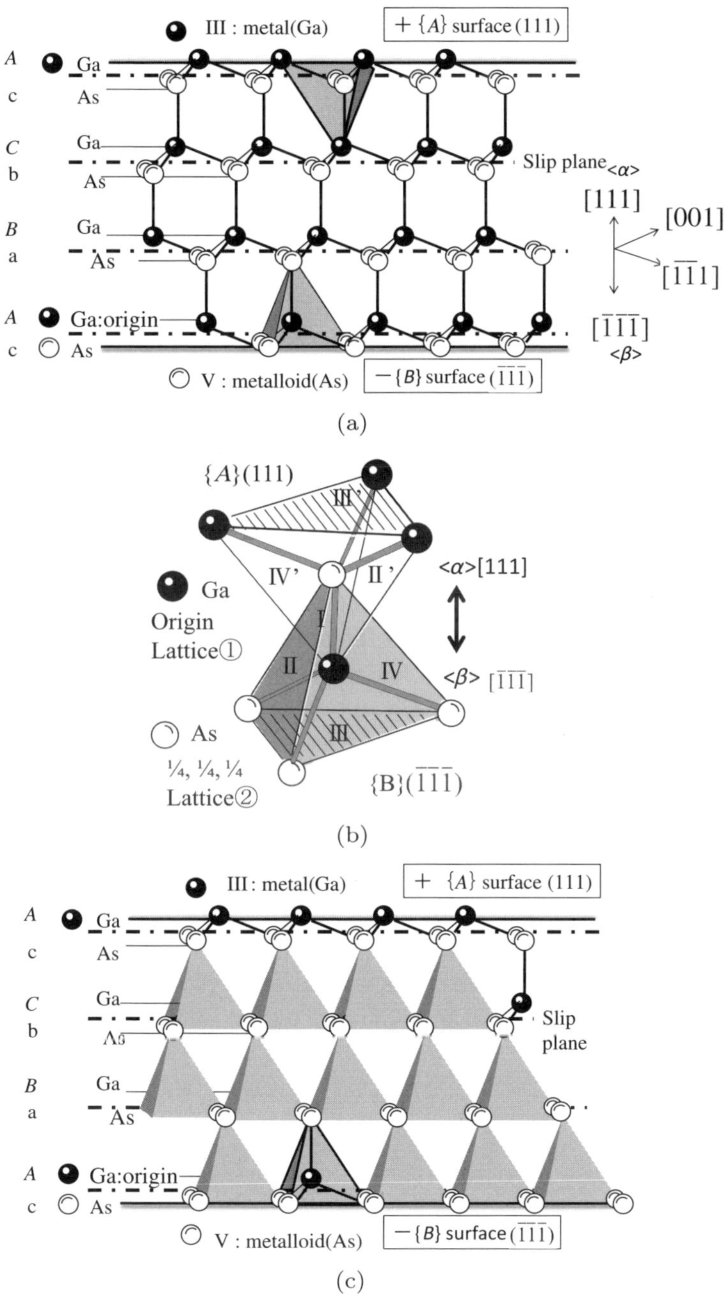

Fig. 11.17 Polarity of {111} plane.

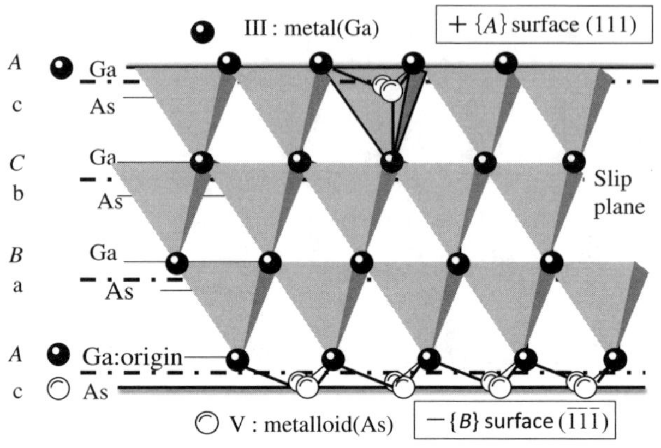

(d)

Fig. 11.17 (*Continued*)

such as Ac three bonds are present (corresponding to II–IV in Fig. 11.17(b)), while in the widely separated space such as cC only one bond is present (corresponding to I in Fig. 11.17(b)). Therefore, bonding across Ac is much stronger than that across cC, resulting in cleavage taking place always at bond I. In Figs. 11.17(a)–11.17(c) the upper surface ((111) plane) is Ga(●)-terminated surface and the lower surface (($\bar{1}\bar{1}\bar{1}$) plane) is As (○)-terminated surface. In other words, the crystal surface has a polarity, with one surface Ga-terminated and the other As-terminated. The surface terminated by elements of group III (A: ●(Ga)) is referred to as A-surface or plus (+) (meaning (111) plane) surface, and the surface terminated by elements of V group (B: ○(As)) as B-surface or minus (−) (meaning ($\bar{1}\bar{1}\bar{1}$) plane) surface. Throughout this book, the symbols A and B will be indicated by $\{A\}$ and $\{B\}$ in order to make clear that they stand for crystal surfaces. Similarly, [111] and [$\bar{1}\bar{1}\bar{1}$] directions will be indicated by $\langle\alpha\rangle$ and $\langle\beta\rangle$.

Shuffle-set dislocations

Let us make a slit from $\{B\}$ surface and remove a region indicated by a square □ as shown in Fig. 11.18(a), then weld the slit as shown in Fig. 11.18(b). An extra half-plane (ehp) is upward, i.e., in the upper half of the crystal. As can be seen from Fig. 11.18(b), in the dislocation core As atoms (○) have dangling bonds (DB). When the ehp is downward, i.e., in the lower half of crystal, the dangling bonds in the dislocation core are Ga (●) (Fig. 11.18(c)).

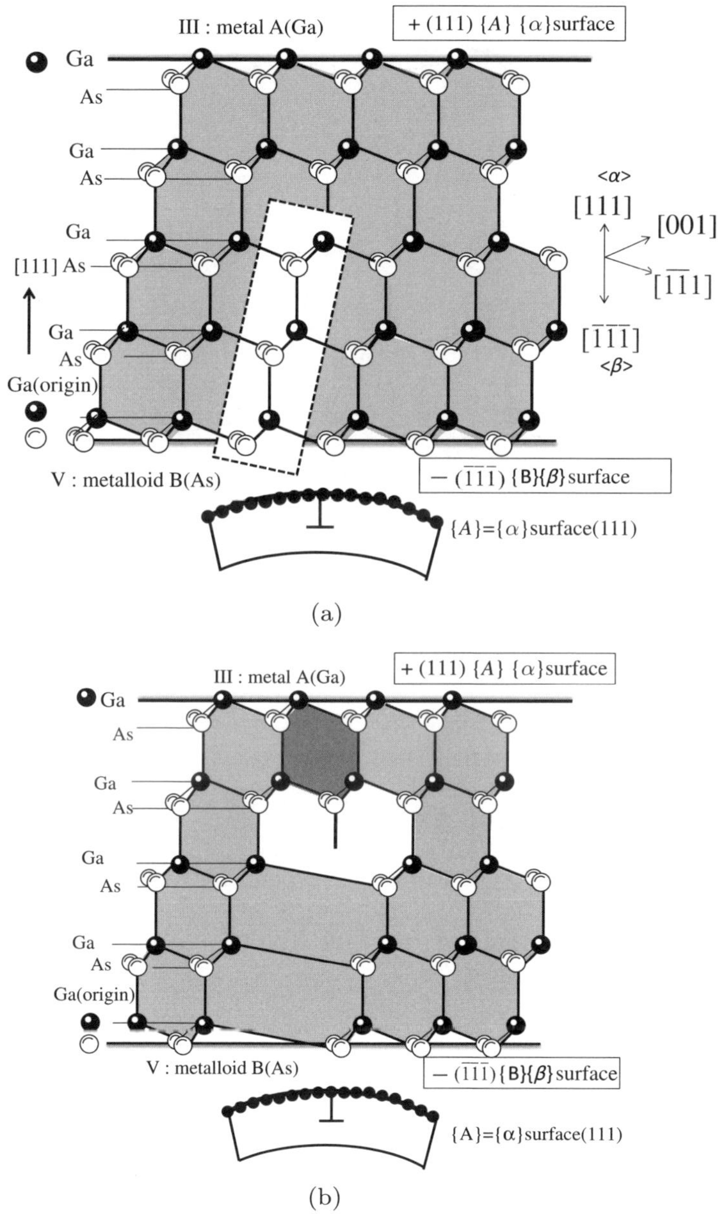

Fig. 11.18 (a)–(c) 6 member rings corresponding to the extra half-plane (ehp) are drawn thicker.

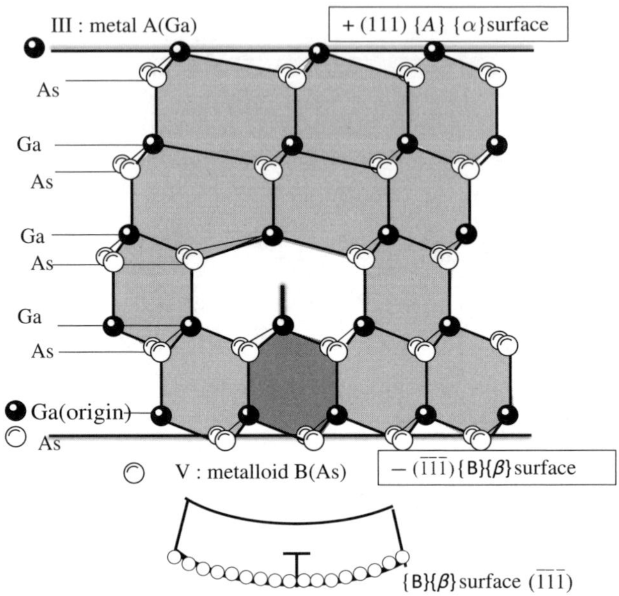

(c)

Fig. 11.18 (*Continued*)

Glide-set dislocations

If an operation similar to Figs. 11.18(a)–11.18(c) is performed to the GD, the results are shown in Figs. 11.19(a) and 11.19(b).

Nomenclature

Edge (perfect) dislocations in the zincblende structure have different core structures depending on whether they are the SPs or the GDs, and also whether positive or negative. In other words, the core has polarity.

In order to define the polarity of dislocation core, there are three possible definitions.

① In the dislocation core whether the DB is A element or B element?

Suppose that you shrink and enter the tunnel of a dislocation core and search DB. If the DB is element A(●) you are in the tunnel (core) of an A-dislocation. If it is element B (○), you are in the tunnel (core) of a B-dislocation.

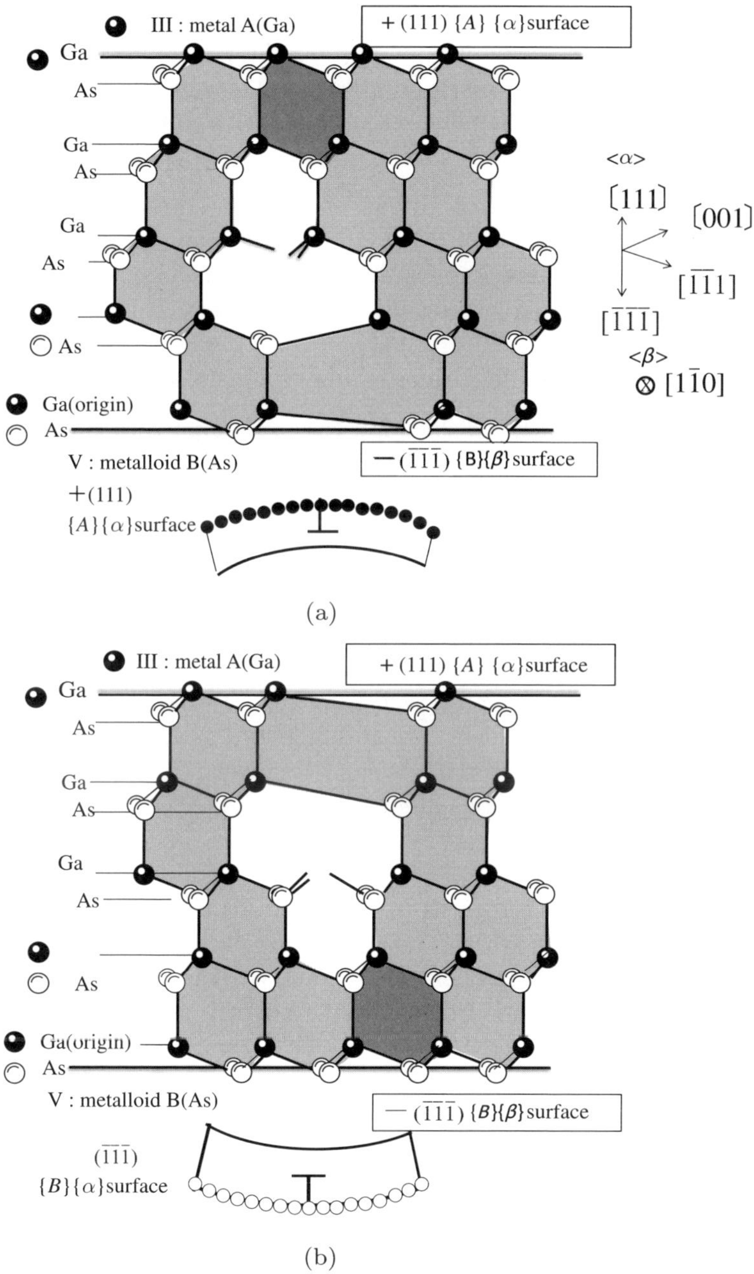

Fig. 11.19 Core structures of GD dislocation.

② Suppose that you are a geographer and walking on surfaces to search a surfaced mine of ehp. If you find an ehp on (111) plane (= $\{A\}$ plane = $\{\alpha\}$ plane), you are on top of $\{\alpha\}$ dislocation. If you find a surfaced mine of ehp on $(\bar{1}\bar{1}\bar{1})$ plane (=$\{B\}$ plane=$\{\beta\}$ plane), then you are on top of $\{\beta\}$ dislocation.

③ Suppose that a wedge ehp is driven by a hammer into a crystal. If the ehp is driven in the direction of $\langle\alpha\rangle$ [111], then the edge dislocation created by driving the ehp is defined as $\langle\alpha\rangle$-dislocation. If the ehp is driven in the direction of $\langle\beta\rangle$ $[\bar{1}\bar{1}\bar{1}]$, then the resultant edge dislocation is defined as $\langle\beta\rangle$ dislocation.

The features of these definitions can be summarized as follows:

① When interaction with a dislocation core and impurities is considered, this definition is most straightforward.

②, ③ When a crystal whose polarity is known is deformed, say by bending, the polarity of the edge dislocations introduced can be determined easily.

Historically, definition ③ was used when it was believed that only the shuffle-set dislocations operated in the diamond and zincblende structures. In reality, it was found that the GD also operated. In Huenfeld Conference in 1978, Haasen *et al.* proposed definition ① in Table 11.2(a). Since then definition ① is used widely. The author feels that definition ③ appears inconsistent with the definition of the ehp.[4] Definition ② appears more consistent. But readers should be careful when reading literatures (especially old ones) [14].

Table 11.2(b) summarizes.

It is noted that with regards to definition for the GD definitions, ① and ② are consistent. Even so, inevitably there are still some ambiguity with regard to the definition of elements A and B, + and − of {111} planes. Therefore, it is recommended to define the polarity of dislocation as Ga-core(g), As-core(g) etc., in the new definition (definition ①).

The foregoing discussion assumes that lattice ① (●) passes through the origin $(0, 0, 0)$ and that lattice ② (○) passes through the coordinate $(1/4,\ 1/4,\ 1/4)$. In this definition, elements ○ sit at the four corners and ● sits at the centroid of the tetrahedron (Fig. 11.17(c)).

What happens when the definition is reversed, that is, lattice ① (As(○)) passes through the origin and lattice ② (Ga(●)) passes through

[4] $\mathbf{e} = \frac{\xi \times b}{|\xi \times b|}$ Eq. (2.5)

Table 11.2a Definitions 1, 2 and 3.

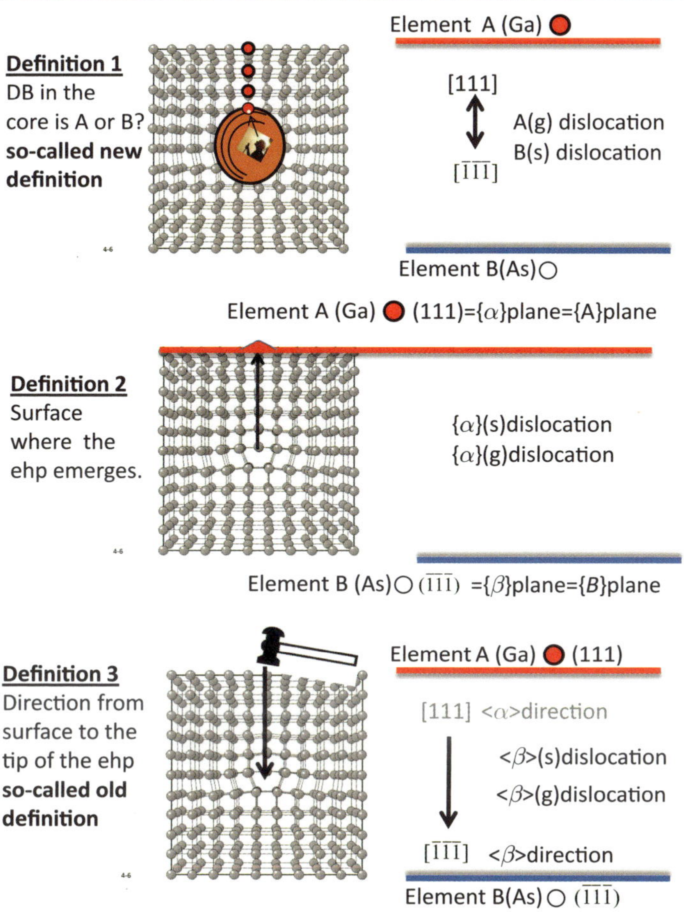

coordinate (1/4, 1/4, 1/4) (Fig. 11.20(b))? Then, As\{B\} plane becomes (111) plane, just opposite to the previous definition. However, Fig. 11.20(b) is obtained by turning Fig. 11.20(a) upside down (Fig. 11.20(c)), where Ga\{A\} plane is defined as (111). In Fig. 11.20(a) Ga tetrahedron ($G0$–$G1$–$G2$–$G3$) is viewed from [111] towards the origin, while in Fig. 11.20(c) it is viewed in the reverse direction, that is from the origin towards [111].

Exercise 11.2 If this is discussed based on the Thompson tetrahedron, Fig. 11.20(a) corresponds to Figs. 8.1(a) and 11.20(c) corresponds to Fig. 8.1(d). Show this.

Table 11.2b Definition of polarity of dislocations [15].

	s/g	Definition ① DB in the core	Definition ② Surface where ehp emerges	Definition ③ Direction of the tip of ehp	Definition ① DB in the core	Definition ② Surface where ehp emerges	Definition ③ Direction of the tip of ehp
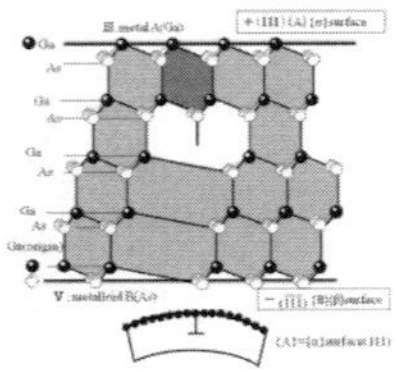	s	○ Element B	$\{\alpha\} = \{A\}$ plane	$\downarrow \langle\beta\rangle$ direction	B(s)	$\{\alpha\}$ (s)	$\langle\beta\rangle$ (s)
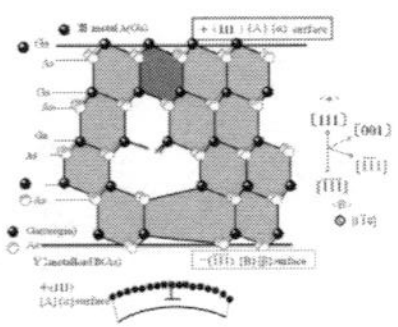	g	● Element A	$\{\alpha\} = \{A\}$ plane	$\downarrow \langle\beta\rangle$ direction	A(g)	$\{\alpha\}$ (s)	$\langle\beta\rangle$ (g)

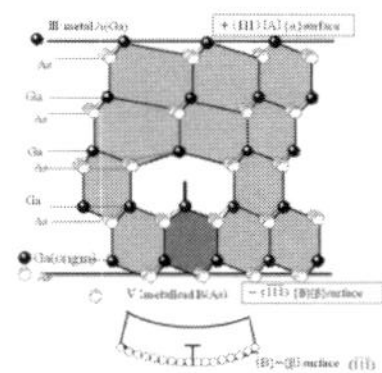	s	● Element A	$\{\beta\} = \{B\}$ plane	$\uparrow < \alpha >$ direction	A(s)	$\{\beta\}$ (s)	$< \alpha >$ (s)
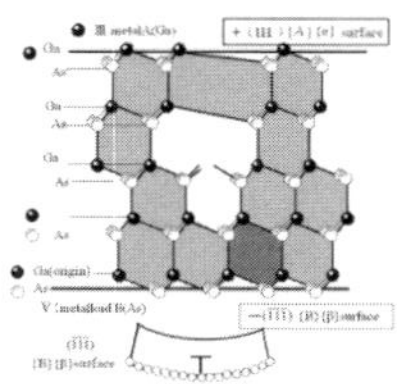	g	○ Element B	$\{\beta\} = \{B\}$ plane	$\uparrow < \alpha >$ direction	B(g)	$\{\beta\}$ (g)	$< \alpha >$ (g)

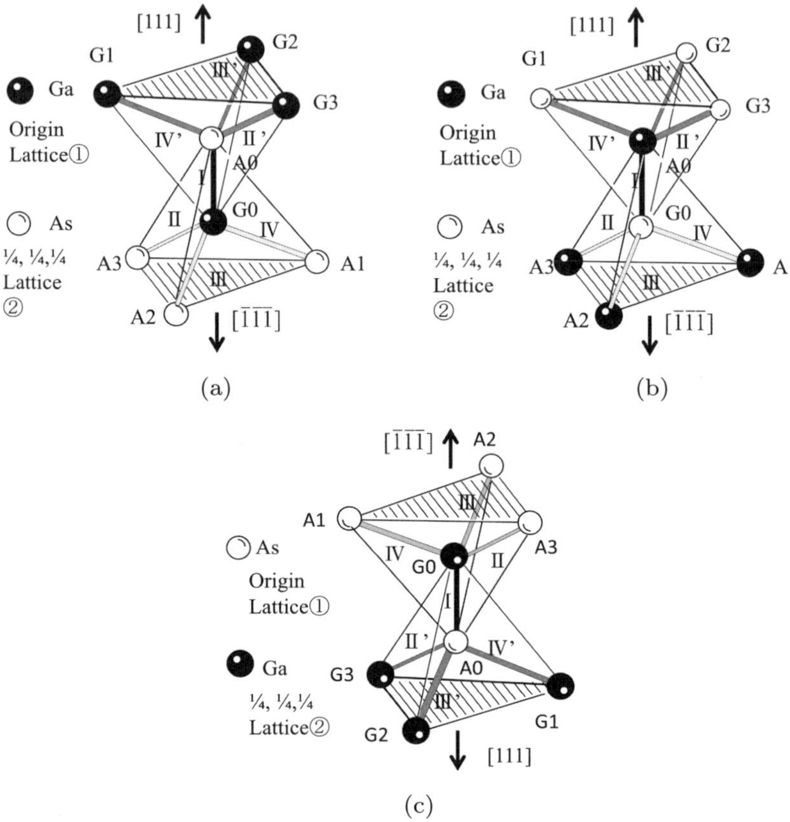

Fig. 11.20 Relationship between Ga tetrahedron (a) and As tetrahedron (b). (a) and (b) are viewed from (111) toward the origin. (c) is viewed in the reverse direction. Note that (b) and (c) are crystallographically equivalent.

11.2.2 *Shockley partial dislocations*

The GD is dissociated on {111} plane in the same manner as in the FCC structure, and hence can be described by the Thompson tetrahedron. That is to say, Hirth–Lothe axiom regarding the Burgers vectors of Shockley partials for FCC can be applied.

The polarity of partial dislocations is defined such as As (B) core or Ga (A) core, as shown in Fig. 11.21. Important is that polarity is defined by the edge component alone of a partial dislocation.

Hirth–Lothe axiom holds for Shockley partials of the GD in the zincblende structure.

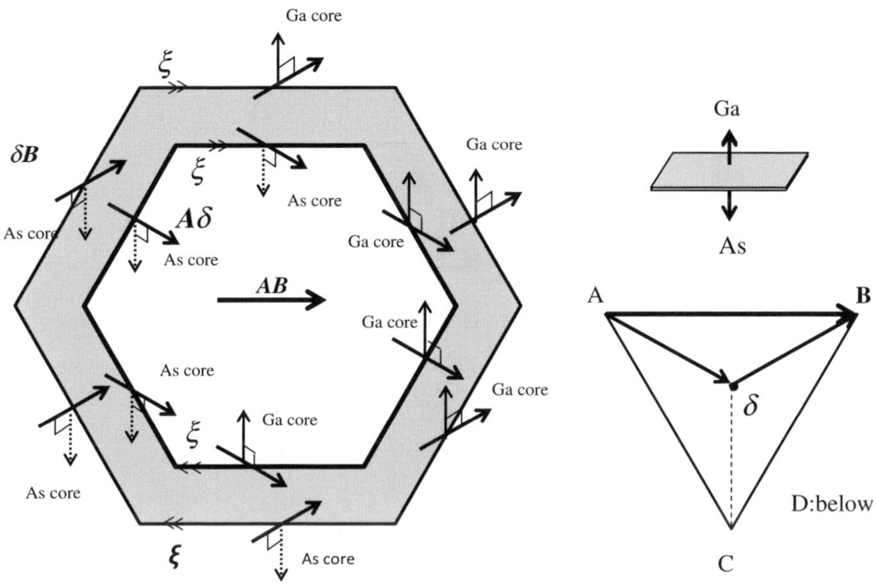

Fig. 11.21 Polarity of Shockley partials.

Exercise 11.3 Discuss similarity and difference between Figs. 11.21 and 8.38.

11.3 Shockley Partial Dislocations in the Wurtzite Structure

Figure 11.22 shows the crystal structure of the wurtzite structure. The diamond and zincblende structures are based on FCC, but the wurtzite structure is based on hexagonal close parking (HCP). Tetrahedra composing the wurtzite structure (for instance GaN[5]) have a twin relationship between the stacking orders AB (△ in Frank symbol) and BA (▽ in Frank symbol).

Consider a tetrahedron with N atoms (◯) at the four corners. These tetrahedra are stacked in the order of ABAB. The tetrahedra A and B in Fig. 11.22 (corresponding to ① and ② in Fig. 9.5) are shown in Figs. 11.23(a) and 11.23(b). Figure 11.23(c) shows the twin relationship between (a) and (b). As shown in Fig. 11.22 the tetrahedra are viewed in

[5]GaN crystallizes both in the wurtzite and zincblende structures. But most of industrially important ones such as blue LED crystallize in the wurtzite structure.

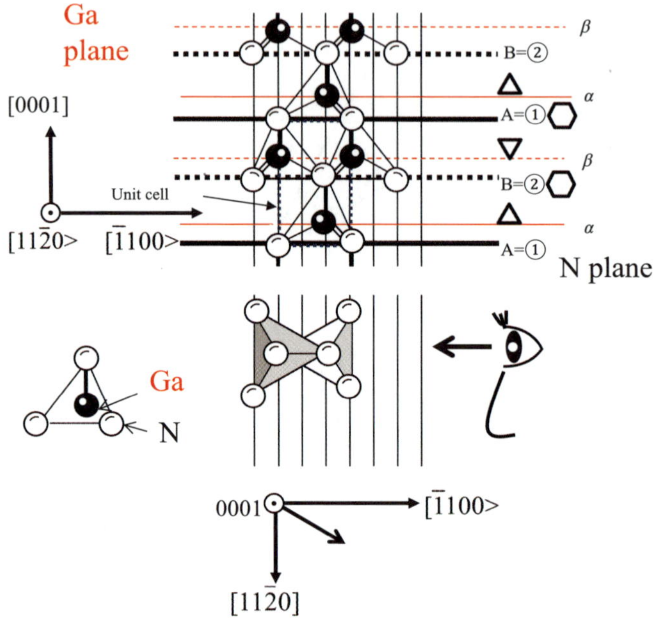

Fig. 11.22 Structure of GaN.

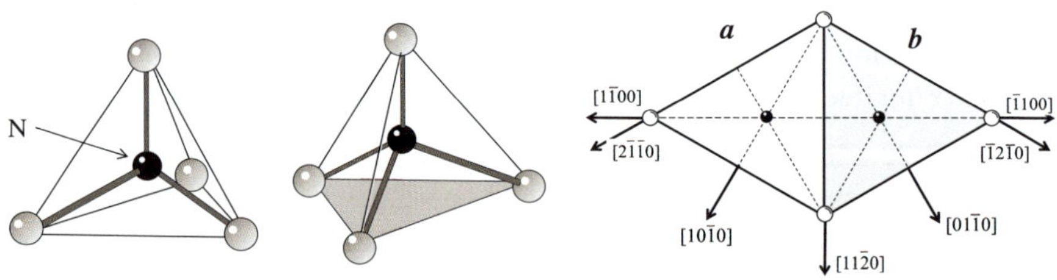

(a) (b) (c)

Fig. 11.23 Twin relationship of tetrahedra.

the direction [$1\bar{1}00$]. Similar consideration can be made with regard to Ga which is in the centroid of the tetrahedron. However, since Shockley partials glide between $A\alpha$ (and $B\beta$), it is not necessary to consider α and β planes.

With regard to Shockley partial dislocations, the axiom for a dissociated dislocation in the HCP structure, described in Chapter 9, is applicable. That is, when the slip plane is in $\triangle$ of Frank symbol (i.e., $A\alpha$), the Greeks are outside (Fig. 11.24(a)), but when it is in $\bigtriangledown$ (i.e., $B\beta$) the Greeks are inside (Fig. 11.24(b)).

11.4 Dislocations in SiC

11.4.1 *Polytypes in SiC*

In SiC crystal, C atoms are at the four corners and Si is in the centroid (or inversely C in the centroid and Si at the corners). Ga (●) is to be replaced by Si and N(○) by C in Figs. 11.23(a)–11.23(c). Thus, a twin relationship is present.

The horizontal plane of the tetrahedron is the close packed plane and corresponds to {111} planes of the FCC structure or (0001) plane of the HCP structure. Therefore, there are three equivalent positions a, b and c for stacking. Now consider the stacking of the tetrahedron of SiC.

First let the close packed plane (shown in Fig. 11.23(a)) be stacked in the order of:

$$a, b, c \text{ (as in Fig. 8.2(e))},$$

or its twin-related plane shown in Fig. 11.23(b) be stacked in the order of:

$$a_T, b_T, c_T, a_T, b_T, c_T.$$

This corresponds to FCC, resulting in the zincblende structure. This is referred to as β-SiC.

In these stacking orders, mixture between a, b, c and its twin related a_T, b_T, c_T is not present. Suppose that there is a mixture like, for instance,

$$c, a,\ b, c_T, b_T, a_T,\ c, a,\ b, c_T, b_T, a_T.$$

In this case, a base consisting of two zincblende structures with twin relationship is stacked. That is, three tetrahedra shown in Fig. 11.23(a) and three tetrahedra shown in Fig. 11.23(b) are stacked alternately to form a unit cell composed of six tetrahedra (Fig. 11.25). This structure is referred to as 6H–SiC. This structure is nothing but two unit cells of the zincblende structures stacked with twin relationship as can be seen in Fig. 11.25.

Instead, if the order of stacking is

$$c, a, b_T, a_T, c, a, b_T, a_T,$$

to form a unit cell with four close packed layers, this results in a structure called 4H–SiC. This structure is nothing but two unit cells of the wurtzite

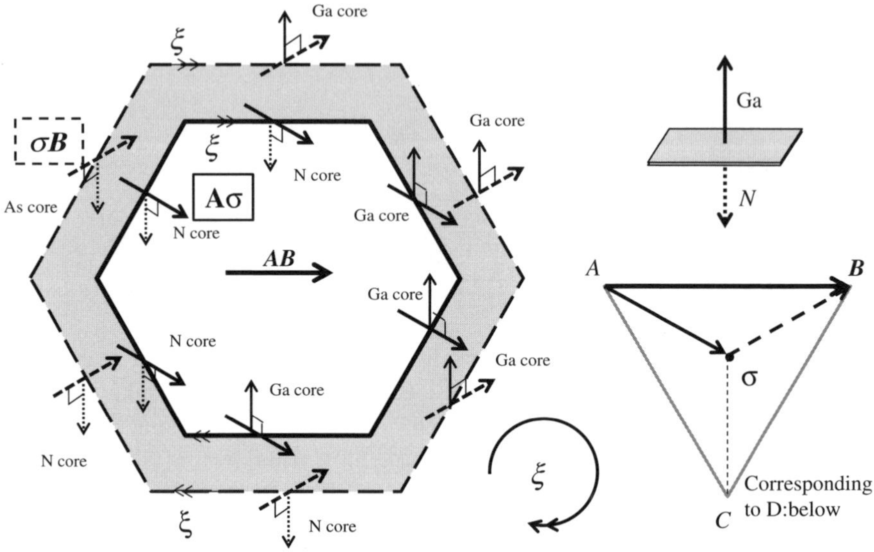

(a)

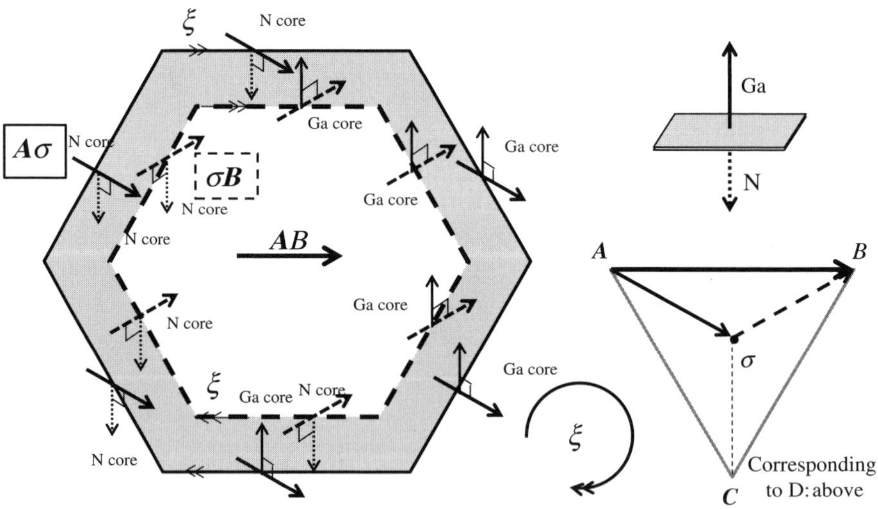

(b)

Fig. 11.24 (a), (b) Polarity of Shockley partials in GaN. Note the altitude of σ.

structure stacked with twin relationship as can be seen in (Figs. 11.26(a) and 11.26(b).

These similar but different structures are referred to as polytype. The reason why polytypes occur is that interaction between atoms extends over the nearest neighbours and the next-nearest neighbours.[6] Interaction energy decreases rapidly with the distance. Differences in interaction energy among polytypes of SiC is quite small. As a result, in a single grain many polytypes can exist.

Exercise 11.4 Show that the stacking order for the wurtzite structure is $a, b_T, a, b_T, a, b_T, \ldots$, or $b, c_T, b, c_T, b, c_T, \ldots$, or $c, a_T, c, a, c, a_T, \ldots$. (Hint): Fig. 11.22

Exercise 11.5 Indicate the stacking of (0001) planes in 4H–SiC shown in Fig. 11.26(b) in symbols ⬡ and ■ discussed in Sec. 9.1.1. Show that in 4H–SiC, the FCC and the HCP structures are stacked alternately.

11.4.2 *Dislocations in 4H–SiC*

Dislocations in β-SiC and 2H–SiC are essentially same as those in the zincblende and wurtzite structures, respectively. Let us consider dislocations in 4H–SiC.

Basal plane dislocation

A plane on which tetrahedra are packed closely is referred to as the basal plane, and dislocations in the basal plane are referred to as BPD. The Burgers vectors of the BPD's are $1/3\langle 11\bar{2}0\rangle$ and correspond to $1/2\langle 110\rangle$ in the FCC structure and $1/3\langle 11\bar{2}0\rangle$ ($\boldsymbol{a}_1$ or $\boldsymbol{a}_2$ or $\boldsymbol{a}_3$ in the unit cell (Fig. 1.5)) in the HCP structure. Difference in free energies of SiC polytypes is quite small, so that the stacking fault energy γ_{SFE} is quite low. Hong *et al.* [16] measured experimentally that $\gamma_{\text{SFE}} = 12.3\text{–}17.2\ \text{mJm}^{-2}$; BPD is dissociated widely into two Shockley partials.

Burgers vectors of Shockley partials follow the same axiom as for the wurtzite structure. That is, if BPD is between C and α, (and equivalently A and β) the Greeks are outside, but if it is between B and α (equivalently A and γ) the Greeks are inside.

[6]In this respect, the difference between FCC and HCP is not to be attributed to the nearest layers but the next-nearest layers.

(a)

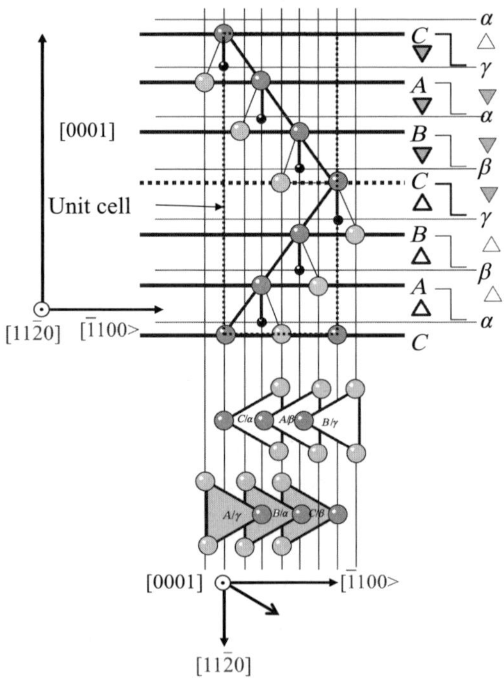

(b)

Fig. 11.25 (a), (b) Structure of 6H–SiC (Reproduced with permission from [16]).

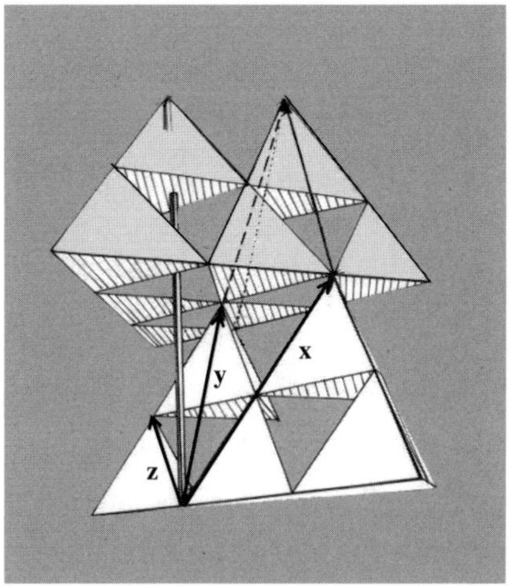

(a)

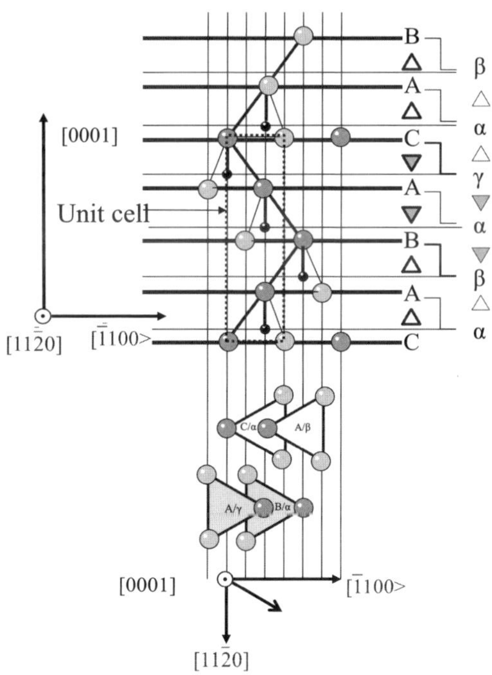

(b)

Fig. 11.26 Structure of 4H–SiC ((a) Reproduced with permission from [17] and (b) Refer to [16]).

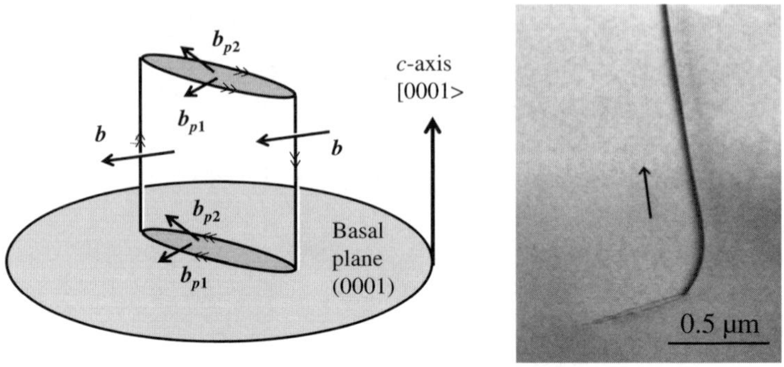

(a) (b)

Fig. 11.27 TED. (a) Schematic illustration, (b) TEM micrograph. The dislocation is dissociated into two Shockley partials on the basal plane [18].

Threading dislocations

Threading edge dislocations

A threading edge dislocation (TED) has $\boldsymbol{b} = 1/3 < 11\bar{2}0 >$ and dissociated into two Shockley partials on the basal plane, but once it lies off the basal plane it is a perfect dislocation.

Threading screw dislocations

Strictly speaking, a threading screw dislocation (TSD) has $\boldsymbol{b}$ parallel to [0001] with the direction of dislocation line parallel to $\xi//[0001]$. However, so-called TSD's can be classified into four, as follows:

① Threading pure screw dislocation (TpSD)
② Threading near screw dislocation(TnSD)
③ Threading mixed dislocation 1(TMD)
④ Threading mixed dislocation 2(TMD')

Dissociation of TSD [17]

All of the so-called TSD's are dissociated into two partial dislocations, an example of which is shown in Fig. 11.28.

As already described, the unit cell of 4H–SiC is composed of two unit cells of the wurtzite structure stacked along $\boldsymbol{c}$-axis with twin relationship. Therefore, if threading dislocations with $\boldsymbol{c}$-component are dissociated at all, they are expected to be dissociated into two partials, p_1 and p_2 with

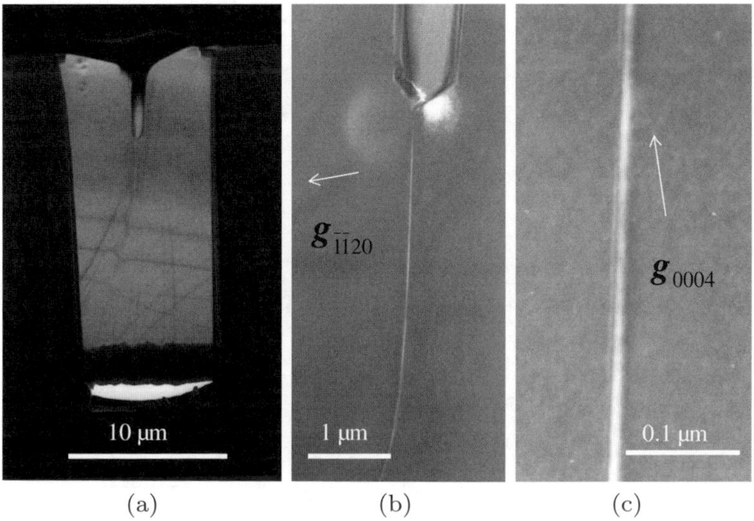

Fig. 11.28 Threading screw dislocation. (a) is a low-maginification micrograph. A dislocation is present at the bottom of an etch pit, (b) and (c) are enlarged images (Reproduced with permission from [19]).

$\boldsymbol{b} = 1/2\boldsymbol{c}$. That is,

$$\boldsymbol{b}_T(\boldsymbol{c}\text{-component} = c) \rightarrow \boldsymbol{b}_{p1}(\boldsymbol{c}\text{-component} = 1/2\boldsymbol{c}) + \boldsymbol{b}_{p2}(\boldsymbol{c}\text{-component} = 1/2\boldsymbol{c}). \quad (11.1)$$

Let us refer to this dissociation mode as mode-1 dissociation. Candidates for $\boldsymbol{b}_{p1}$ and $\boldsymbol{b}_{p2}$ are $\boldsymbol{x}$ and $\boldsymbol{y}$ in Fig. 11.26(a). All of the crystallographical equivalents will be indicated by x and y.

Now let us construct a TSD dislocation according to the mode-1 dissociation model (Fig. 11.29). Let us start with $b_{p1} = x$. The starting point of $b_{p1} = x$ is indicated by ●, and the ending point by ■. b_{p1} threads the lower part of the twinned structure of 4H–SiC. b_{p1} is connected with b_{p2}, which in turn threads the upper part of the twinned structure of 4H–SiC. Thus the starting point of b_{p2} is ■. As for the ending point of b_{p2} there are six candidates, three of which are x and the remaining three are y.

The total Burgers vectors of the resultant perfect dislocations are summarized as follows:

$$1_1, \text{TpSD}; b_T = x + x \quad 1_2, \text{TnSD}; b_T = x + y, \quad 1_3, \text{TnSD}; b_T = x + y,$$
$$1_4, \text{TMD}; b_T = x + y, \quad 1_5, \text{TM'D}; b_T = x + x, \quad 1_6, \text{TM'D}; b_T = x + x. \quad (11.2)$$

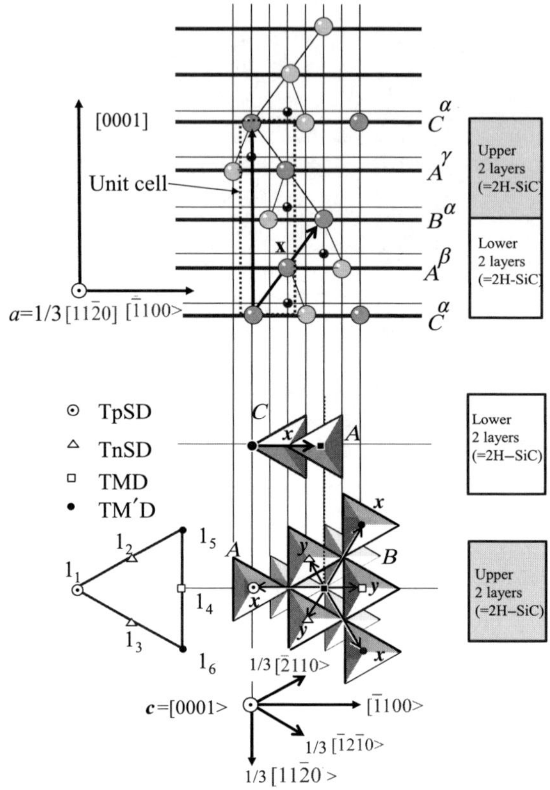

Fig. 11.29 Dissociation of a TSD in 4H–SiC (I) (Reproduced with permission from [17]).

where the first term (x) in the extreme right sides stands for b_{p1} and the second term (x or y) b_{p2}.

Similarly, if we start with $b_{p1} = y$, the Burgers vectors of the resultant total dislocations are summarized as follows (Fig. 11.30),

$$2_1, \text{TnSD}; b_T = y + x, \quad 2_2, \text{TpSD}; b_T = y + y, \quad 2_3, \text{TnSD}; b_T = y + y,$$
$$2_4, \text{TnSD}; b_T = y + y, \quad 2_5, \text{TnSD}; b_T = y + x, \quad 2_6.\text{TMD}; b_T = y + x. \tag{11.3}$$

where the first term (y) in the extreme right sides stands for b_{p1} and the second term (x or y) b_{p2}.

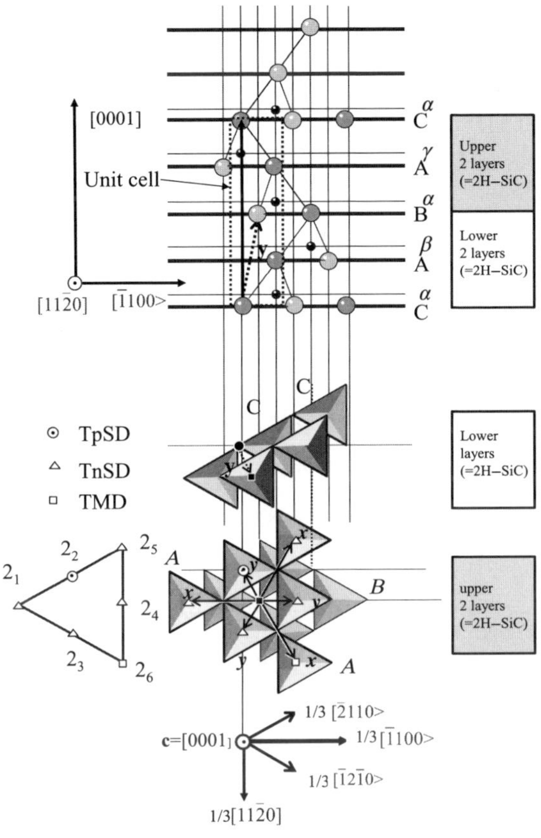

Fig. 11.30 Dissociation of a TSD in 4H–SiC (II) (Reproduced with permission from [17]).

The energies of the respective dislocations are expressed as follows[7]:

$$
\begin{aligned}
(b_T^{pS})^2 &= c^2 = 10.7a^2,\\
(b_T^{nS})^2 &= c^2 + a^2 = 10.7a^2 + a^2 = 11.7a^2,\\
(b_T^{M})^2 &= c^2 + 3a^2 = 10.7a^2 + 3a^2 = 13.7a^2,\\
(b_p^{\alpha})^2 &= 4a^2,\\
(b_p^{\beta})^2 &= 3a^2,
\end{aligned}
\tag{11.4}
$$

[7] $\frac{\mu}{4\pi}\ln\frac{R}{r_0}$ in Eq. (4.5) is replaced by 1, because we are interested only in the relative amount of elastic energy and not the absolute value.

where b_T^{pS}, b_T^{nS}, b_T^{M}, b_p^x and b_p^y are Burgers vectors of TpSD, TnSD, TMD, partial x and partial y, respectively.[8]

Obviously the energy of the total dislocations increases in the order of TpSD, TnSD, TMD, as expected.

Next, we apply the Frank criterion to judge whether or not the dissociation takes place.

$$b_T^2 \geq b_{p1}^2 + b_{p2}^2. \tag{11.5}$$

The left and right sides of Eq. (11.5) for dislocations 1_1, 1_2, 1_3, 1_4, 1_5 and 1_6 are expressed as follows:

$$\begin{aligned}
&1_1(b_T^{pS})^2 = 10.7a^2, x^2 + x^2 = 8a^2,\\
&1_2(b_T^{nS})^2 = 11.7a^2, x^2 + y^2 = 7a^2,\\
&1_3(b_T^{nS})^2 = 11.7a^2, x^2 + y^2 = 7a^2,\\
&1_4(b_T^{M})^2 = 13.7a^2, x^2 + y^2 = 7a^2,\\
&1_5(b_T^{M'})^2 = c^2 + 4a^2 = 10.7a^2 + 4a^2 = 14.7a^2, x^2 + x^2 = 8a^2,\\
&1_6(b_T^{M'})^2 = c^2 + 4a^2 = 10.7a^2 + 4a^2 = 14.7a^2, x^2 + x^2 = 8a^2.
\end{aligned} \tag{11.5'}$$

It is obvious that the Frank judgment holds.

Similarly, for dislocations 2_1, 2_2, 2_3, 2_4, 2_5 and 2_6 are expressed as follows:

$$\begin{aligned}
2_1(b_T^{nS})^2 &= 11.7a^2, y^2 + x^2 = 7a^2,\\
2_2(b_T^{pS})^2 &= 10.7a^2, y^2 + y^2 = 6a^2,\\
2_3(b_T^{nS})^2 &= 11.7a^2, y^2 + y^2 = 6a^2,\\
2_4(b_T^{nS})^2 &= 11.7a^2, y^2 + y^2 = 6a^2,\\
2_5(b_T^{nS})^2 &= 11.7a^2, y^2 + x^2 = 7a^2,\\
2_6(b_T^{M})^2 &= 13.7a^2, y^2 + x^2 = 7a^2.
\end{aligned} \tag{11.6}$$

Again, the Frank's judgment holds.

Thus, the energy of mode-1-dissociated dislocations may be categorized into three groups in terms of energy, that is, $6a^2$, $7a^2$ and $8a^2$. Thus, it is reasonable to conclude that the mode-1 dissociation can take place.

[8] a = 0.308 nmm, c = 1.008 nm.

Next, coming back to Fig. 11.26(a), it is noted that both x and y may be dissociated into $2z$. That is,

$$\begin{aligned} x &\to z + z, \\ y &\to z + z. \end{aligned} \tag{11.7}$$

Let us refer to this dissociation mode as mode-2 dissociation. In this case, the energies of mode-1 partial dislocations x and y are reduced further by mode-2 dissociation as follows:

$$\begin{aligned} (b_p^x)^2 &= 4a^2 \to 2a^2, \\ (b_p^y)^2 &= 3a^2 \to 2a^2. \end{aligned} \tag{11.8}$$

If the mode-2 dissociation takes place, the aforementioned mode-1 dissociations expressed by Eqs. (11.5′) and (11.6) are to be modified as follows:

$$\begin{aligned}
&1_1 : (b_T^{pS})^2 = 10.7a^2, x^2 + x^2 = 8a^2 \to 4a^2, \\
&1_2 : (b_T^{nS})^2 = 11.7a^2, x^2 + y^2 = 7a^2 \to 4a^2, \\
&1_3 : (b_T^{nS})^2 = 11.7a^2, x^2 + y^2 = 7a^2 \to 4a^2, \\
&1_4 : (b_T^{M})^2 = 13.7a^2, x^2 + y^2 = 7a^2 \to 4a^2, \\
&1_5 : (b_T^{M'})^2 = c^2 + 4a^2 = 10.7a^2 + 4a^2 = 14.7a^2, x^2 + x^2 = 8a^2 \to 4a^2, \\
&1_6 : (b_T^{M'})^2 = c^2 + 4a^2 = 10.7a^2 + 4a^2 = 14.7a^2, x^2 + x^2 = 8a^2 \to 4a^2.
\end{aligned} \tag{11.9}$$

and

$$\begin{aligned}
&2_1 : (b_T^{nS})^2 = 11.7a^2, y^2 + x^2 = 7a^2 \to 4a^2, \\
&2_2 : (b_T^{pS})^2 = 10.7a^2, y^2 + y^2 = 6a^2 \to 4a^2, \\
&2_3 : (b_T^{nS})^2 = 11.7a^2, y^2 + y^2 = 6a^2 \to 4a^2, \\
&2_4 : (b_T^{nS})^2 = 11.7a^2, y^2 + y^2 = 6a^2 \to 4a^2, \\
&2_5 : (b_T^{nS})^2 = 11.7a^2, y^2 + x^2 = 7a^2 \to 4a^2, \\
&2_6 : (b_T^{M})^2 = 13.7a^2, y^2 + x^2 = 7a^2 \to 4a^2.
\end{aligned} \tag{11.10}$$

For the mode-2 dissociation Frank judgment holds.

All the so-called TSD's have the same level of energy if the mode-1 and mode-2 dissociations take place. This may appear surprising, however, it is quite reasonable if the unit cell of 4H–SiC is considered to be composed of four layers of the close packed basal plane of SiC.

Needless to say, the above discussion is very approximate. The stacking fault energy, interaction energies, orientation effect of the self-energy of the dissociated dislocations involved etc., are neglected. Furthermore, while adequate experimental evidence for the mode-1 dissociation has been obtained [19], no evidence exists for the mode-2 dissociation. There are two possibilities: The dissociation does not take place at all, or the dissociation is too narrow to be resolved by the dark-field weak beam electron microscope technique.

Even taking such ambiguities into account, it is noted that the energy of TpSD is very similar to those of TnSD, TMD and TM'D to the first approximation. This may explain abundance of the TnSD, TMD and TM'D in 4H–SiC [20].

References

1. Hornstra, J. *J. Phys. Chemi. Solids* **5** (1958), 129.
2. Roberts, S. G. and Hirsch, P. B. *Phil. Mag.* **86** (2006), 4099.
3. Nishino, Y., Saka, H. and Imura, T. *Phy. Stat. Sol.* A **70** (1982), 729.
4. Hirsch, P. B., Roberts, S. G. and Samuels, J. *Scripta Met.* **21** (1987), 1523.
5. Wessel, K. and Alexander, H. *Phil. Mag.* **35** (1977), 1523.
6. Nishino, Y., Saka, H. and Imura, T. *J. Materi. Sci.* **19** (1984), 245.
7. Izumi, S., Ohta, H., Takahashi, C., Suzuki, T. and Saka, H. *Phil. Mag. Lett.* **90** (2010), 707.
8. Castaing, J., Veyssiere, P., Kubin, L. P. and Rabier, J. *Phil. Mag.* **44** (1981), 1497.
9. Okuno, T. and Saka, H. *J. Materi. Sci.* **48** (2013), 115.
10. Asaoka, K., Umeda, T., Arai, S. and Saka, H. *Materi. Sci. Engg.* **A400–401** (2005), 92.
11. Ray, I. L. F. and Cockayne, D. J. H. *Proc. Roy. Soc., Ser. A* **325** (1971), 543.
12. Rabier, J. *Phy. Stat. Sol* A **204** (2007), 2248.
13. Saka, H. Yamamoto, Y., Arai, S., Kuroda, K. *Phil. Mag.* **86** (2006), 4841.
14. Holt, D. B., Yacobi, B. G. *Extended Defects in Semiconductors*, Cambridge Univ. Press., 2007, p. 193.
15. Saka, H. and Yonenaga, I. in preparation.
16. Hong, M. H., Samant, A. S. V. and Pirouz, P. *Phil. Mag. A* **80** (2000), 919.
17. Onda, S. *et al. Phil. Mag. Lett.* **93** (2013), 591.
18. Kitou, Y. *et al.* Oral presentation at the 70th meeting of the Japanese Society of Applied Physics (at Toyama University), 2009, September 11.
19. Onda, S. *et al. Phil. Mag. Lett.* **93** (2013), 439.
20. Onda, S. *et al. Phil. Mag. Lett.* **95** (2015), 489.

Chapter 12

Dislocations and Macroscopic Strength

12.1 Geometry of Yielding of a Single Crystal

12.1.1 *Critical resolved shear stress — Schmid law*

Let us consider simple tension of a single crystal. As shown in Fig. 2.2, a glide takes place on a plane (slip plane) inclined approximately 45° with respect to the tensile axis. This glide is caused by a shear stress. The axial tensile force F applied externally is resolved into a shear force G on the slip plane and a normal force N perpendicular to the slip plane (Fig. 12.1(a)).

$$F = G + N$$

G is the resolved shear force and N is the resolved normal force. The relationship between the tensile stress and the resolved shear stress is obtained as follows. Let the cross-sectional area of the specimen be S and suppose that the only external force is the axial force F along the x_1-axis.

The only stress is σ_{11} along the x_1-axis (Fig.12.1(b)), so that

$$\sigma_{ij} = \begin{pmatrix} \sigma_{11} & 0 & 0 \\ 0 & 0 & 0 \\ 0 & 0 & 0 \end{pmatrix}.$$

This stress σ_{ij} is transformed into a shear stress (τ'_{12})[1] acting in the slip direction on the slip plane. By referring to Eq. 3.22 τ'_{12} is given by,

$$\tau'_{12} = T_{11}T_{21}\sigma_{11} = \cos\lambda\cos\phi\sigma_{11} = m\sigma_{11} = m\left(\frac{F}{S}\right), \tag{12.1}$$

[1]In Chapter 3, a stress was indicated by σ in general. However, a shear stress $(\sigma_{ij}, i \neq j)$ is occasionally denoted by $\tau_{ij}(\sigma_{ij}, i \neq j)$ in order to distinguish clearly from tensile and compressive stresses $(\sigma_{ij}, i = j)$.

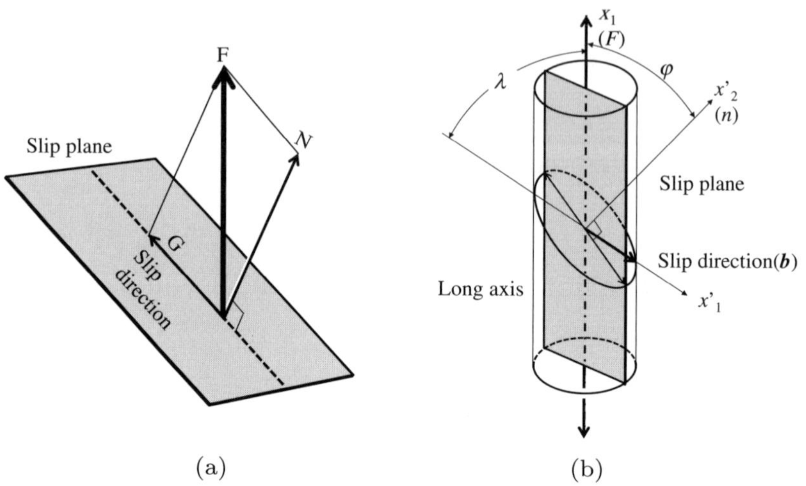

Fig. 12.1 (a) Resolved shear force (G) and normal force (N), (b) Calculation of Schmid factor.

where λ is an angle between the tensile axis and the slip direction $x_1'(\boldsymbol{b})$, φ is an angle between x_1 and the normal of the slip plane $x_2'(\boldsymbol{n})$ (Fig. 12.1(b)). m is referred to as Schmid factor [1].

Exercise 12.1 Derive Eq. (12.1).

When the resolved shear stress τ'_{ij} reaches a critical stress, the crystal yields. This critical stress is referred to as the critical resolved shear stress (CRSS), and this phenomenon is referred to as law of CRSS or Schmid law. From the viewpoint of dislocation theory, it is the resolved shear force (G) alone that governs the slip, i.e., motion of dislocations, irrespective of the amount of the normal force (N). Figure 12.2 shows the axial yield stress (σ_{11}) of single crystals Zn as a function of Schmid factor ($\chi = 90° - \varphi$ in Fig. 12.1(b)). By putting CRSS $= 18.4\text{g/mm}^2$, it is clear that Schmid law holds.[2]

Exercise 12.2 The maximum value of Schmid factor is 0.5. Prove this.

[2] Zn crystallizes in the HCP structure, where the basal slip is preferred. It is possible to measure CRSS in a wide range of low m. This is difficult in the FCC and the BCC structures which have high symmetry. This is true for some of HCP metals where the prismatic or pyramidal slip occurs.

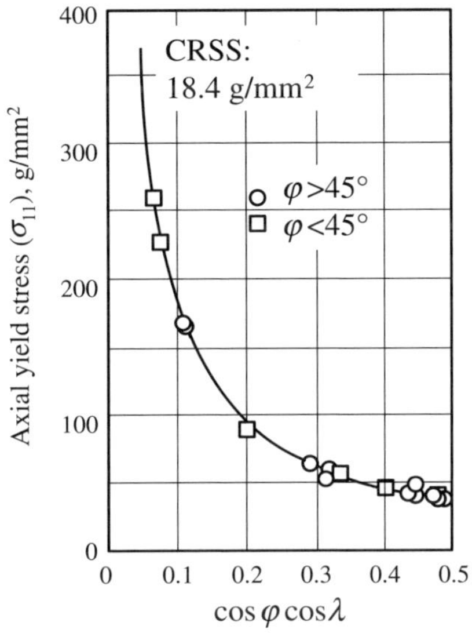

Fig. 12.2 (Reproduced from [2]).

12.1.2 *Derivation of shear stress and shear strain in the uniaxial deformation*

Equation (12.1) holds within the framework of elasticity theory. On the onset of plastic deformation, a plastic strain ε is generated. Therefore, a correction is necessary to Eq. (12.1). Relationship between a plastic tensile strain and a plastic shear strain is given as follows. For the sake of simplicity, let the cross-section of the specimen be circular.

In a triangle $\Delta AB'B$ (weakly hatched in Fig. 12.3(b)), let $AB = l_o, AB' = l, AM = L$. Then, because $\angle AB'B = \lambda, \angle ABM = \lambda_o$,

$$\ell_o \sin \lambda_o = \ell \sin \lambda = L. \tag{12.2}$$

(see Fig. 12.3(b)).

Similarly in triangles $\Delta ANB'$ and ΔANB,

$$\ell_o \sin \chi_o = \ell \sin \chi = AN, \tag{12.3}$$

where AN is perpendicular to the slip plane.

Since $\chi = 90° - \phi$,

$$\ell_o \cos \phi_o = \ell \cos \phi. \tag{12.3'}$$

(a)

(b)

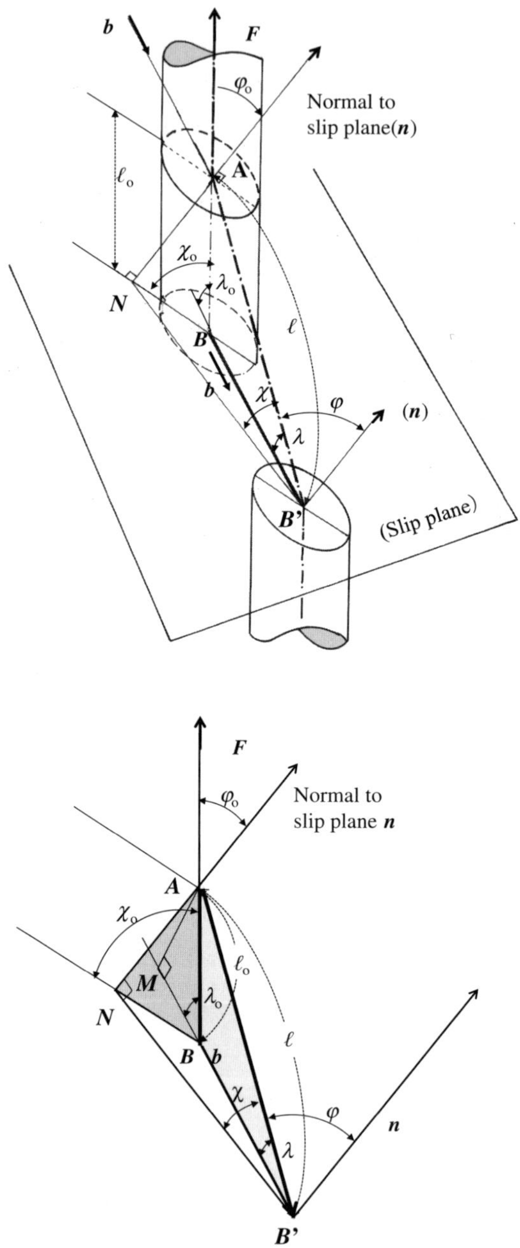

Fig. 12.3 Shear strain (Reproduced from [1]).

Coming back to Fig. 12.1(b), let the area of the slip plane be S,

$$\frac{S_o}{\cos\phi_o} = \frac{S}{\cos\phi}. \tag{12.4}$$

Substituting these into Eq. (12.1),

$$\begin{aligned}\sigma'_{12} &= \left(\frac{F}{S}\right)\cos\lambda\cos\phi = \left(\frac{F\cos\phi_o}{S_o\cos\phi}\right)\cos\lambda\cos\phi = \left(\frac{F}{S_o}\right)\cos\phi_o\cos\lambda \\ &= \left(\frac{F}{S_o}\right)\cos\phi_o\sqrt{1-\left(\frac{\ell_o}{\ell}\right)^2\sin^2\lambda_o}.\end{aligned} \tag{12.5}$$

Next, let us consider the plastic shear strain. Slip takes place on the adjacent planes separated by AN by an amount of shear displacement BB'. Therefore, the resultant shear strain ε is written as follows.

$$\varepsilon = \frac{BB'}{AN} = \frac{\ell\cos\lambda - \ell_o\cos\lambda_o}{\ell_o\sin\chi_o} = \frac{\cos\lambda}{\sin\chi} - \frac{\cos\lambda_o}{\sin\chi_o} \tag{12.6}$$

Using Eqs. (12.2) and (12.3), ε is given by,

$$\begin{aligned}\varepsilon &= \frac{1}{\sin\chi}\sqrt{1-\left(\frac{\ell_o}{\ell}\right)^2\sin^2\chi_o} - \frac{\cos\lambda_o}{\sin\chi_o} \\ &= \frac{1}{\ell_o\sin\chi_o}\sqrt{1-\left(\frac{\ell_o}{\ell}\right)^2\sin^2\chi_o} - \frac{\cos\lambda_o}{\sin\chi_o} \\ &= \frac{1}{\cos\phi_o}\left(\sqrt{\left(\frac{\ell_o}{\ell}\right)^2 - \sin^2\lambda_o} - \cos\lambda_o\right).\end{aligned} \tag{12.7}$$

Exercise 12.3 Derive Eq. (12.7).

As can be seen from Fig. 12.4, the tensile axis rotates towards the slip direction during tensile deformation. In compressive deformation, the slip plane rotates towards the compressive plane as shown in Fig. 12.5.

The rotation of the tensile axis during a tensile test of a single crystal provides the most reliable information on the slip direction. Figures 12.6(a) and 12.6(b) show the change in the axial orientation of single crystals of β'-CuZn, a BCC-based long-range ordered alloy described in Sec. 10.4.2 (same as Fig. 10.28). (a) shows results obtained at room temperature and (b) shows the results at 200°C. The slip direction at room temperature is [111], while that at 200°C is definitely not <111>. Especially crystal $A2$

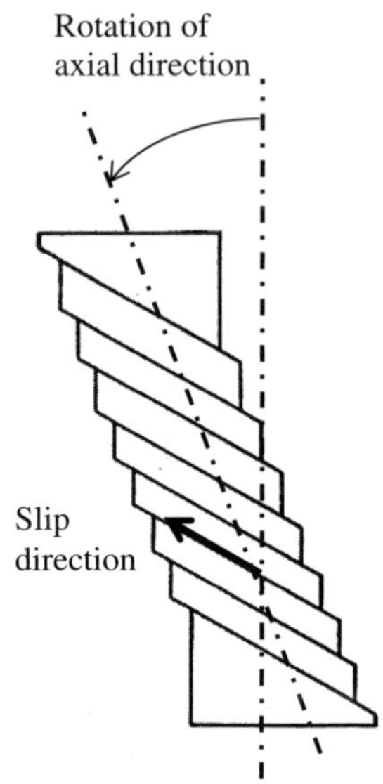

Fig. 12.4 Rotation of the tensile axis in tension.

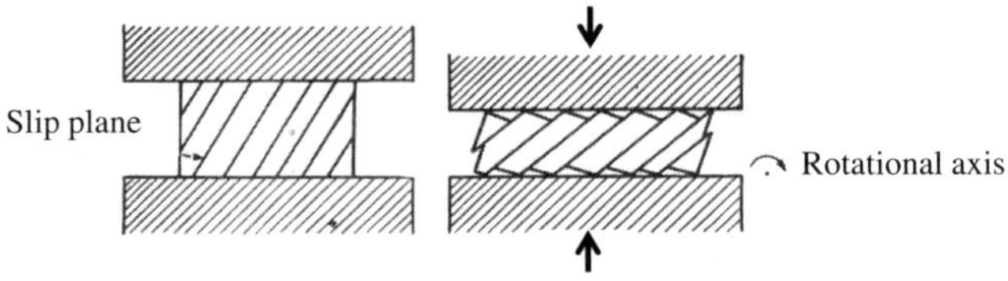

Fig. 12.5 Rotation of slip plane in compression (Reproduced from [2]).

remains virtually unchanged during deformation, suggesting that multiple slips take place.

Face-centered Cubic (FCC) In FCC crystals the slip direction is 1/2<110> and the slip plane is {111}. As the Thompson tetrahedron shows, there are four types of {111} slip plane, each of which has three slip directions.

Therefore, the number of the <111>/{101} systems is $4 \times 3 = 12$. In Fig. 12.7 the Thompson vectors are shown on the 001 standard stereographic projection. Now, let the tensile (or compressive) axis be in the standard triangle bounded by 001, 011 and $\bar{1}11$ (hatched area in Fig. 12.8). All the twelve slip systems are summarized in Table 12.1.

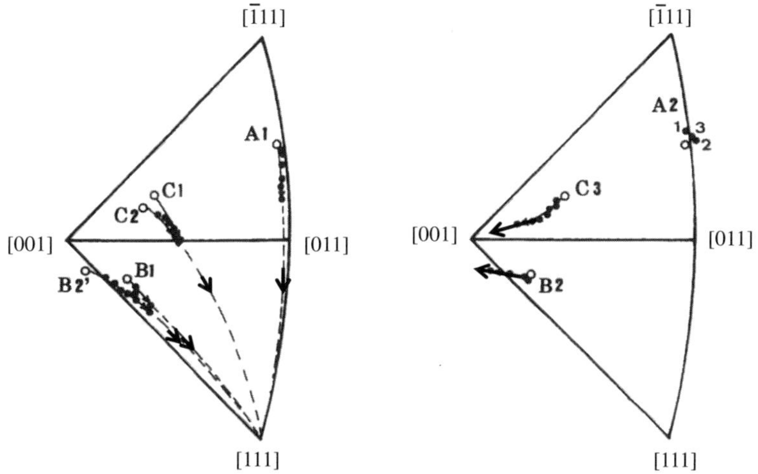

(a) (b)

Fig. 12.6 (a),(b) Rotation of tensile axis during tensile tests in β'-CuZn (same as Fig. 10.28) (Reproduced with permission from [3]). ○ stands for the starting point.

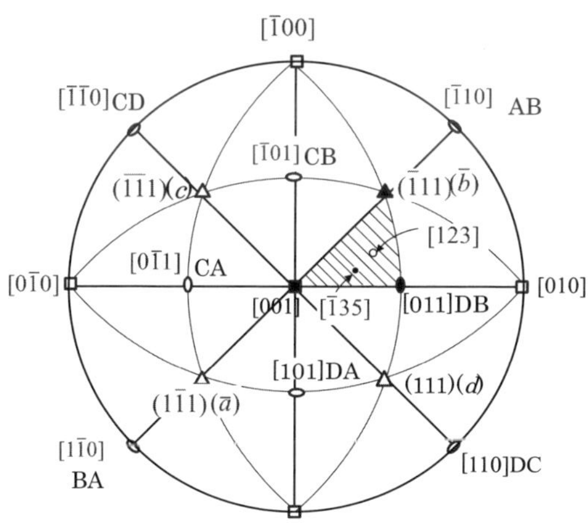

Fig. 12.7 Slip systems in FCC.

Among the 12 slip systems, some of the most important ones have their own names as shown in Fig. 12.8. The coplanar slip system has the same slip plane as the primary slip system but with a different slip direction. The cross slip system has the same slip direction as the primary slip system

Table 12.1 Slip systems in FCC.

Slip plane	Slip direction		Name	Schmid factor for ([$\bar{1}$23])	Schmid factor for ([$\bar{1}$35])
(d)(111)	CB	[$\bar{1}$01]	Primary		
	AB	[$\bar{1}$10]	Coplanar		
	CA	[0$\bar{1}$1]	Coplanar		
(a) ($\bar{1}$1$\bar{1}$)	CB	[$\bar{1}$01]	Cross		
	DC	[110]			
	DB	[011]			
(b) (1$\bar{1}\bar{1}$)	CA	[0$\bar{1}$1]			
	DC	[110]			
	DA	[101]	Critical		
(c) ($\bar{1}\bar{1}$1)	AB	[$\bar{1}$10]			
	DB	[011]	Conjugate		
	DA	[101]			

but with a different slip plane. The conjugate slip system is a slip system which operates when, during a tensile test, the tensile axis reaches the boundary $001-\bar{1}11$ of the standard triangle. On the other hand, the critical slip system is a slip system which operates when, during a compressive test, the compressive axis reaches the boundary 001–011 of the standard triangle.

Exercise 12.4 Fill the columns of Schmid factor in Table 12.1.

Hint: Let the slip direction be $[uvw]$, the slip plane be (hkl), the tensile (or compressive) axis be $[UVW]$, then Schmid factor is given by

$$m = \cos\lambda\cos\phi = \cos[uvw]^\wedge[UVW] \times \cos[hk\ell]^\wedge[UVW]$$
$$= \frac{uU+vV+wW}{\sqrt{u^2+v^2+w^2}\sqrt{U^2+V^2+W^2}} \times \frac{hU+kV+\ell W}{\sqrt{h^2+k^2+\ell^2}\sqrt{U^2+V^2+W^2}} \tag{12.8}$$

Body-centered Cubic (BCC) The slip direction of BCC crystals is 1/2<111>, but the slip plane is not unique. This will be described in detail in Sec. 12.4.

12.2 Phenomenology of Yielding of a Single Crystal

Yielding is, by definition, the transition from elastic deformation to plastic deformation. Except for special cases such as deformation twinning,

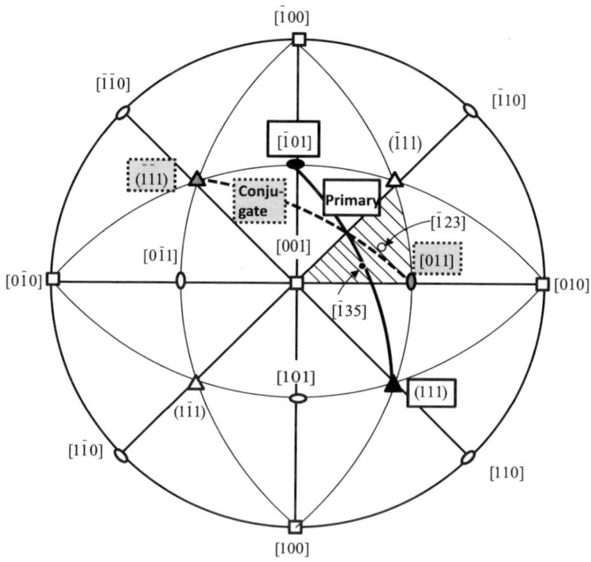

(a)

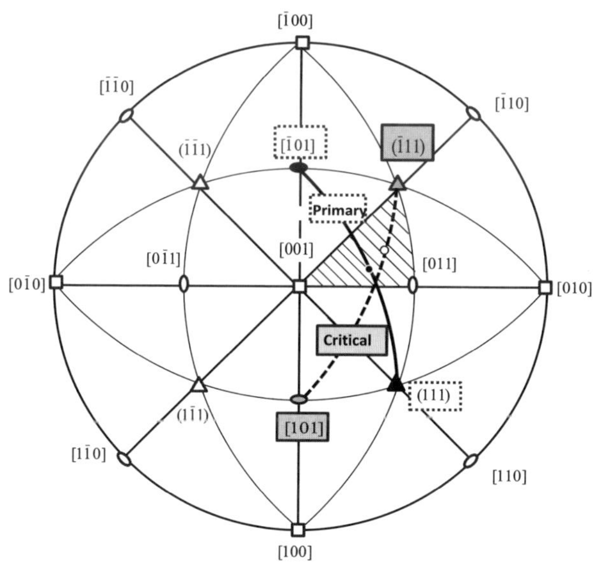

(b)

Fig. 12.8 (a) Primary slip system and conjugate slip system, (b) Critical slip system, (c) Coplanar slip system, (d) Cross slip system.

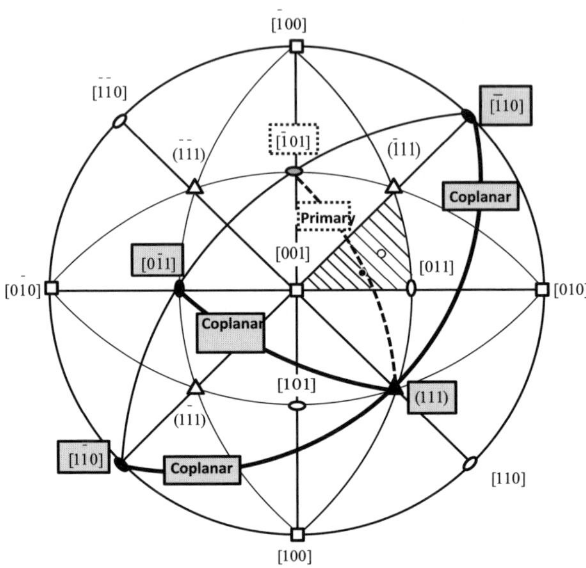

(c)

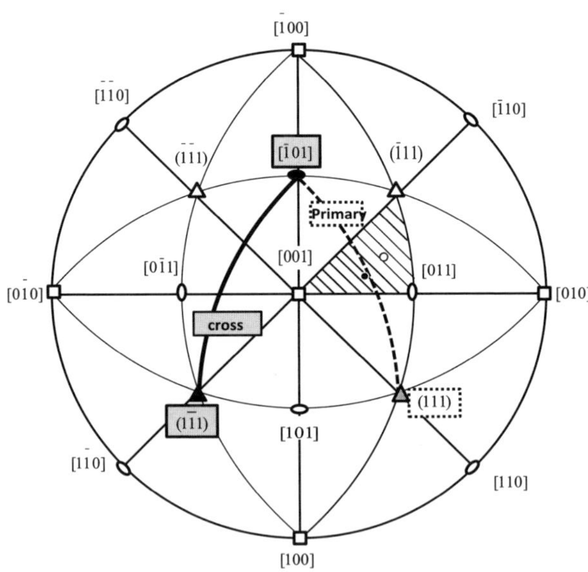

(d)

Fig. 12.8 (*Continued*)

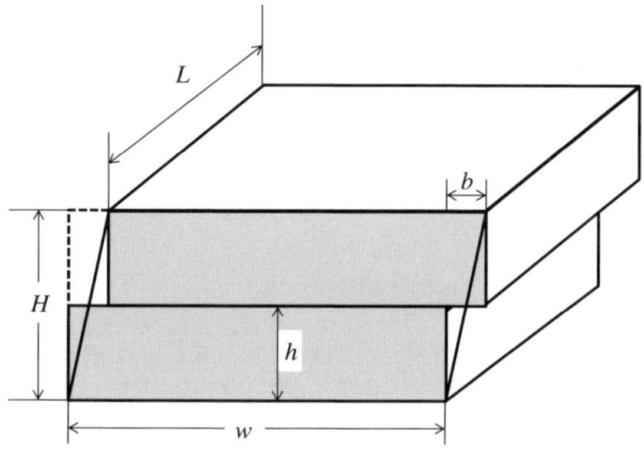

Fig. 12.9

this corresponds to the onset of dislocation motions. The practical plastic deformation is a result of motion of a vast amount of dislocations.

Let us suppose that, under the action of a shear stress σ_{yx}, a single edge dislocation sweeps completely a slip plane located at the height of $y = h$ of a single crystal with a dimension of width W, height H and depth L (Fig. 12.9). Then an atomic step with the length of the Burgers vector b will be created at the height $y = h$. The overall plastic shear strain is given by $\gamma_{yx} = b/H$ (Note that the strain is not b/h and is independent to h). If the dislocation does not sweep the slip plane completely but moves only Δx, then the shear strain is given by $\Delta\gamma = b(\Delta x/W)/H$. When n dislocations move $\Delta x, \Delta\gamma = bn(\Delta x/W)/H$ is given by

$$\Delta\gamma = b\left(\frac{n}{WH}\right)\Delta x = b\rho\Delta x, \tag{12.9}$$

where n/WH is the number of dislocations per unit area, and was replaced by a dislocation density ρ (strictly speaking, the density of mobile dislocations). In order for a macroscopic plastic deformation to take place, many dislocations must move. Now, let us consider two extremes. In one, it is assumed that the mobile dislocation density ρ is virtually constant. Differentiating Eq. (12.9) with respect to time, the strain rate is given by

$$\begin{aligned} \dot{\gamma} &= b\rho\bar{v} \\ \bar{v} &= \left(\frac{\Delta x}{\Delta t}\right). \end{aligned} \tag{12.10}$$

In the other case, dislocations are initially locked and cannot move, but once they are unpinned, they move at a high velocity and after sweeping certain areas, they become immobile again or get out of the crystal. Assuming that, on average, ΔN dislocations are unlocked and sweep an area S per Δt, then

$$\dot{\gamma} = b\left(\frac{S}{V}\right)\dot{N}, \tag{12.11}$$

where $V = LWH$ is the volume of the crystal.

12.2.1 *Mobile dislocation density is constant (Johnston–Gilman theory)*

Let us consider the first case where Eq. (12.10) is applicable. Usually tensile or compression tests are performed on an Instron-type testing machine. Two screws rotate to displace a crosshead up and down at a constant speed. The crystal to be tested is inserted between the crosshead and the fixed floor. The force applied to the crystal is measured by a spring (a load cell) connected in series (Fig. 12.10(a)). Let the speed of the crosshead be S_c and the displacement be Δy (Fig. 12.10(b)),

$$\Delta y \equiv S_c t = \Delta y_{el} + \Delta L_p, \tag{12.12}$$

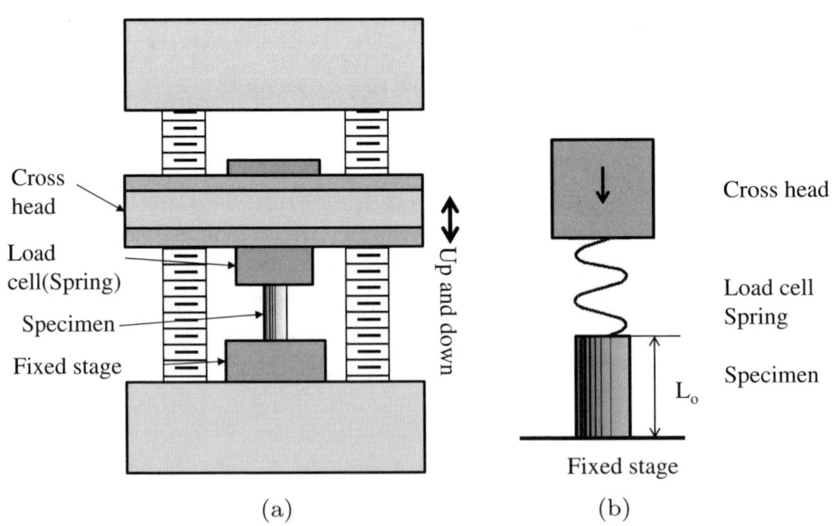

Fig. 12.10 (a),(b)

$\Delta L_{\rm p}$ is the plastic displacement of the crystal. $\Delta y_{\rm el}$ is the elastic displacement of the overall system and given by,

$$\Delta y_{\rm el} = \Delta L_{\rm el} + F/K + F/K^{\rm CH} \approx F/K,$$

where $\Delta L_{\rm el}$ is elastic deformation of the crystal, F/K is the elastic deformation of the load cell, F is the applied load and K is the spring constant of the load cell. $F/K^{\rm CH}$ is the elastic deformation of the crosshead and $K^{\rm CH}$ is the spring constant of the crosshead.

If the initial length of the crystal is $L_o, \Delta L_p = \varepsilon L_o$ (ε is a longitude strain). ε and shear strain γ is related by Schmid law, $\varepsilon = f_s\gamma$ (f_s: Schmid factor), so that Eq. (12.12) is given by,

$$\gamma = \frac{\Delta L_{\rm p}}{f_{\rm s}L_0} = \frac{S_{\rm c}t - F/K}{f_{\rm s}L_{\rm o}}. \qquad (12.13)$$

Therefore, the strain rate imposed externally ($\dot{\gamma}_{\rm ext}$) is given by

$$\dot{\gamma}_{\rm ext} = \left\{S_{\rm c} - \frac{1}{K}\left(\frac{dF}{dt}\right)\right\}\Big/ f_{\rm s}L_{\rm o}. \qquad (12.14)$$

On the other hand, the internal strain rate of the crystal from the microscopic viewpoint of dislocation motion ($\dot{\gamma}_{\rm int}$) is given by Eq. 12.10.

These two should be equal, so that

$$b\rho\bar{v} = \left(S_{\rm c} - \frac{1}{K}\frac{dF}{dt}\right)\Big/ (f_{\rm s}L_{\rm o}). \qquad (12.15)$$

Johnston and Gilman [4],[5] succeeded in measuring velocity of an individual dislocation in a single crystal of LiF. First, the surface of a crystal is etched chemically to form an etch pit at the emerging point of a dislocation. Then, a pulsed stress (σ) is applied for a short time (Δt) and the surface of the crystal is again etched. If the dislocation moved during the pulse stress, a new etch pit is formed at the new emerging point ②. Now, since no dislocation is present at the original position ①, the original etch pit flattens.

Suppose that the distance between ② and ① is Δl, then $\Delta l/\Delta t$ is the average velocity ($\bar{v}$) with which the dislocation moved under the stress of σ. In Fig. 12.11(a) a pulsed stress was applied twice consecutively, and the dislocation moved ① → ② → ③. Dislocation indicated by B did not move during the pulses, and the etch pit grew by the consecutive etching. In

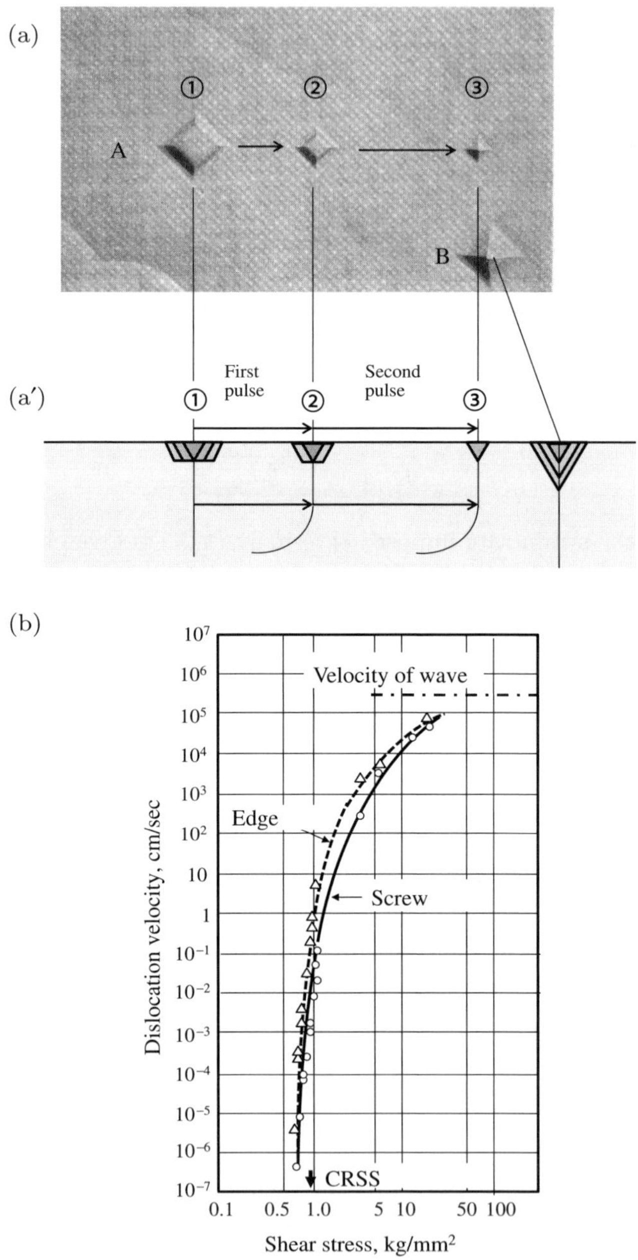

Fig. 12.11 (a) Observation of motion of a dislocation by etch pits (Reproduced with permission from[4]), (a′) Schematic illustration, (b) Mobility of dislocations in LiF (Reproduced with permission from [5]).

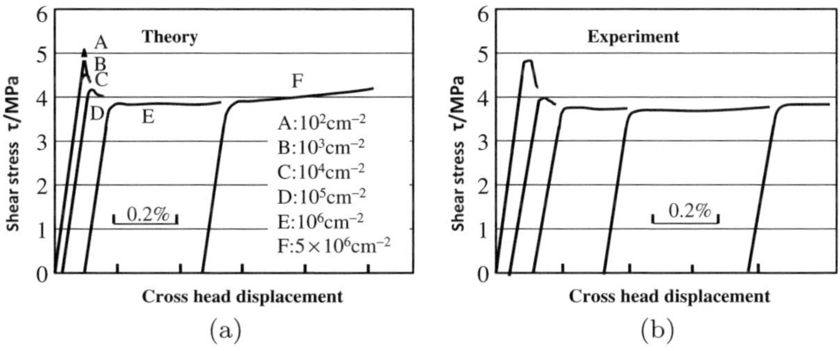

Fig. 12.12 (a) Theoretical stress-strain curves (Reproduced with permission from [4]), (b) Experimental stress-strain curve (Reproduced with permission from [5]).

this way Johnston and Gilman measured experimentally, the mobility of dislocations in LiF and obtained the following relation;

$$\overline{v} = (\tau/\tau_0)^m, \tag{12.16}$$

$$\rho = \alpha\gamma, \tag{12.17}$$

(see Fig. 12.11(b)). From Eqs. (12.13)–(12.16), the following differential equation was obtained (assuming $f_s = 0.5$),

$$\begin{aligned} \frac{d\tau}{dy} &= C - B\left(\frac{\tau}{\tau_o}\right)^m (Cy - \tau), \\ C &= \frac{K}{2A_\mathrm{o}}, \quad B = \frac{2b\alpha}{S_\mathrm{c}}. \end{aligned} \tag{12.18}$$

Johnston and Gilman obtained stress-strain curves shown in Fig. 12.12(a) by solving Eq. (12.18) (numerically). Figure 12.12(b) shows the corresponding experimental stress-strain curves. Agreement between theory and experiment is excellent. A–F in Fig. 12.12(a) correspond to varying the initial dislocation density (ρ), and as can be seen from Eq. (12.17), correspond to a prestrain being given in experiment.

12.2.2 *Modification to Johnston–Gilman theory*

Next consider the case where Eq. (12.11) is applicable, that is, in the initial stage of deformation, dislocations do not move. Figure 12.13 shows mobility of the dislocations $v(\sigma)$ in a variety of crystals. Mobility is different from one crystal to another, but in general it is much higher in FCC metals and in basal slip in HCP metals than in BCC metals and ionic crystals.

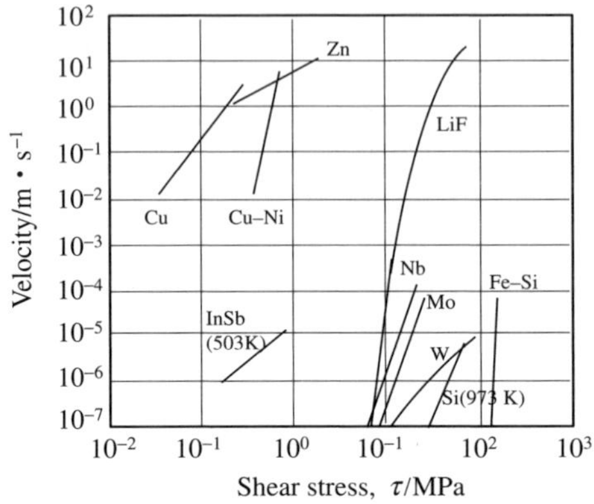

Fig. 12.13 Mobility of dislocations in a variety of crystals (Reproduced with permission from [6]).

Let us consider the initial stage of plasticity, where dislocations start to move under the action of an external force. Usually even a well annealed crystal contains dislocations of a density as high as $10^5 \sim 10^6\,\mathrm{cm}^{-2}$, most of which comprise dislocation networks (see for instance, Fig. 6.5(a)). Under the action of an external force, the dislocation segment whose Schmid factor is high will start to move (Fig. 6.5(b)). In order for a dislocation segment with the length of l to move freely, the stress must be higher than the back stress given by the following relation

$$\tau_c = \frac{\mu b}{l}, \tag{12.19}$$

(see Eq. (6.4)). In other words, below τ_c the density of mobile dislocations is zero.

Dislocation mobility such as shown in Fig. 12.13 has been measured in nearly perfect crystals with a dislocation density less than $10^5\,\mathrm{cm}^{-2}$: It represents the intrinsic velocity of dislocation. However, in a usual crystal whose dislocation density is not so low, dislocation velocity is virtually zero below τ_c. It is only above τ_c that the velocity approaches the intrinsic velocity. Therefore, the relation between $\bar{v}$ and stress is to be modified as represented by solid line in Fig. 12.14. In FCC metals the dislocation mobility is high because of low P-N force and a cross slip is difficult because of dissociation. Therefore, once the firstly activated dislocation source comes to halt, another new source is unlikely to be activated, so that the dislocation motion diminishes. This is exactly what Eq. (12.11) describes. Furthermore,

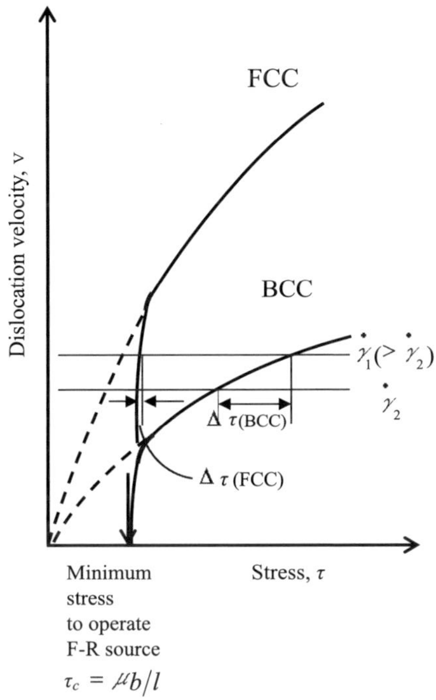

Fig. 12.14 (Reproduced with permission from [7]).

the microscopic strain rate due to the dislocation motion ($\dot{\gamma}_{\text{int}}$) exceeds the macroscopic strain rate imposed by the testing machine ($\dot{\gamma}_{\text{ext}}$) (right side of Eq. (12.14)).

By contrast, in BCC metals the intrinsic velocity of dislocations is low, so that the microscopic strain rate due to dislocation motion ($\dot{\gamma}_{\text{int}}$) cannot follow $\dot{\gamma}_{\text{ext}}$ at τ_c; it is necessary to increase the dislocation velocity by increasing τ. On top of this, screw dislocations in BCC metals can cross slip easily: Once screw dislocations start to move, they generate new dislocations continuously in such a way that they cancel those dislocations which are locked or disappear out of crystal. The density of mobile dislocation remains virtually constant. This is the situation to which Johnston–Gilman theory can be applied.

This mechanism explains difference between FCC and BCC metals in $\dot{\gamma}$ dependence of τ_{o}. In FCC metals $\dot{\gamma}_{\text{int}}$ at the yield point is much larger than $\dot{\gamma}_{\text{ext}}$, so that an increment of stress $\Delta\tau_{\text{o}}$ caused by changing $\dot{\gamma}_{\text{ext}}$ is negligibly small, while in BCC metals $\Delta\tau$ should be larger as can be seen from Fig. 12.14.

12.3 Thermally Activated Process of Dislocation Motion — Internal Stress and Effective Stress

12.3.1 *General theory*

In the foregoing discussion on Johnston–Gilman theory it is assumed that a stress imposed externally (external stress or applied stress) is directly transferred to dislocations. However, in real crystals, barriers are present which hinder dislocation motion. The barriers can be classified into two, i.e., long-range barrier and short-range one. Long-range barriers are referred to as athermal barriers, and a large external force is required for dislocations to overcome them. This is the origin of the so-called internal stress. On the other hand, short-range barriers can be overcome by a thermally activated process. That is,

Long range barrier=internal stress=athermal stress (cannot be overcome by a thermally activated process) Short range barrier=effective stress=thermal stress(can be overcome by a thermally activated process)

These two overlap to become barriers to the dislocation motion (Fig. 12.15). Figure 12.16(a) shows yield stress (or flow stress) of a crystal as a function of temperature. The flow stress increases monotonously with decreasing temperature below a critical temperature (T_c), but above T_c the flow stress

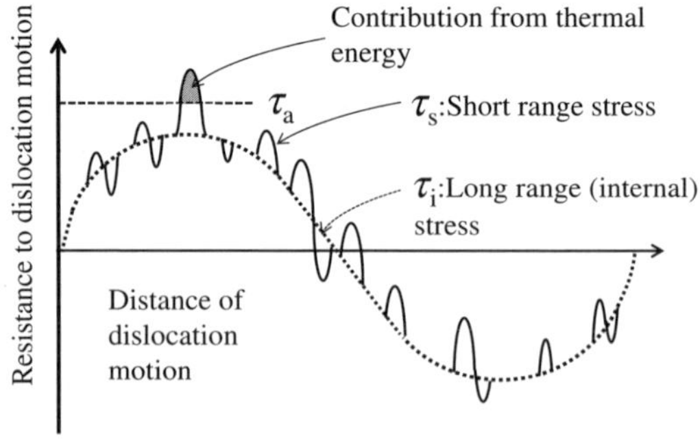

Fig. 12.15 Long-range stress and short-range stress.

remains virtually unchanged.[3] The stress which is independent to temperature above T_c is the internal stress caused by athermal long-range barriers and the stress increment which increases with decreasing temperature below T_c is thermal stress caused by short-range barriers. As will be discussed later, a thermally activated process is time-dependent, so that, at a high strain rate, a higher stress is required for dislocations to be activated because dislocations have the shorter waiting time under the higher strain rate. Thus, as shown in Fig. 12.16(b), stress-strain curves shift to the higher-temperature side.

One of the most important origins of the internal stress is dislocation cell walls. Formation process of the dislocation cell will be described later (see Fig. 12.53). Here, it is pointed out how the cell walls hinder dislocation motions. Figure 12.17 shows motions of dislocations in pure Ni which was deformed *in situ* in a HVEM. Figure 12.17(h) shows a macroscopic stress-strain (load-displacement) curve measured concurrently with the *in situ* deformation. The specimen had been prestrained in the bulk state, so that dislocation cells had been introduced prior to the *in-situ* deformation, as can be seen in (f). Now, let us focus on dislocations A and B. In Fig. 12.17(a) the load is at a maximum and dislocations A and B bow out rightward. From the radius (r) of the bowing-out dislocations an effective stress acting on these dislocation is given by $\tau_{\text{eff}} = \mu b/2r$ (see Eq. (12.19)). On reducing the load, both A and B shrink (Fig. 12.17(b)), eventually bowing out to the opposite diection (i.e., leftward) (Fig. 12.17(c)).

In other words, the effective stress acting on dislocations A and B is now opposite to the externally applied stress. Since $\tau_a = \tau_i + \tau_{\text{eff}}$, $\tau_{\text{eff}} = \tau_a - \tau_i < 0$, that is, the internal stress τ_i is larger than the applied stress τ_a. On reloading, A and B bow out rightward again. Below the point (a) (Fig. 12.17(e)) A and B move forward and backward reversibly, but as soon as the load exceeds point (a) (Fig. 12.17(f)), A and B move large distances. At the same time, the macroscopic stress-strain curve is bent, indicating the crystal yields (Fig. 12.17(h)).

The region a→b→c→d→e in Fig. 12.17(h) corresponds to the elastic region and the macroscopic stress-strain curve is apparently straight. However, dislocations A and B did move actually. One can draw a microscopic stress-strain curve by taking the area swept by dislocations A and B as a microscopic strain, as is shown in Fig. 12.17(i). At

[3]Strictly speaking, above T_c athermal flow stress decreases slightly through the temperature dependence of μ.

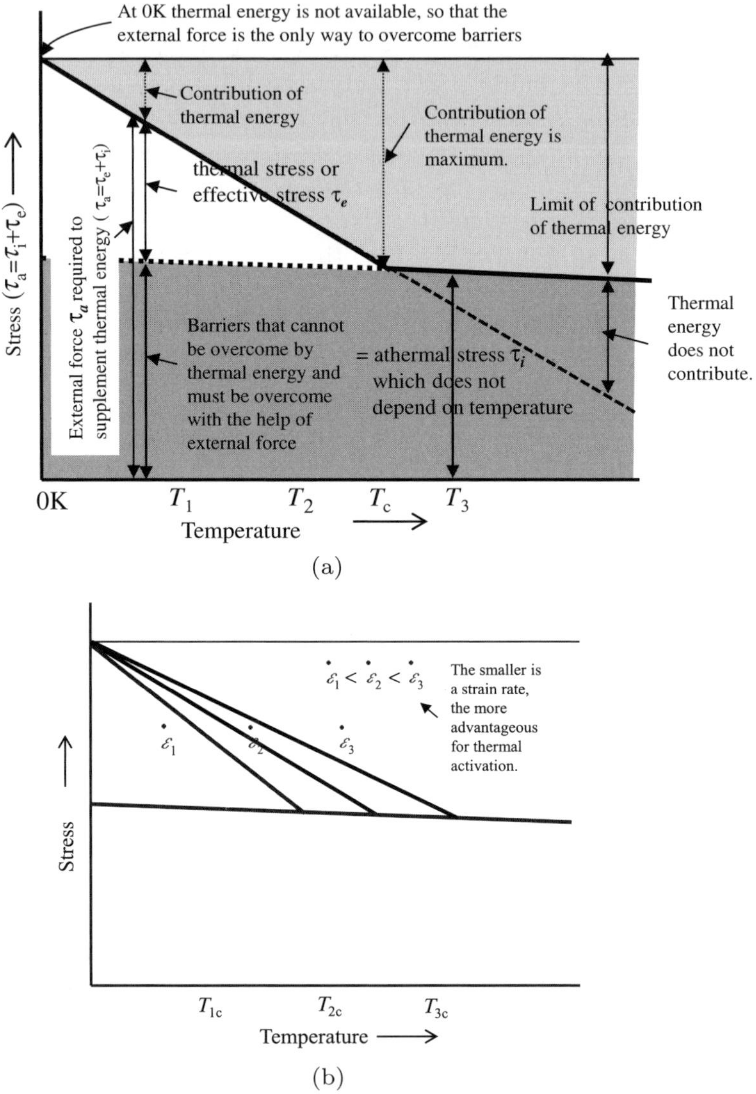

Fig. 12.16 (a) Thermal stress and athermal stress, (b) Effect of strain rate on thermal stress.

the macroscopic yielding, dislocations A and B moved large distances (Fig. 12.17(f)), and many slip traces (intersections of the slip planes on the crystal surfaces, indicated by a pair of arrows (black and white)) appeared (Fig. 12.17(g)). Dislocation motions before a macroscopic yielding

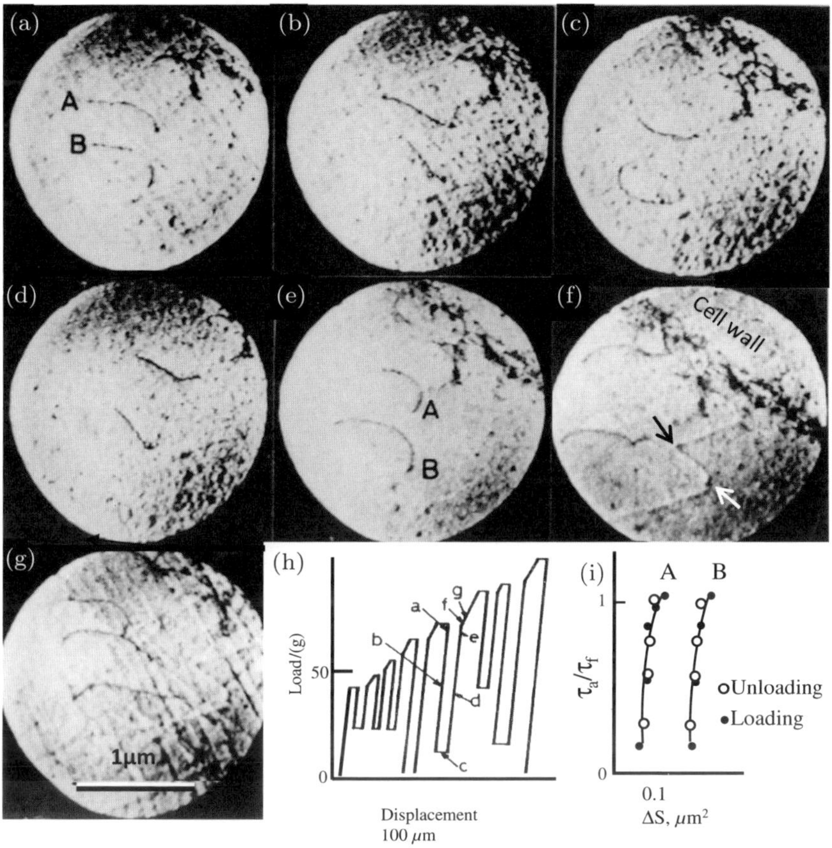

Fig. 12.17 Motion of dislocations during loading-unloading cycle in Ni (Reproduced with permission from [8]).

is referred to as microplasticity. From experimental point of view, energy dissipated during the reversible motion of a dislocation is measured as an internal friction to detect microplasticity.

In addition to dislocation cell walls, sources of the internal stresses include precipitates and grain boundaries. Typical short-range obstacles which can be overcome by thermally activated processes are listed in Figs. 12.18(a)–12.18(d).

Now let us consider a process by which a short–range barrier (hatched peak of a short-range obstacle in Fig. 12.15) is overcome by a dislocation. The potential of the barrier is shown in Fig. 12.19(a).

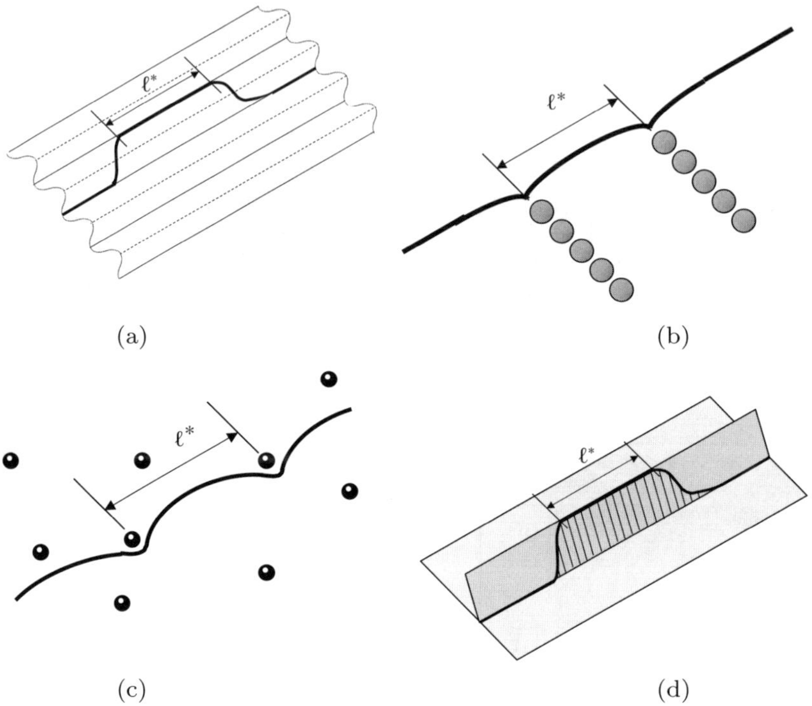

Fig. 12.18 (a)–(d) (a) P–N potential, (b) non-conservative motion of a jog, (c) cutting of solute atoms, fine precipitates and forest dislocation, (d) cross slip.

Overcoming a barrier with the help of thermal energy alone (Fig. 12.19) In order for a dislocation to move from position A to B, it must overcome a potential peak C. To achieve this, first of all, the dislocation must reach C. Dislocations are incessantly vibrating due to thermal energy. Supposing that the height of the potential peak is G^*, the probability of the dislocation to reach C is given by $\mathrm{e}^{-\mathrm{G}*/\mathrm{kT}}$. The actual frequency is obtained by multiplying this probability by the number of trial ν.

However, the dislocation that managed to reach B may return to A with the same probability $\mathrm{e}^{-\mathrm{G}*/\mathrm{kT}}$. As a result, the net probability of the dislocation moving forward (A→B) is given by

$$\mathrm{e}^{-\mathrm{G}*/\mathrm{kT}} - \mathrm{e}^{-\mathrm{G}*/\mathrm{kT}} = 0.$$

That is, the dislocation stays at the initial position A on average.

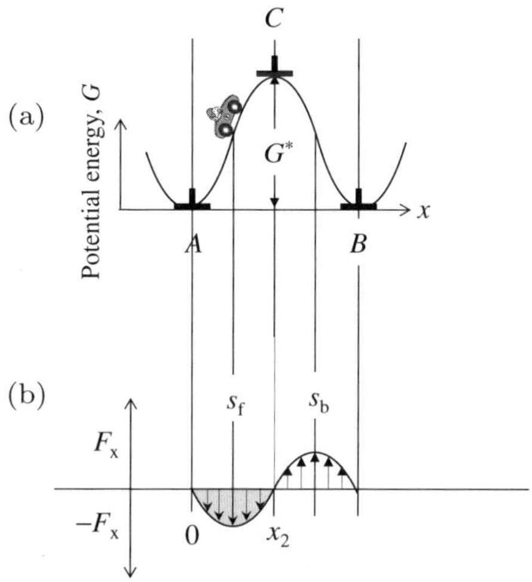

Fig. 12.19 (a),(b)

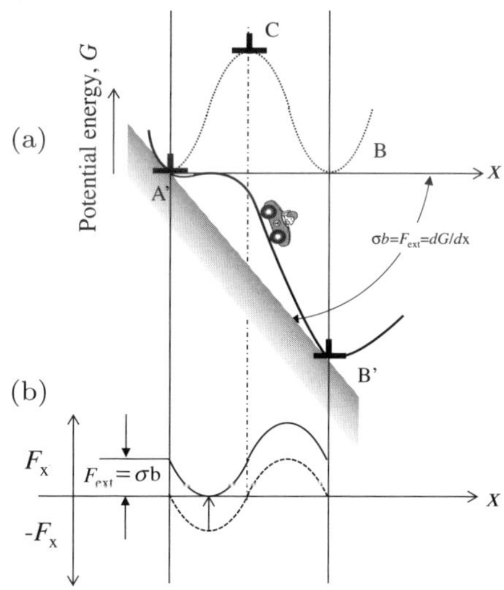

Fig. 12.20 (a),(b)

Overcoming a barrier by an external force without help of thermal energy (Fig. 12.20) The dislocation must overcome the peak of the potential energy. The resistive force exerted on the dislocation from the slope of the peak is given by $F = -dG^*/dx$ (x is distance). Thus, the steepest slope (S_f) is the critical position. Supposing that you are climbing a slope by a car, you must push the accelerator pedal to the maximum. Once the peak(C) is passed, the force from the slope becomes positive (the car is accelerated). The $F - x$ curve is schematically shown in Fig. 12.19(b).

Now let us suppose that an external force $F_{\text{ext}} = \sigma^* b$ pushes the dislocation toward the summit of the potential. This corresponds to tilting the $G^* - x$ curve clockwise (Fig. 12.20(a)). Equivalently, the force $-x$ curve for the dislocation is escalated by the external force from zero to F_{ex} (see Fig. 12.20(b)). When $F_{\text{ext}} (= \sigma^* b)$ exceeds the maximum resistive force $(dG^*/dx)_{\max}$ that exerts on the dislocation at the steepest point, then the dislocation moves from A' to B'.

Both external force and thermal energy assist (Fig. 12.21) However, when $F_{\text{ext}} < (dG^*/dx)_{\max}$, the dislocation comes to a halt halfway on the slope. Now let us consider effects of thermal energy. The hatched area (G^*) in Fig. 12.21(b) corresponds to what must be assisted by thermal energy.

Tilting $G^* - x$ curve clockwise by F_{ext} results in making the probability of the forward and backward motions of a dislocation asymmetric. Now, the probability of the forward motion (A"→B") is $e^{-G_{f^*}/kT}$, while that for the backward motion (B'→A") is $e^{-G_b^*/kT}$. The net probability of the forward

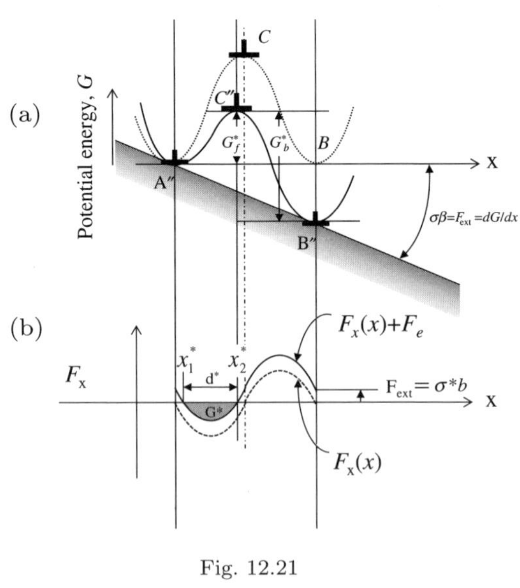

Fig. 12.21

motion is given by,

$$e^{-G_{f*}/kT} - e^{-G_b^*/kT}.$$

Since the second term is much smaller than the first one, it can be neglected. The motion of a dislocation under the action of an external force of $F_{\mathrm{ext}} = \sigma^* b$ at temperature T is given by $\nu e^{-G_{f*}/kT}$.

Contribution of thermal energy G^* (Fig. 12.21(b)) is given by

$$\begin{aligned} G^* &= \int_{x_1^*}^{x_2^*} \{0 - (F_x(x) + F_{ext})\}\, dx \\ &= \int_{x_1^*}^{x_2^*} (-F_{ext} + |F_x(x)|)\, dx = \int_{x_1^*}^{x_2^*} (-\sigma b l^* + |F_x(x)|)\, dx \\ &= \int_{x_1^*}^{x_2^*} |F_x(x)| dx - \sigma b l^* (x_2^* - x_1^*) \cong G^0 - \sigma b l^* d^* = G^0 - \sigma V^*, \end{aligned} \tag{12.20}$$

where approximated as follows:

$$G^0 = \int_0^{x_2} |F_x(x)|\, dx \cong \int_{x_1^*}^{x_2^*} |F_x(x)| dx.$$

x_2 is indicated in Fig. 12.19(a). Furthermore, in

$$V^* = bl^* d^*, \tag{12.21}$$

$d^* (= x_2^* - x_1^*)$ corresponds to the lateral motion of a dislocation, and l^* is the length of a dislocation segment which succeeds in overcoming the peak of the potential barrier (Fig. 12.18). V^* has a dimension of $[\mathrm{L}^3]$ and for this reason, referred to as the activation volume. However, b corresponds to the Burgers vector of the dislocation and has nothing with the thermally activated process. Since $l^* d^*$ is a value characterizing the thermally activated process, it is appropriate to refer to $l^* d^* = A^*$ as the activation area.

d^* is a spacing between short-range obstacles. P–N barrier will be discussed later. G^* is the thermal energy supplied during the thermally activated process. Thermal activation is absent at absolute zero, that is, contribution from thermal energy in Fig. 12.21 is absent, either. Therefore, unless an external force reaches $F_{\max}$, the barrier cannot be overcome. This is what Fig. 12.20 indicates. At T (below T_c) thanks to contribution of G^* by thermal energy, the external force required to overcome the barrier is reduced to as small as F_{ext}. In other words, the hatched area in

Fig. 12.21 is contribution from thermal energy which increases with temperature. Coming back to Fig. 12.16(a), above T_c, whatever barriers that can be overcome by thermal energy have already been overcome by thermal activated process. The remaining barriers are only long-range barriers that cannot be overcome by thermal energy and can be overcome entirely by the assistance of the external force. This is σ_μ.

The motion of an individual dislocation below T_c is given by an Arrhenius equation as follows:

$$\begin{aligned}\dot{\varepsilon} &= bn\dot{v} \\ &= bn\nu^* \exp\left(-\frac{G^*}{kT}\right),\end{aligned}$$

where n is the number of sites where a thermal activation can occur. Thermal activations can occur at different sites on a dislocation, so that n is not the number of dislocations.

$$\begin{aligned}G^* &= -k\left(\frac{\partial \ln \dot{\varepsilon}}{\partial\left(\frac{1}{T}\right)}\right)_{\sigma^*} \cong -k\left(\frac{\ln \dot{\varepsilon_2} - \ln \dot{\varepsilon_1}}{\frac{1}{T_2} - \frac{1}{T_1}}\right)_{\sigma^*} \\ &= -8.62 \times 10^{-5}[\text{eV}][\text{K}^{-1}]\left(\frac{2.3\Delta \log \dot{\varepsilon}}{\Delta \frac{1}{T}}\right)_{\sigma^*}[\text{K}] \\ &= -19.8 \times \left(\frac{\Delta \log \dot{\varepsilon}}{\Delta \frac{1}{T}}\right)_{\sigma^*} \times 10^{-5}[\text{eV}] \qquad (12.22)\end{aligned}$$

The stress under discussion here is an effective stress (τ_{eff}) associated with thermally activated process described in Figs. 12.19 or 12.21. Putting $\tau_{\text{eff}} = \sigma^*, \sigma_a = \sigma_i + \sigma^*$. Here, σ_i is the internal stress (which has nothing with thermal activation) and σ_a is an externally imposed stress, which can be measured experimentally. Also,

$$\begin{aligned}V^* &= -\frac{dG*}{d\sigma^*} = kT\left(\frac{\partial \ln \dot{\varepsilon}}{\partial \sigma}\right)_T \\ &= 1.38 \times 10^{-16}[\text{erg}][\text{K}^{-1}]T\,[\text{K}^{-1}]\left(\frac{\partial \ln \dot{\varepsilon}}{\partial \sigma}\right)_T [10^{-7}\text{cm}^2]\,[\text{dyn}^{-1}] \\ &= 1.38 \times 10^{-23}\,[\text{cm}^3]\,T\left(\frac{\partial \ln \dot{\varepsilon}}{\partial \sigma}\right)_T\end{aligned}$$

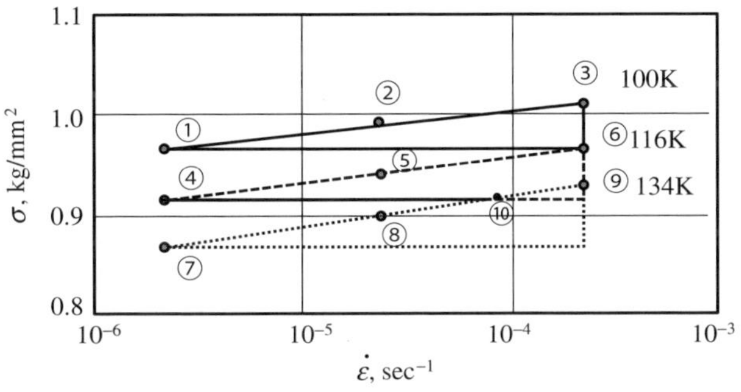

Fig. 12.22 Calculation of activation parameters [9].

$$= 31.8 \times [\text{Å}]^3 T \left[\frac{\Delta \log \dot{\varepsilon}}{\Delta \sigma}\right]_T$$
$$(\sigma \quad \text{in} \quad \text{MPa}). \tag{12.23}$$

Here, σ is indicated in MPa.

The point is that G^* and V^* can be deduced from stress-strain curves obtained at different strain rates at different temperatures.[4]

Exercise 12.5 Derive Eqs. (12.22) and (12.23).

Table 12.2 shows yield stresses of a Cu–2at.%Al alloy (σ:kg/mm^2) [9]. Figure 12.22 shows σ-$\log_{10}\dot{\varepsilon}$ curves redrawn from these data.

Let us calculate V^* according to Eq. (12.23). Connecting ① and ⑥,

$$\triangle \log_{10} \dot{\varepsilon} = 2,$$
$$\triangle(1/T) = 1/116 - 1/100 = 1.38\text{x}10^{-3}[\text{K}^{-1}]$$

Table 12.2 (Reproduced from [9]).

Strain rate $\dot{\varepsilon}$	2.2×10^{-6}	2.2×10^{-5}	2.2×10^{-4}	Deformation	
$\log_{10} \dot{\varepsilon}$	−5.66	−4.66	−3.66	Temperature (T)	1/T
	0.963①	0.988②	1.010③	100 K	0.0010 K^{-1}
	0.916④	0.938⑤	0.9608⑥	116 K	0.00862 K^{-1}
	0.8628⑦	0.896⑧	0.9263⑨	134 K	0.00746 K^{-1}

[4]One must be careful not to make a mistake in taking unit in practical computation.

$$\therefore G^* = 19.8 \times 10^{-5} \times 2 \div (1.38 \times 10^{-3})[\text{eV}]$$
$$= 0.28\ eV(@\sigma = 0.96\ \text{kg/mm}^2).$$

Similarly connecting ④ and ⑩,

$\triangle \log_{10}\dot{\varepsilon} = 1.57$, $\triangle(1/T) = 1/134 - 1/116 = 1.158 \times 10^{-3}[\text{K}^{-1}]$, so that

$G^* = 19.8 \times 10^{-5} \times 1.57 \div (1.158 \times 10^{-3}) = 0.268\ eV(@\sigma = 0.916\,\text{kg/mm}^2)$

On the other hand, the activation volume is given by,

$$V^* = 31.8\text{T}(\triangle \log_{10}\dot{\varepsilon}/\triangle\sigma)(\text{unit of } \sigma \text{ is MPa}).$$

Connecting ①, ② and ③ (at 100 K),

$$\triangle \log_{10}\dot{\varepsilon} = 2, \quad \triangle\sigma = (1.010 - 0.963)\ \text{kg/mm}^2 = 0.07\ \text{kg/mm}^2$$
$$= 0.686\,\text{MPa}.$$
$$\therefore \quad V^* = 31.8 \times 100 \times 2 \div 0.686 = 9271\mathring{\text{A}}^3(@\sigma = 0.988\ \text{kg/mm}^2)$$
$$= 561b^3(b = 2.5\text{A}).$$

Similarly, connecting ⑦, ⑧ and ⑨ (at 134 K),

$\triangle\sigma = 0.0635\,\text{kg/mm}^2 = 0.622\text{MPa}$, so that

$V^* = 31.8 \times 134 \times 2 \div 0.622 = 13700\ \mathring{\text{A}}^3(@\sigma = 0.896\,\text{kg/mm}^2) = 876\,b^3$.

Exercise 12.6 Calculate V^* from the line connecting ④⑤⑥ (134 K).

As shown in Fig. 12.18, short-range obstacles include cutting solute atoms (or very tiny precipitates and forest dislocations), non-conservative motion of a jog, cross slip and overcoming P–N potential barrier. The corresponding activation volumes (areas) are shown in Table 12.3.

In the traditional dislocation theory, thermal activation parameters such as V* and G^* have been measured to obtain information on the motion

Table 12.3 Short-range barriers.

Mechanism	V*(A*)	Fig. 12.18
P–N barrier	$10 - 10^2\text{b}^3(\text{b}^2)$	(a)
non conservative motion of a jog	$10^2 - 10^4\text{b}^3(\text{b}^2)$	(b)
solute atoms, tiny precipitates	$10^2 - 10^4\text{b}^3(\text{b}^2)$	(c)
cutting forest dislocations	$10^2 - 10^4\text{b}^3(\text{b}^2)$	(c)
cross slip	$10 - 10^2\text{b}^3(b^2)$	(d)

Table 12.4 Activation volume of some functional materials.

	Γ[10]	Γ_1[11]	ζ[12]	γ_1[13]	LaN_5[14]	Fe_2Al_5[15]	FeDy[16]
V* (Å^3)	~ 200	~ 200	~ 160	~ 200	200–300	~ 300	300–600

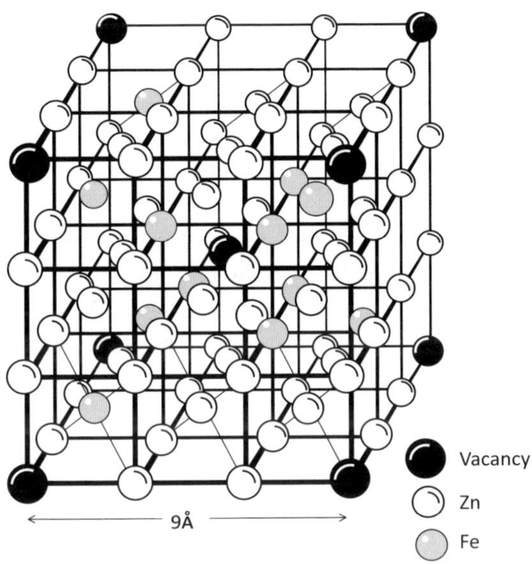

Fig. 12.23 Unit cell of Γ. It crystallizes in the BCC structure with $a = 9$Å. Vacancies (indicated by ●) form the BCC lattice. If it is assumed that $\boldsymbol{b} = a/2\langle 111\rangle$ as in conventional BCC structure, $b = 7.794$Å and hence $b^3 = 473$Å^3.

of dislocations of the known $\boldsymbol{b}$ in a well controlled crystal with a simple crystal structure. However, cutting-edge functional materials often have complex crystal structures and giant unit cells, so that $\boldsymbol{b}$ is unknown. Measuring activation parameters (especially activation volume) in these materials provides information on the absolute values of $\boldsymbol{b}$.

Table 12.4 summarizes values of V^* measured in some functional materials. For instance, Γ (capital gamma) is an Fe–Zn intermetallic compound and has V^* of ~ 200Å^3. However, the structure of Γ is composed of 27(3×3×3) small BCC lattices each having $a = 3$Å(Fig. 12.23). The Burgers vector of a perfect dislocation b_{T} is a translational vector of the unit cell of Γ, which is quite large. The energy of the dislocation (proportional to b_{T}^2) is also quite large and the dislocation is likely dissociated into partial dislocations. If one assumes that the dislocation is dissociated into smaller BCC lattices with $a = 3$Å, one obtains $b_p = 2.6$Å, so that $V^* \sim 10b_p^3$. Then, as can be seen

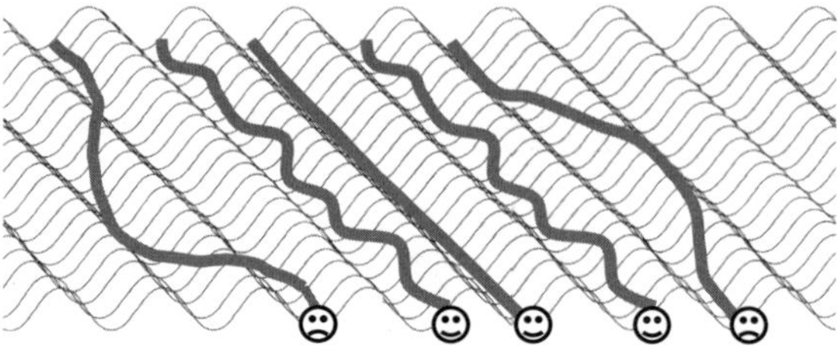

(a)

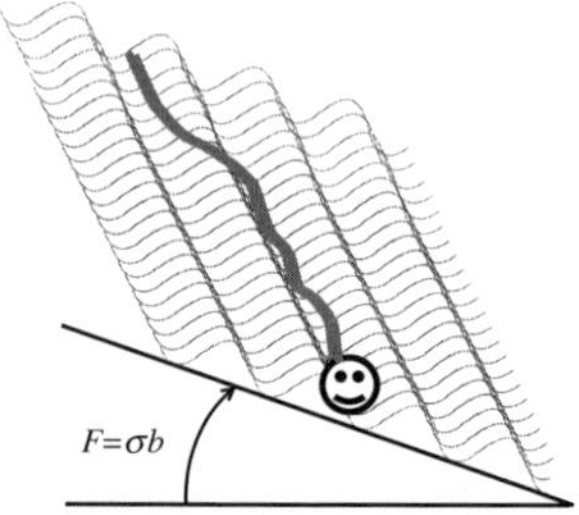

(b)

Fig. 12.24 (a), (b)

from Table 12.3, the motion of the dislocation is controlled by the P–N force. Measurement of V^* in functional materials may provide otherwise unobtainable information concerning mechanism of deformation of these materials (usually deformed at high temperature during processing).

12.3.2 *Peierls–Nabarro force*

The P–N potential can be compared to a corrugated galvanized iron roof with alternating valleys and hills along a crystallographic orientation. A dislocation lies in one of the valleys. Let us suppose that a snake lies (Fig. 12.24). At low temperature the snake is still and lies entirely in a valley. On warming, it will wriggle and part of it may cross the hill to reach the adjacent valley. However, when the roof is horizontal the snake wriggles equally to the right and left, so that on average the position remains unchanged (Fig. 12.24(a)). However, on a tilted roof the snake will drop from the roof (Fig. 12.24(b)).

This corresponds to the motion of a dislocation in a crystal with a high P–N force. The corrugation corresponds to the P–N potential. The snake wriggling on a warm roof corresponds to thermally activated process. Tilting the roof corresponds to imposing an external force; the larger tilting angle corresponds to the larger external force.

12.4 Yielding of BCC Metals

12.4.1 *Peierls force in BCC metals*

Strength of BCC metals is characterized as follows:

(1) It is much stronger than those of FCC and HCP metals.
(2) Small amount of impurity increases the strength significantly.

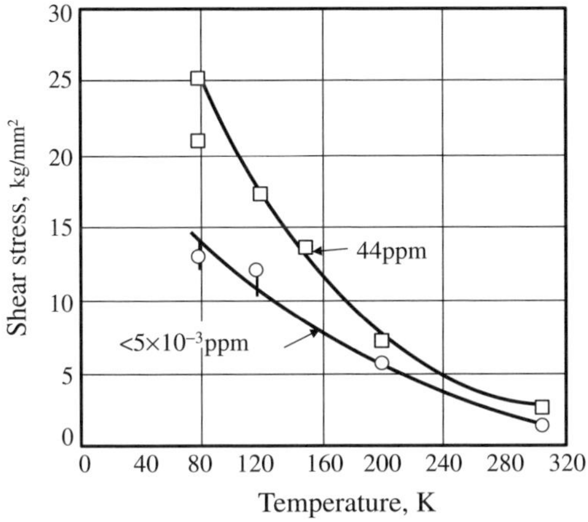

Fig. 12.25 Temperature dependence of yield stress in highly pure Fe and in less pure Fe (Reproduced with permission from [17]).

(3) The yield stress depends on temperature significantly.
(4) Slip planes are often not crystallographic planes.

Whether these features of strength of BCC metals is of extrinsic nature influenced by impurity or of intrinsic one inherent to the BCC lattice was discussed in 1960s and 1970s. Figure 12.25 shows temperature dependence of the yield stresses of a conventional pure Fe and an ultra high pure Fe. It is true that Fe with impurity is stronger and more temperature sensitive than the ultra-pure Fe. However, even the strength of the ultra-pure Fe is definitely temperature sensitive. This led to a conclusion that the strength of BCC metals is of intrinsic nature inherent to the BCC structure itself.

Mobility of edge and screw dislocations in BCC metals is quite different. Figure 12.26 shows mobility of edge and screw dislocations in Fe measured by means of an *in situ* deformation experiment in a HVEM [18]. The mobility of edge dislocations is higher than that of screw dislocation by one order. Figure 12.27 shows a process of a dislocation loop expanding under the action of an external force in Mo. The dislocation loop is elongated along the Burgers vector $\boldsymbol{b}$. In other words, the velocity of edge component is much larger than that of screw one. This tendency is enhanced with decreasing temperature. Figures 12.28(a)–12.28(c) show dislocations under stress in Fe at various temperature. On increasing temperature from 25 K (a) to 153 K (b) and to room temperature (c), difference in mobility between edge and screw dislocation decreases.

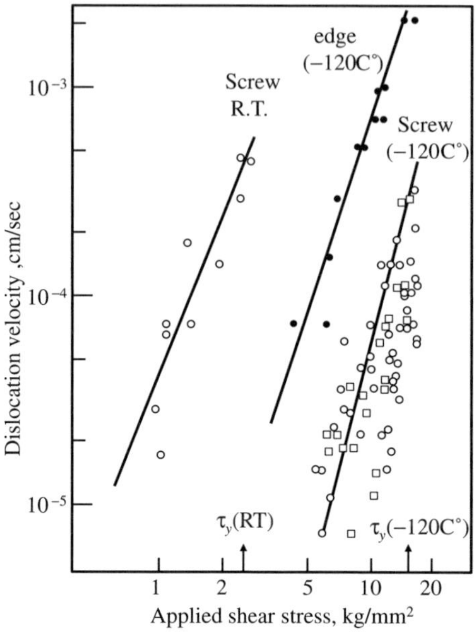

Fig. 12.26 Mobility of dislocations in Fe (Reproduced from [18]).

Low mobility of screw dislocations in BCC crystals is to be attributed to the crystallographic symmetry of the screw dislocation. The Burgers vector $\boldsymbol{b}$ in the BCC structure is 1/2<111>. Now let us view the BCC crystal in the direction of [111]. It is obvious from Figs. 2.28(a) and 2.28(b) that with respect to <111> direction a threefold symmetry is present. In accordance with this, a <111> screw dislocation assumes the threefold symmetry configuration. Furthermore, as shown in Fig. 12.29(a), a strain associated with $\boldsymbol{b}$ is distributed in threefold symmetrical planes like,

$$\frac{1}{2}[111] = \frac{1}{6}[111] + \frac{1}{6}[111] + \frac{1}{6}[111].$$

Figure 12.30(b) [21] shows a computer simulation and suggests that $\boldsymbol{b}$ is extended on three planes. It is to be noted here that this does not mean that a dislocation is dissociated definitely into partial dislocations like Shockley partials in FCC crystal. Displacement associated with $\boldsymbol{b}$ is distributed in a threefold symmetrical manner.

Now let us consider the process by which this screw dislocation glides. Glide motion is essentially a 2D phenomenon in which a dislocation glides on a planar glide plane. Therefore, it is necessary that the threefold

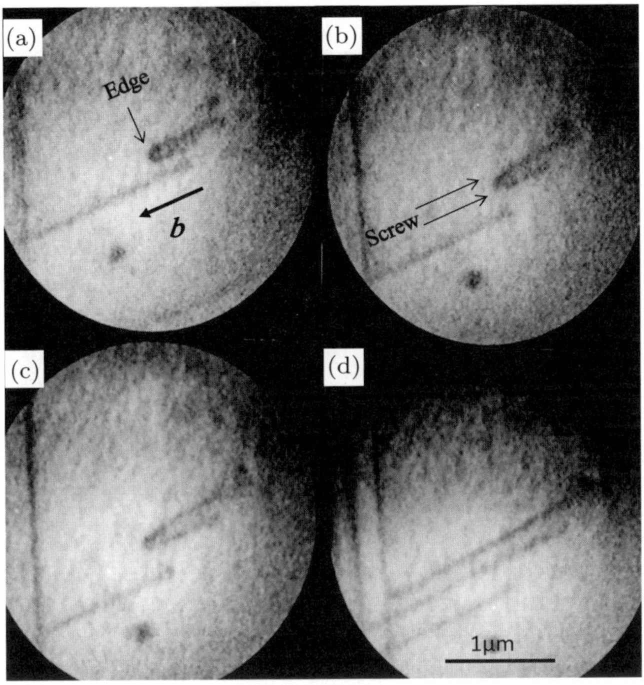

Fig. 12.27 (Reproduced from [19]).

symmetrical configuration must transform into a planar configuration. The necessity of transformation in configuration from threefold to twofold symmetry is responsible for the high P–N force in BCC crystals. This is Hirsch model of screw dislocations in BCC crystals [22].

Suzuki noticed configurations of atoms aligning along <111> direction. He pointed out that the sites in which a screw dislocation can get in are only two, A and B shown in Fig. 12.31(a). In site A, atoms assume a right-handed (RH) screw (b), while in site B, atoms assume a left-handed (LH) screw (c). Now let us suppose that a screw dislocation sitting in site A is to move. In order for this dislocation to move it must pass site B. In other words, the screw dislocation must go through $A \to B \to A \to B \cdots$ by changing its chirality. This is Suzuki model to explain why the P–N force in BCC crystal is high [23].

In either model, the characteristic configuration of a 1/2<111> screw dislocation in a BCC crystal is inherent to the BCC crystal structure itself, and explains various anomalous features of the strength of BCC metals, as will be described in the following.

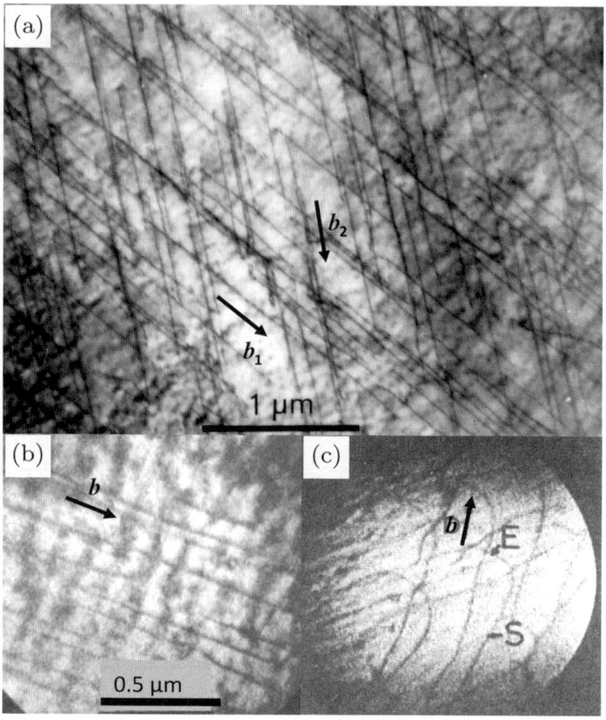

Fig. 12.28 Configuration of screw dislocations at (a) 25 K [20], at (b) 183 K [18] and at (c) room temperature [18].

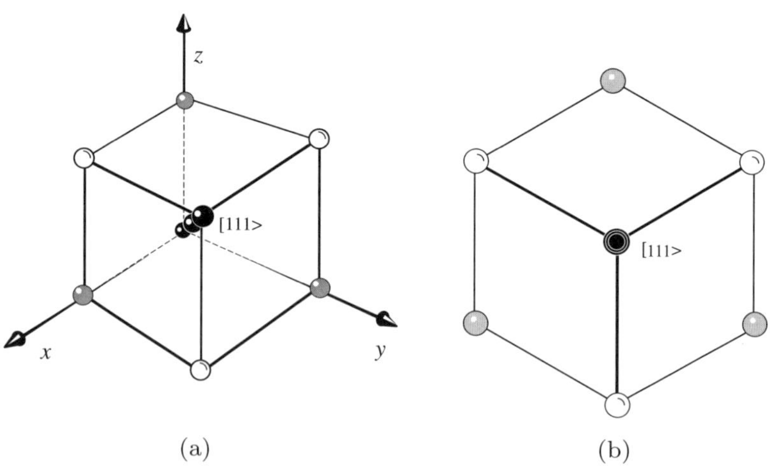

Fig. 12.29 (a),(b)

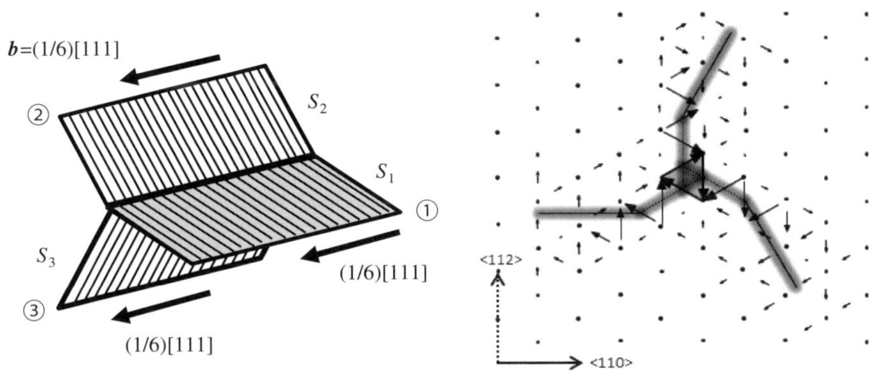

(a) (b)

Fig. 12.30 Models for extension of a screw dislocation in the BCC structure. (a) Schematic illustration, (b) Computer simulation (Reproduced with permission from [21]).

12.4.2 $\psi - \chi$ *curve*

Slip direction in the BCC structure is 1/2<111>, but it is often difficult to identify slip planes. Figure 12.32 shows an optical micrograph of slip lines observed on a Li–70at.%Mg alloy (BCC).[5] The slip lines are not straight, so that their crystallographic orientations cannot be identified. For the purpose of describing slip planes in the BCC structure, $\psi - \chi$ curve is used (Fig. 12.33) [25].

Now let the slip direction be [111] and the axial (tensile or compression) direction be S on the standard stereographic projection. λ is the angle between S and [111] and is $\sim 45°$. ψ and χ are defined as follows.

χ: The point where the great circle connecting the slip direction (S) and [111] crosses the great circle connecting $\bar{1}\bar{1}2$-$\bar{1}01$-$\bar{2}11$. That is, χ indicates the axial (tensile or compression) direction.

ψ: The observed slip plane on the great circle connecting $\bar{1}\bar{1}2$-$\bar{1}0\bar{1}$-$\bar{2}11$.

Both χ and ψ are expressed by angles from the pole $\bar{1}01$. Right side is positive and the left side is negative. That is, ($\bar{1}\bar{1}2$) is $-30°$ and ($\bar{2}11$) is $+30°$.

Exercise 12.7 Show that, for $\chi < 0$, S is within standard triangle of 001-011-111. (Hint): Consider $\lambda = 45°$ as a typical example.

[5]In the binary Li–Mg system the solubility limit of the primary solid solution of Li (BCC) is ~70at.%Mg.

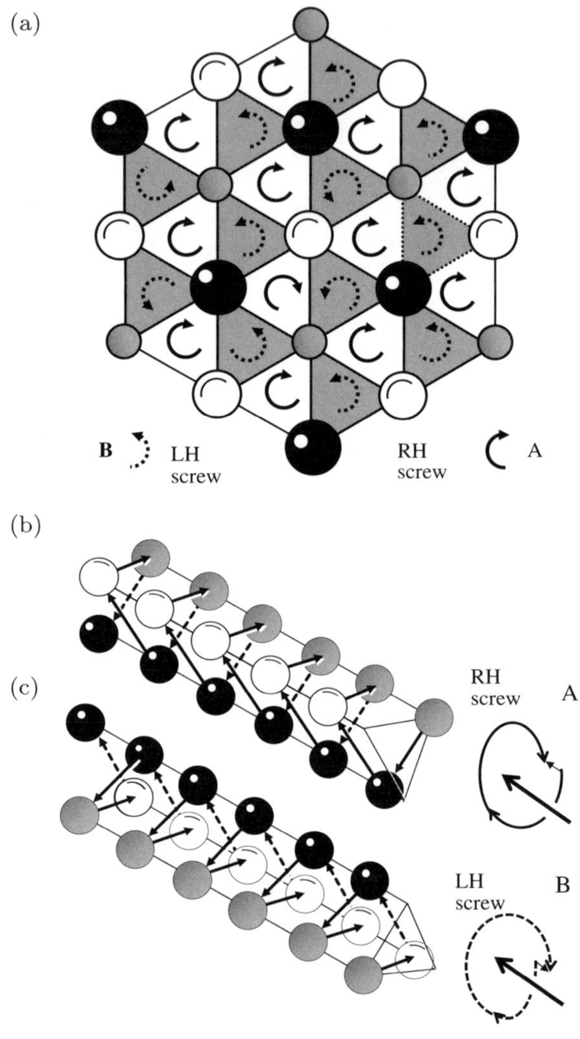

Fig. 12.31 Suzuki model of a screw dislocation in BCC (redrawn based on [23]).

Some examples of $\psi-\chi$ curves of a variety of BCC metals are shown in Figs. 12.34–12.36. To summarize, the ψ—χ curves are classified into three categories as shown in Figs. 12.36(a)–12.36(c).

(a) No difference between {110} and {112} slips.
(b) {110} slip is preferred to {112} slip.
(c) {112} slip is preferred to {110} slip.

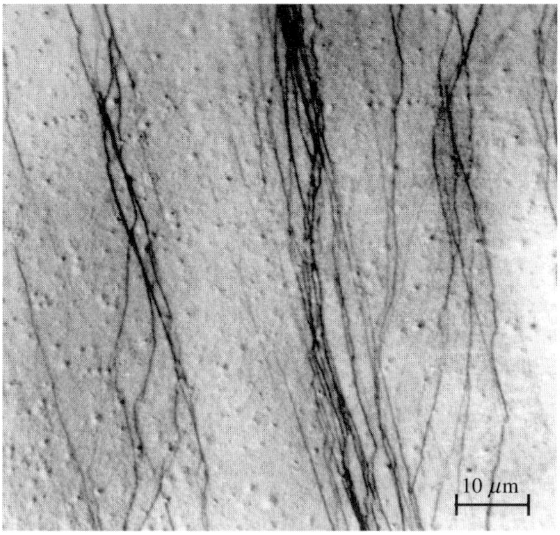

Fig. 12.32 Slip lines in Li–70at.%Mg alloy (BCC) (Reproduced with permission from [24]).

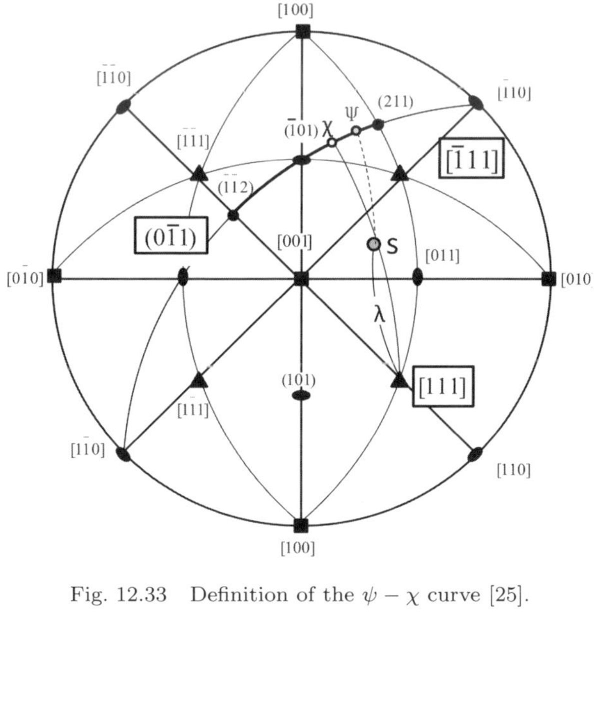

Fig. 12.33 Definition of the $\psi - \chi$ curve [25].

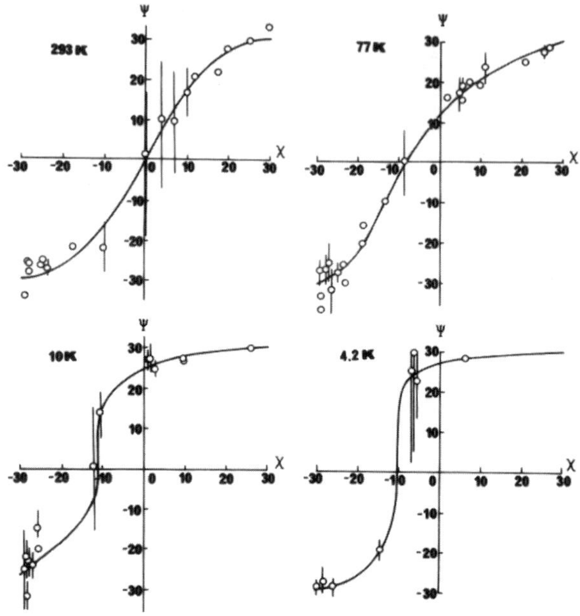

Fig. 12.34 Li–70at.%Mg alloy(BCC) (Reproduced with permission from [24]).

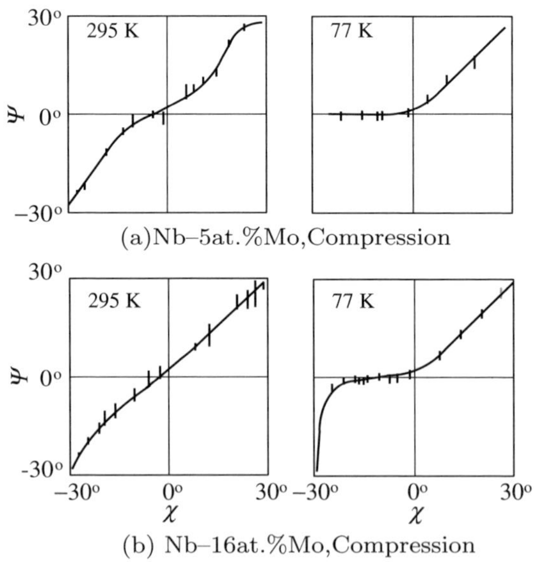

(a)Nb–5at.%Mo,Compression

(b) Nb–16at.%Mo,Compression

Fig. 12.35 Nb–Mo alloy (Reproduced with permission from [26]).

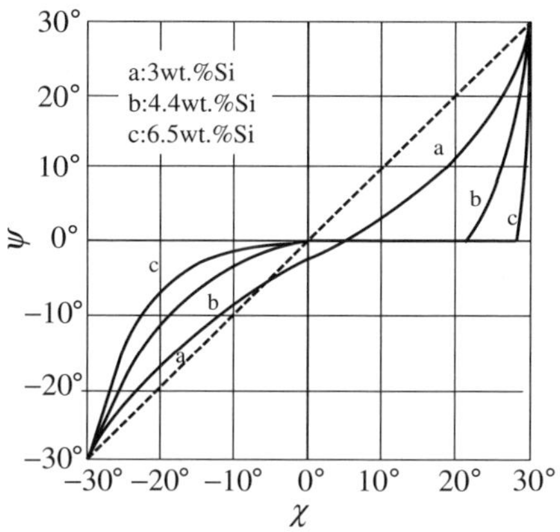

Fig. 12.36 (a) Fe–Si alloy (Redrawn based on [27]–[29]).

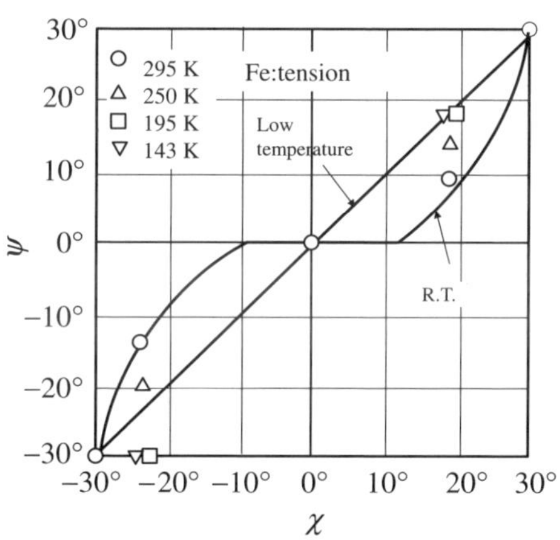

Fig. 12.36 (b) Fe (Reproduced with permission from [30])

These differences are to be attributed to subtle difference in the core structures shown in Fig. 12.30.

12.4.3 *Failure of Schmid law*

It is well known that in BCC metals Schmid law does not hold. Schmid law, or the law of critical resolved shear stress (CRRS), signifies that, once the resolved shear stress onto the slip system as described by Eq. (12.1) reaches a critical value, the crystal yields. The failure of Schmid law is

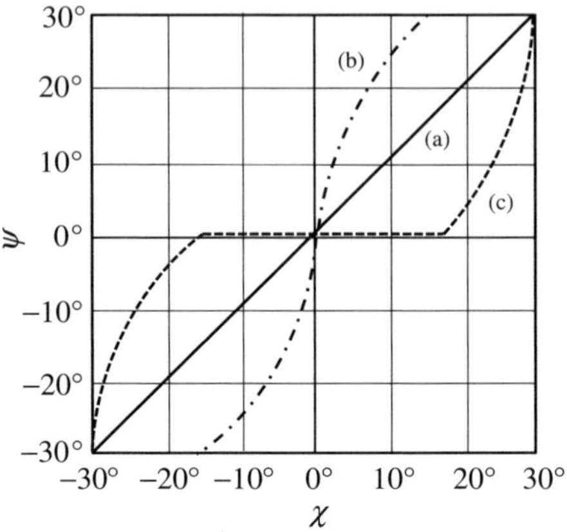

Fig. 12.37

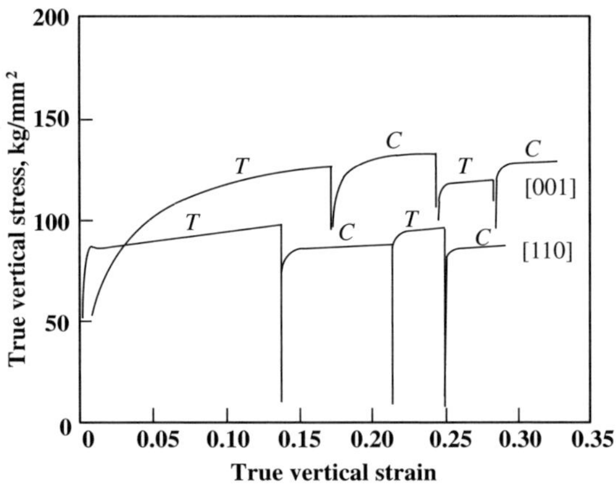

Fig. 12.38 Stress–strain curves of Mo single crystals deformed in tension (T) and compression (C) (Reproduced with permission from [31]).

demonstrated typically by the difference in yield stress between tensile and compression tests.

Figure 12.38 [31] shows stress-strain curves obtained on Mo single crystals with two axial orientations ([001] and [110]); during the tests, the crystals are deformed in tension and compression alternately (T stands for tension and C compression). When deformed along [110], flow stress in tension is higher than that in compression, while when deformed along

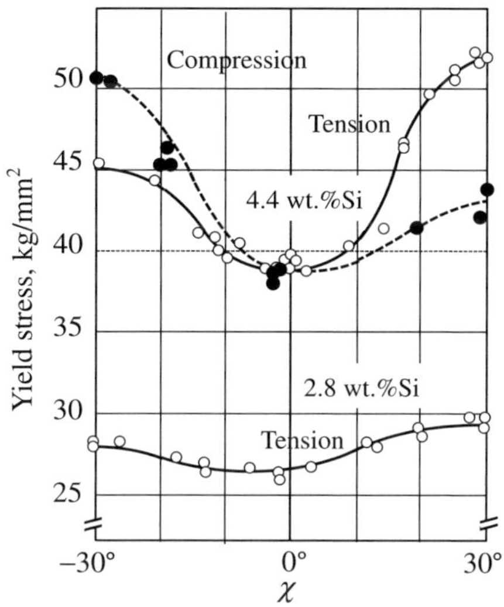

Fig. 12.39 Yield stress versus χ curves in Fe–Si alloys (Reproduced with permission from [29]).

[001], the reverse is true. Since the Schmid factor should be the same both in tension and compression, this indicates that CRSS is different between tension and compression.

Figure 12.39 [29] shows yield stresses of single crystals of Fe–Si alloys with the BCC structure measured in tension and compression as functions of χ. If the Schmid law holds, the curves should be symmetrical with respect to the reversal of the sign of χ, but it is not the case; the curves are asymmetrical. Furthermore, at $\chi = 30°$ the yield stress (τ_o) is higher in tension than in compression. Inversely, at $\chi = -30°$ the yield stress in compression is higher than in tension. Generally speaking, reversing the $\tau_o(\chi)_\mathrm{t}$ curve in tension with respect to χ is coincident to the $\tau_\mathrm{o}(\chi)_\mathrm{c}$ curves in compression and vice versa. That is, $\tau_\mathrm{o}(\chi)_\mathrm{t} = \tau_\mathrm{o}(-\chi)_\mathrm{c}$ (Subscripts t and c stand for tension and compression, respectively). This is attributed to the core structure of a screw dislocation in BCC crystals.

Directions where the yield stress is lower ($\chi = 30°$ in compression and $\chi = -30°$ in tension) is referred to as the easy sense and the opposite direction ($\chi = 30°$ in tension and $\chi = -30°$ in compression) as the hard sense.

The easy and hard senses are also referred to as the twinning and anti-twinning senses, respectively. The reason for this is as follows. Figure 12.40

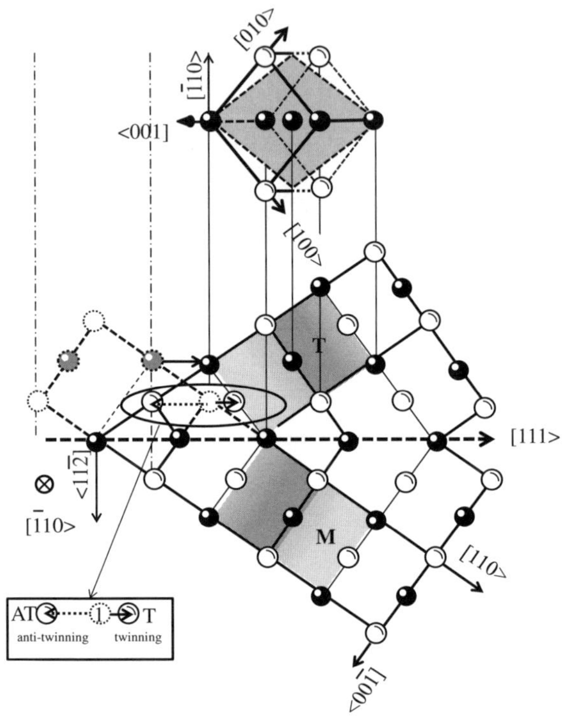

Fig. 12.40 Deformation twin in BCC.

shows orientation relationship of a deformation twin[6] in BCC crystal. Twins in BCC crystal are faced with 180° rotated each other on a {112} plane as a mirror plane. Let the upper crystal be displaced, with the lower crystal fixed. In order for this to be reached, atom 1 must be displaced to position T rightward as indicated by a solid arrow. The same geometry is reached by displacing atom 1 leftward to position AT (indicated by a broken line), however, the displacement vector $1 \rightarrow AT$ is twice as large as the displacement vector $1 \rightarrow T$, so that the displacement $1 \rightarrow T$ is much easier to take place. The displacement $1 \rightarrow T$ (solid line) is referred to as the twinning sense and $1 \rightarrow AT$ (broken line) as the anti-twinning sense.

[6]Deformation twin: a twin formed by an external force. Twinning is a competing mode of deformation against dislocation motion. In the stress–strain curve, sharp drops in stress are observed. Anneal twins are formed during annealing.

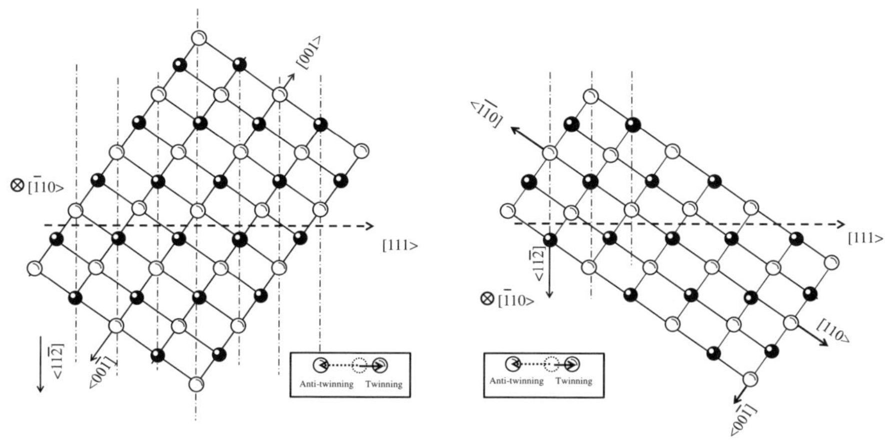

Fig. 12.41 (a) (b) (Redrawn based on [31]).

Table 12.3 Orientation dependence of yield stress of BCC metals.

Axis	Tension	Compression
$\chi = +30°$ (near <110>)	hard (anti-twinning) sense	easy (twinning) sense
$\chi = -30°$ (near <001>)	easy (twinning) sense	hard (anti-twinning) sense

As shown in Fig. 12.38, in tension, <001> direction ($\chi = -30°$) is the easy sense and <011> direction ($\chi = +30°$) is the hard sense. In compression the reverse is true. Figures 12.41(a) and 12.41(b) show the lower crystal in Fig.12-40 once again. Suppose that this crystal is stretched along [001] direction (Fig. 12.41a). In tension, the upper crystal is displaced rightward, that is, in the twinning sense. When stretched along $[\bar{1}\bar{1}0]$ direction (Fig. 12.41(b)) the upper crystal is displaced leftward, that is, in the anti-twinning sense.

Exercise 12.8 Show that, in compression, <100> is the anti-twinning and <110> is the twinning sense.

The asymmetry in the $\tau_o - \chi$ curves can be summarized as follows:

Glide in the twinning sense on {112} plane is easier than glide in the anti-twinning sense. That is, $\tau_{0_{\text{twinning}}} < \tau_{0_{\text{anti-twinning}}}$. Table 12.3 summarizes the results.

The asymmetry in {112} slip is attributed to the core structure of a screw dislocation. As already discussed in Sec. 12.4.1, a screw dislocation in BCC crystals has threefold symmetry along <111> direction (Figs. 12.29(a) and 12.29(b)). Qualitatively the screw dislocation is dissociated into three partials as follows.

$$\frac{1}{2}[111] = \frac{1}{6}[111] + \frac{1}{6}[111] + \frac{1}{6}[111], \tag{12.24}$$

(see Figs. 12.30(a) and 12.30(b)).

In other words, a screw dislocation in BCC crystals is dissociated on three slip planes, not on two planes, so that it is in a sessile configuration. This is the reason why the P–N force is large in BCC crystals. It contrasts sharply with dislocations in FCC and HCP crystals where dislocations are dissociated on one slip plane.

Now let us suppose that such a screw dislocation moves under the action of an external stress. In order for a screw dislocation shown in Fig. 12.42 to move on S_1 plane, partial dislocations ② and ③ must slide on S_1 plane. When the overall dislocation moves rightward as a whole, the external force assists ② and ③ to shrink, while when it moves leftward, ② and ③ must shrink against the external force. Needless to say, the latter motion is much harder than the former. This is the reason why asymmetry appears in $\tau_0 - \chi$ and $\psi - \chi$ curves. In the above explanation the rightward glide corresponds to the easy (twinning) sense and the leftward glide corresponds to the hard (anti-twinning) sense. The core structure explains why, compared with FCC crystals where dislocations are dissociated on one slip plane, screw dislocations in BCC crystals are much easier to cross slip.

12.4.4 $(0\bar{1}1)$ *anomalous slip*

In ultra high pure BCC metals another type of failure of Schmid law occurs, that is $(0\bar{1}1)$ anomalous slip. When an ultra-pure BCC crystal is elongated along direction ○ in Fig. 12.43(a), a slip on $(0\bar{1}1)$ takes place. The Schmid factor for this slip is as small as ~ 0.2. Figure 12.43(b) shows an optical micrograph of the anomalous slip in an ultra high pure Nb. The $(0\bar{1}1)$ slip is very coarse.

In ultra high pure BCC metals, mobility of an edge dislocation is much higher than that of a screw dislocation. For the sake of simplicity, let the lip system be <111>/{110}. First an edge dislocation belonging to the primary slip system [111]/(011) starts to glide. Before a screw dislocation

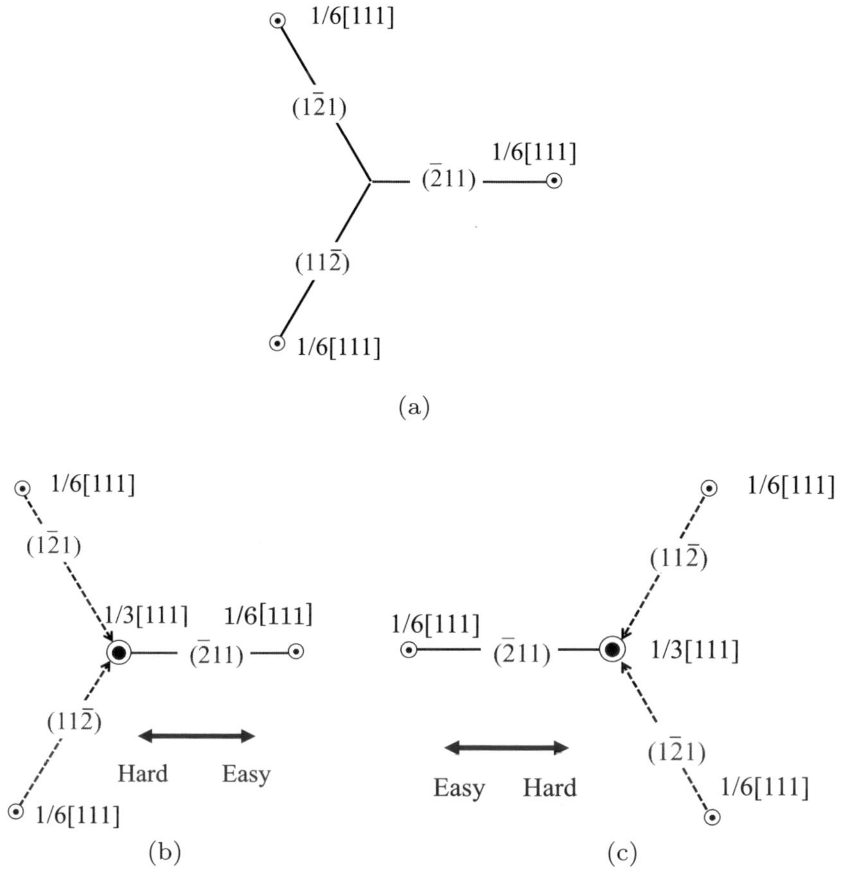

Fig. 12.42 (a)(b)(c)

in the primary system starts to glide, an edge dislocation in the secondary $[\bar{1}11]/(101)$ slip system starts to glide.

This stage is the preyielding stage before the macroscopic yield. As a result, screw dislocations, which were left behind edge dislocations, both in the primary and secondary slip systems, hinder the motion mutually (Fig. 12.44(a)). The only way for the glide to continue to reach the macroscopic yielding is to glide on $(0\bar{1}1)$ plane which contains both the primary and the secondary slip directions. Figures 12.44(b) and 12.44(c) [34] show motion of networks of the primary [111] screw dislocation and the secondary $[\bar{1}11]$ screw dislocation on the coplanar $(0\bar{1}1)$ in an ultra high pure Mo as observed by an *in situ* deformation experiment in a HVEM.

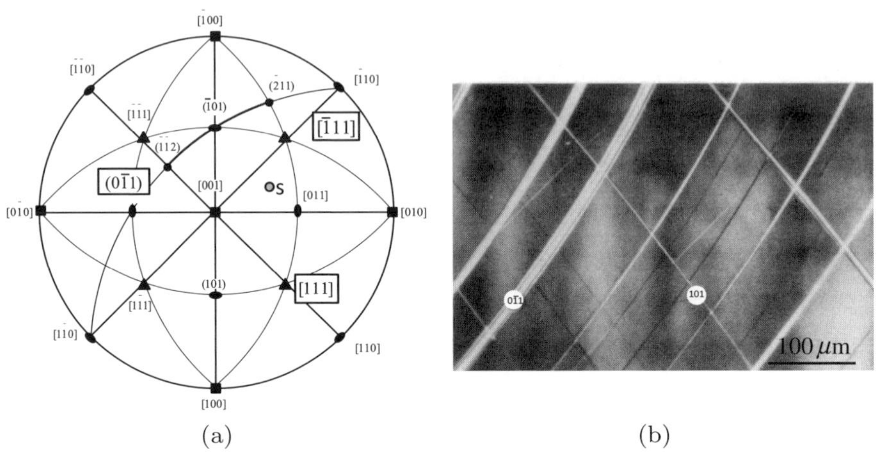

Fig. 12.43 (a), (b) (0$\bar{1}$1) anomalous slip (Reproduced with permission from [33]).

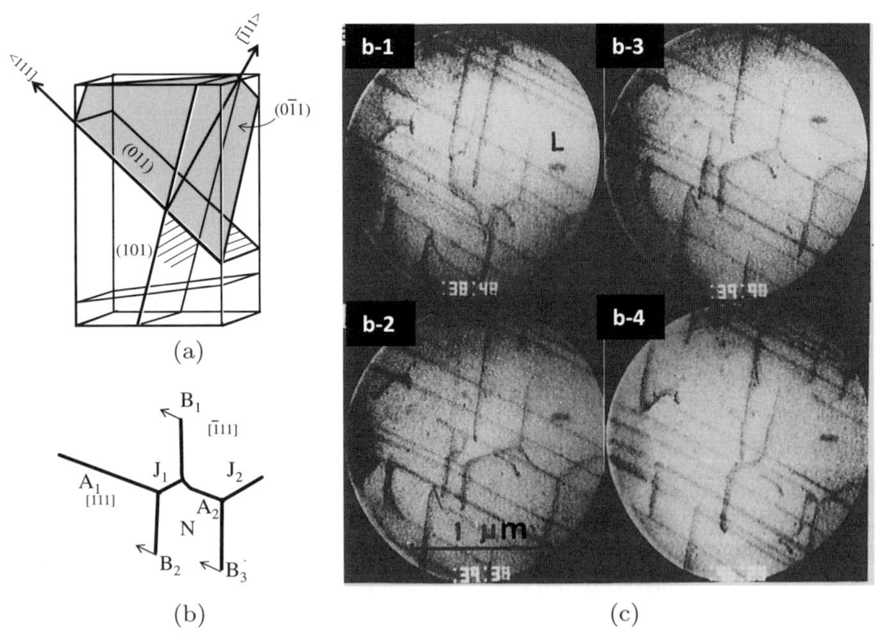

Fig. 12.44 (a)–(c) Anomalous slip in Mo. In (c) A_1 is the primary dislocations, B_1, B_2 and B_3 are secondary dislocations. A_1 and B_1 react to form J_1. Similarly A_2 and B_3 react to form J_2 (Reproduced with permission from [34]).

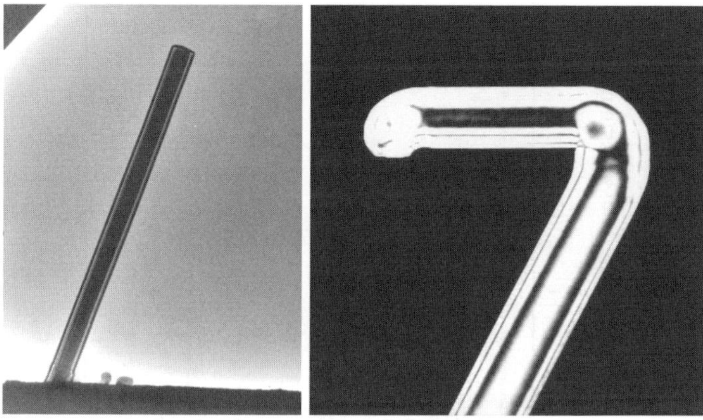

(a) (b)

Fig. 12.45 TEM of whiskers (a), (b) [35].

12.5 Initial Mobile Dislocation Density is Zero

The foregoing discussion considers the case where a small density of dislocations move, in a sense, freely under a small external stress. There are two cases indeed where the mobile dislocation density is zero or virtually zero. One is a dislocation free perfect crystal called a whisker and the other is the case where dislocations are locked perfectly by solute atoms.

12.5.1 *Whisker*

Figures 12.45(a) and 12.45(b) show two examples of TEM micrographs of whiskers. Whisker in (a) is straight, while that in (b) is bent. However, this bent whisker in (b) is a single crystal, free from dislocations since this micrograph was taken in the dark-field imaging mode.[7] When such dislocation-free perfect (straight) whiskers are tested in tension, profound yield drops are observed as shown in Fig. 12.46. Here, a straight line in the very initial stage is in the perfect elastic region. At the upper yield point, dislocations are nucleated and the stress drops precipitously down to the lower yield point, followed by a steady plastic deformation at a constant stress. That is, the upper yield point corresponds to the stress required to nucleate dislocations and the lower yield point to the stress required to move the dislocations. On increasing the diameter of the whisker, the

[7]Dark-field mode: Those areas which have the same orientation appear bright. Therefore, a single crystal appears bright over all.

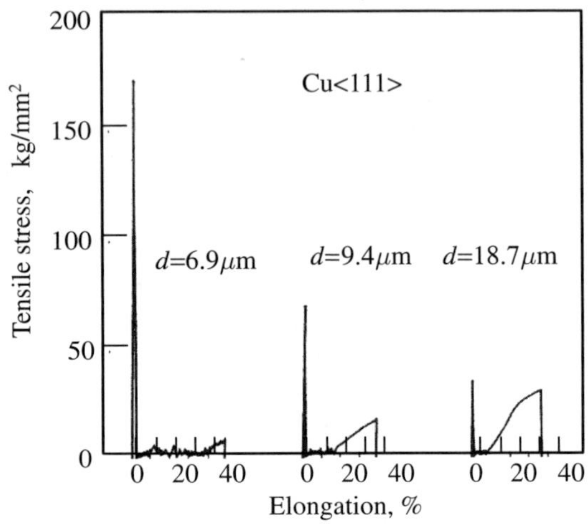

Fig. 12.46 Stress-strain curves of whiskers with different diameter (Reproduced with permission from [36]).

surface area increases, resulting in an increase in probability of existence of sites where stress concentration takes place. As a result the upper yield point decreases with increasing the diameter.

12.5.2 *Yielding of a mild steel*

Steel is an alloy of Fe with carbon (C). C is an interstitial solute and as mentioned in Chapter 5 segregates just beneath an edge dislocation. This is referred to as the Cottrell effect or atmosphere. C can interact with a screw dislocation. Thus, dislocations in an annealed mild steel are locked by C. On imposing an external stress, dislocations are relieved from these locks at the upper yield point. The external stress required to keep the relieved dislocation moving corresponds to the lower yield stress. Figure 12.47 shows some examples of stress–strain curves obtained on mild steels. (a) was obtained on an annealed steel, and a yield point was clearly observed. (b) was obtained after the load was removed to zero, followed by an immediate reloading: The yield point disappeared. However, if the specimen was annealed after unloading and loaded again the yield point recovered (c). The explanation for this is as follows. In (b) dislocations that had been relieved and multiplied by the first run of tensile test were not locked yet because the second run was carried out immediately after unloading, so that the yield point disappeared. In (c) those dislocations introduced in the first run were

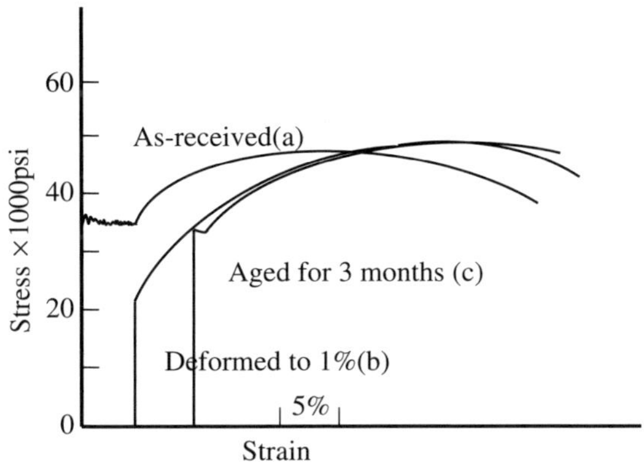

Fig. 12.47 Stress–strain curves of a mild steel after various treatments (Reconstructed based on Figs. 8.36 and 8.37 in [37]).

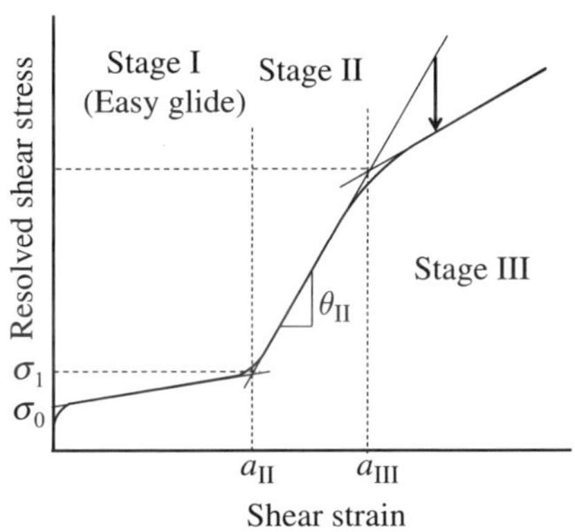

Fig. 12.48 Three-stage hardening.

locked again during the intermediate annealing, resulting in the recovery of the yield point.

12.6 Work Hardening of a Single Crystal

12.6.1 *General description of work hardening of a single crystal*

Work hardening is a phenomenon where metals harden by plastic deformation. It is an everyday experience to bend a metal wire repeatedly to fracture. Figure 12.48 is a schematic illustration of a stress–strain curve of

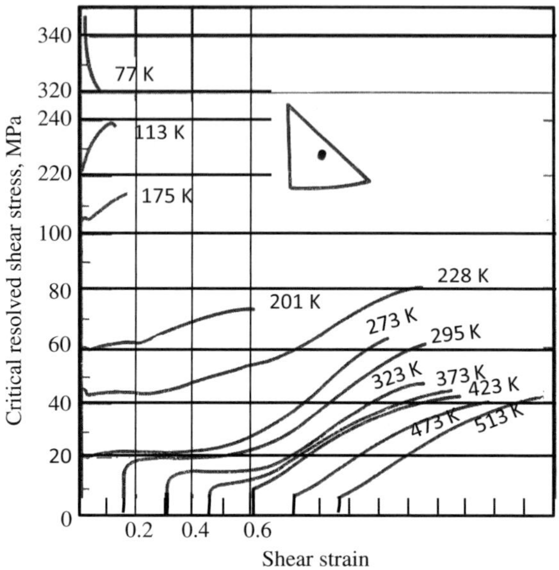

Fig. 12.49 Stress–strain curves of Nb single crystals at different temperatures (Reproduced with permission from [38]).

a single crystal of FCC metals. The flow stress σ increases with increasing strain ε. In other words, the crystal hardens with plastic deformation, and in order for the deformation to proceed, stress σ must be increased. This is work hardening and $\mathrm{d}\sigma/\mathrm{d}\varepsilon = \theta$ is referred to as a work hardening rate. Work hardening consists of three stages; stages I and III with small θ and stage II with the highest θ. Stage I is referred to as the easy glide region and stage II, the linear hardening region.

This three-stage hardening is not characteristic to FCC crystals but observed frequently in BCC metals and covalent crystals deformed at high temperatures. Figure 12.49 shows stress–strain curves on highly pure Nb single crystals at different temperatures: Above room temperature three stage hardening is obvious. Figure 11.6(b) shows stress–stain curves of Si and except for a profound yielding, the curves are quite analogous to those of FCC crystals.

Yield stress is quite sensitive to material and its crystal structure but the work hardening rate in stage II (θ_{II}) is approximately $1/200 - 1/300$ of the shear modulus μ and rather insensitive to material and its crystal structure.

12.6.2 *Mechanism of work hardening*

Stage I Friedel [39] called stage I a laminar flow. Dislocation density in a well annealed crystal ranges from 10^5 to $10^6/\text{cm}^2$. These dislocations form networks. One of the dislocation segments comprising the network happens to start to operate as a Frank–Read source and multiplies itself. The dislocation has (1) the highest Schmid factor (i.e., belongs to the primary slip system) and (2) the length of the segment l is the longest, since $\sigma_{\text{max}} = \frac{\mu b}{l}$.

From these most preferred F–R sources, dislocations are emitted one after another. Those dislocations, which are emitted before other sources get in operation, are compared to, in a sense, cars driving freely on a highway. However, soon these dislocations encounter those emitted from other F–R sources. These two groups of dislocations move in the opposite directions, so that the signs of their $\boldsymbol{b}$'s are opposite. Edge dislocations with $\pm\boldsymbol{b}$ constitute a dislocation dipole which is stable when the component edge dislocations make an angle of 45°, so that the two groups of dislocations restrain each other (Fig. 2.50). Under such a circumstance, the F–R sources continue to emit dislocations, and as a result, many dipoles are formed as shown in Fig. 12.51. These dipoles are referred to as a multiple dipole. Figure 12.52 shows a TEM of a multiple dipole. The individual dipoles in a multiple pole restrain motions of the counterparts. That is, work hardening, albeit small, takes place. Then, F–R source which did not operate because its l was slightly smaller than the critical value, starts to operate. In order for the individual dislocations trapped as a dipole to be released from the

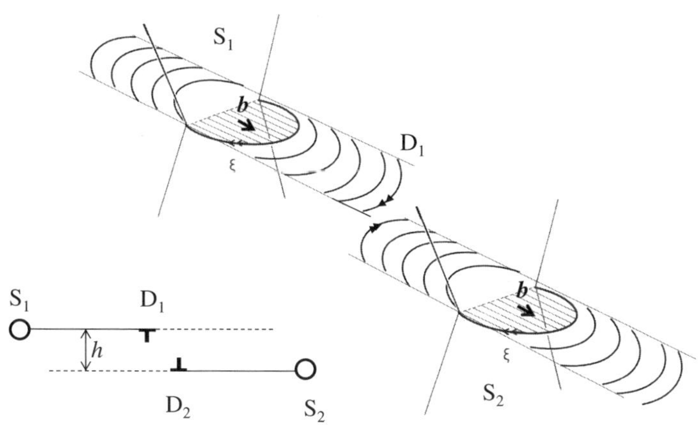

Fig. 12.50 Formation of multiple dipoles (I).

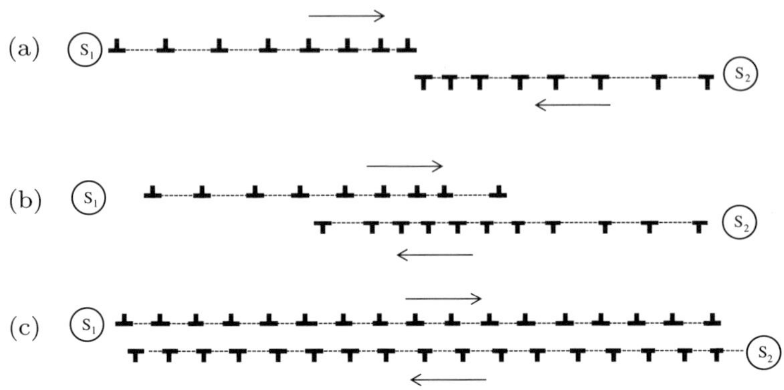

Fig. 12.51 Formation of multiple dipoles (II).

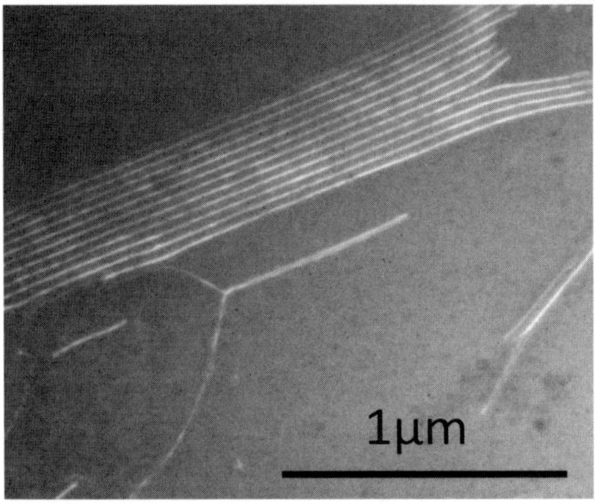

Fig. 12.52 Multiple dipoles (Reproduced with permission from [32]).

dipole, a stress of the order of $\mu b/2\pi h$ is required, where h is the spacing between the slip planes (see 4.4.2). With increasing density of active F–R sources, h decreases, resulting in more hardening. This is the main reason of work hardening in the easy-glide stage. Only interactions of dislocations in the primary slip planes are responsible.

Stage II Friedel [39] called stage II a turbulence flow of dislocations. At this stage $\theta_{\rm II}$ is the maximum, i.e., the dislocation motion becomes suddenly much more difficult. At this stage, secondary slip systems other than the primary slip system come into operation. The dislocations in the primary and secondary slip systems react and restrain mutual motions and tangle

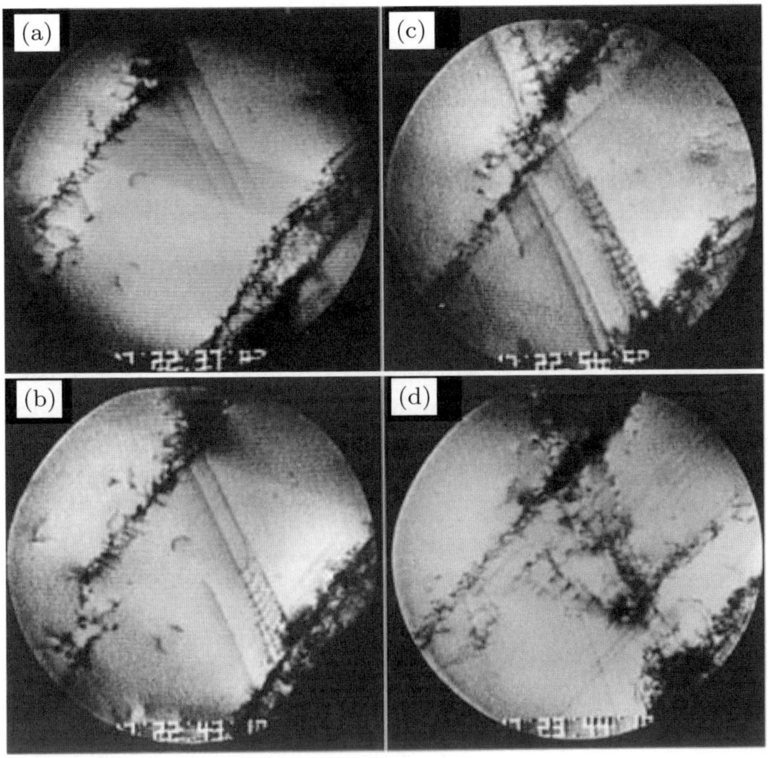

Fig. 12.53 Formation of a CW in Mo as observed by an *in situ* deformation experiment in a HVEM [40].

themselves in a complex way. Regions with a high density of dislocations are formed. This structure is referred to as cell structure and the area with a high density of dislocation is referred to as a cell wall.

Figure 12.53 [40] reproduces a series of TEM micrographs showing the initial stage of cell formation as observed by an *in situ* deformation experiment in a HVEM. The primary slip band running from the upper right to the lower left encounters the secondary slip bands running from the upper left to the lower right; the dislocations in these slip bands tangle and form a cell wall.

Figure 12.17 shows an *in situ* deformation experiment in a HVEM. Under loading (a) those dislocations bowing out rightward are blocked by a cell wall (CW). (b) On removing the load, the dislocations shrink backward, and eventually bow out leftward, i.e., in the opposite direction (d). It is obvious that the back stress from the CW is large. In this way, a CW exerts a back stress to dislocations. This is the essence of the work hardening in stage II.

Stage III The work hardening rate decreases again, indicating that the accumulation of dislocations is not so effective as in stage II. Those dislocations that are piled up in stage II now start leaving the barriers. This is referred to as dynamic recovery.

12.7 Solid-solution Hardening, Precipitation Hardening and Dispersion Hardening

Solid–solution hardening, precipitation hardening and dispersion hardening are strengthening methods by dispersing isolated barriers to dislocation motion. Depending on the size and strength of the dispersed obstacles they are classified into solid–solution hardening, precipitation hardening and dispersion hardening.

12.7.1 *Solid-solution hardening*

In a solid solution, a solute atom replaces a solvent atom on a lattice site or intrudes in between lattice sites while the crystal structure of the matrix remains unchanged. The former is referred to as a substitutional solid solution and the latter as an interstitial solid solution. In metallic materials, interstitial solid solutions are restricted to alloys of BCC (such as Fe) with small atoms like C or N. Substitutional solid solutions are universally formed in materials with all kinds of crystal structures.

Figure 12.54 shows temperature dependence of CRSS in Cu–Ni system which forms a complete solid solution over the entire composition.[8]

(1) On alloying CRSS increases, reaching a peak at Cu–50at%Ni.
(2) Temperature sensitivity of CRSS increases with increasing alloying element.

There have been two competing theories concerning the origin of solid–solution hardening. One is the pinning theory and the other is the friction theory (see Sec. 12.2). In the pinning theory, dislocations are pinned by solute atoms which segregate in the vicinity of dislocations, so that an additional stress is required to move the segregated dislocations. On the

[8]Complete solid solution: Elements A and B (in this case Cu and Ni) are solved completely over the range from 0 to100%.

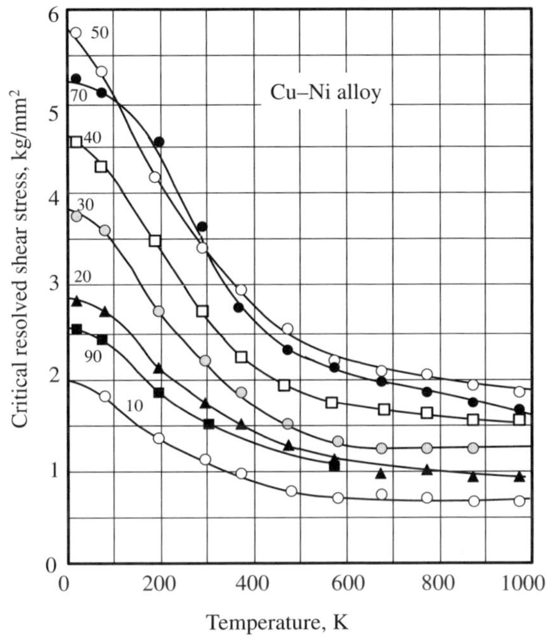

Fig. 12.54 Solid solution hardening in Cu–Ni alloys (Reproduced from [41]).

other hand, in the friction theory, solute atoms are distributed randomly in the matrix and retain motions of individual dislocations.

Pinning theory As a result of interactions between dislocations and solute atoms, solute atoms segregate in the vicinity of dislocations.

Typical examples include the followings.

(1) Interstitial solute atoms such as C and N in BCC metals form Cottrell atmosphere around edge dislocations due to elastic interaction. (Screw dislocations also interact with C and N (see Chapter 5).)
(2) In a substitutional solid solution with the FCC (and possibly HCP) structures Suzuki atmosphere is formed through chemical interaction (see Chapter 8).

When Cottrell atmosphere or Suzuki atmosphere is formed, an additional stress is required for the pinned dislocation to be released for the following reason.

Let us consider why the atmosphere is formed. It is simply because formation of the atmosphere leads to a decrease in the free energy of the system containing dislocations and solute atoms. In other words, both dislocations and solute atoms become happier. That dislocations leave the (segregated) atmosphere means to transplant the system to a higher energy state. To do so, the stress is required, and this is the reason of strengthening by the pinning effect.

Friction theory In a substitutional solid solution, solute atoms are distributed randomly on lattice sites of the matrix (solvent) atoms. Solute atoms are different from solvent atoms, so that, naturally lattices in the vicinity of a solute atom are disturbed. In pure metals, the dislocations are moving as if gliding on a smooth ice apart from the P–N potential. In a solid solution, the road is a little bit modulated because of interaction with solute atoms. Interactions with a solute atom can be repulsive or attractive (Fig. 12.55). The interaction, irrespective of whether repulsive or attractive, does hinder the dislocation motion. The case of repulsive interaction may be easier to understand intuitively. But even in the case of attractive interaction a dislocation is pulled back when leaving.

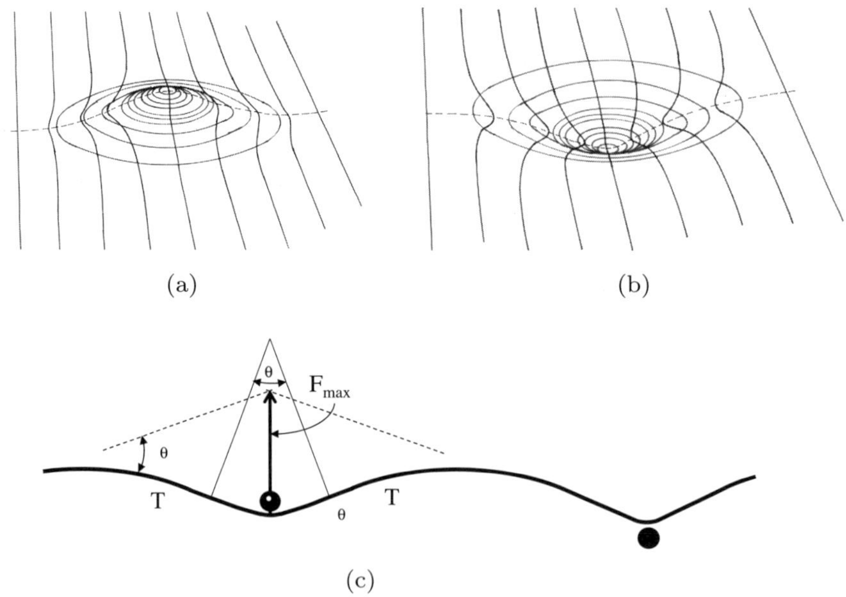

Fig. 12.55 Interaction of a dislocation with a solute atom, (a) Repulsive interaction, (b) Attracive interaction, (c) Plan view of a repulsive interaction (left) and an attractive one (right).

12.7.2 *Precipitation hardening*

Precipitation hardening is analogous to solid solution hardening (as explained by friction theory) in that barriers are distributed randomly in the matrix. However, the biggest difference is that, in solid solution hardening, the crystal lattice of the matrix is preserved while, in precipitation hardening, the crystal structure of the barriers (precipitates) is no longer the same as that of the matrix.

Initial stage of precipitation In the initial stage of precipitation, precipitates (often called G.P.[9] zones) are fine and the spacing is small and, above all, is coherent (or nearly coherent) with the matrix. Therefore, a dislocation can cut the G.P. zones. In that sense, precipitation hardening at this stage can be regarded as an extension of solid solution hardening. The main reason for strengthening is elastic interactions between dislocations and G.P. zones. Essential point is that a dislocation can proceed by cutting G.P. zone without changing significantly its Burgers vector $\boldsymbol{b}$.

However, on further aging, a precipitate transforms into a metastable phase and eventually the equilibrium phase. On coarsening of the precipitates,

(1) Elastic interaction becomes larger.
(2) Spacing between the precipitates (L_o) becomes larger.

Overaging — Orowan's by-passing mechanism On further aging, precipitates coarsen and become incoherent to the matrix; it becomes more and more difficult for a dislocation to overcome them. However, the spacing between the precipitates becomes larger as well. As a result, a dislocation can meander freely avoiding the precipitates. Dislocations now can by-pass the precipitates without overcoming them. This is referred to as Orowan's by-pass mechanism. Figure 12.56 illustrates schematically the Orowan's mechanism.

Exercise 12.9 Assign the sense of the direction to the Orowan's dislocations in Figure 12.56 and apply Kirchhoff's law.

Figure 12.57 shows a TEM micrograph of Orowan's mechanism. Note the dislocation loops left behind around precipitates. The precipitates at this stage are no longer effective barriers to dislocation motion, that is,

[9]G.P. zone: Guinier–Preston zone.

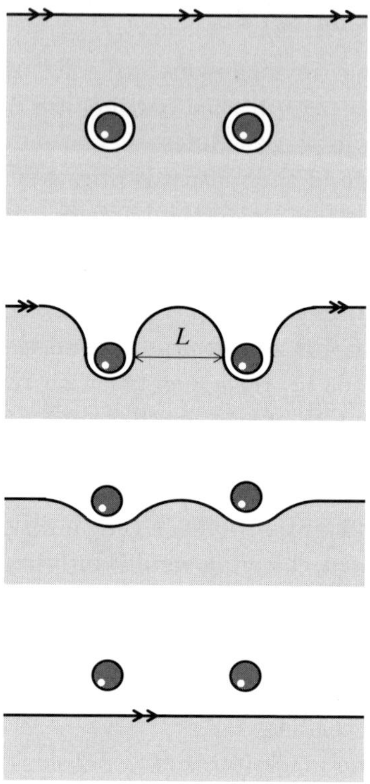

Fig. 12.56 Orowan's by-pass mechanism.

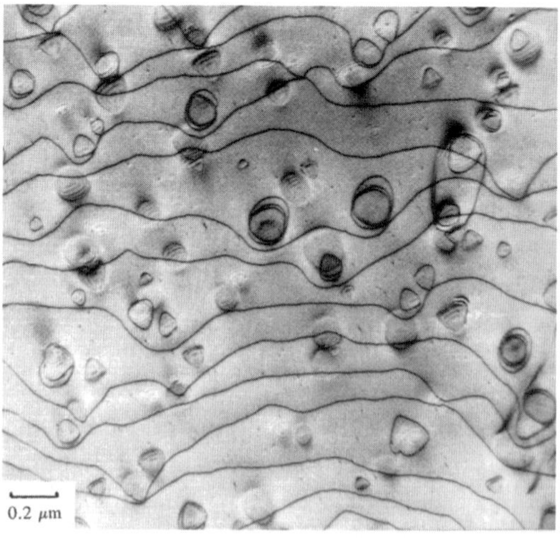

Fig. 12.57 Orowan's by-pass mechanism (Reproduced from [42]).

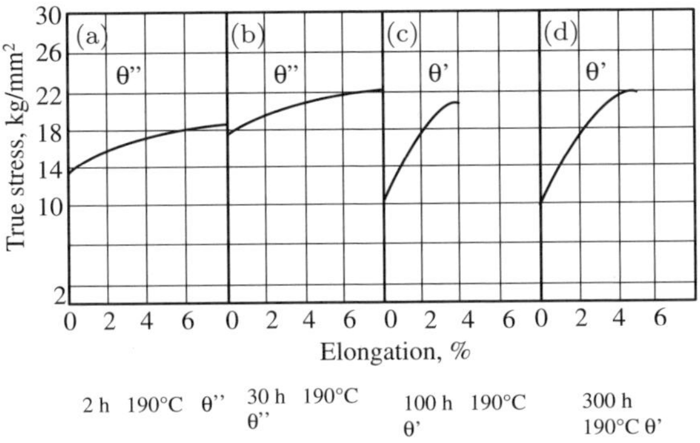

Fig. 12.58 Stress–strain curves at different stages of precipitation. (a), (b) Early stage, (c), (d) Overaged stage (Reproduced with permission from [43]).

Fig. 12.59 Orowan's loops left behind around large precipitates (Reproduced with permission from [44]).

precipitation hardening disappears. This is referred to as overaging softening. Figure 12.58 shows stress–strain curves of an Al alloy aged at 190°C for different times. The yield stress shows a peak at 30 h, while the work hardening rate is much larger after 100 h aging. Orowan's loops left behind exert a strong back stress to the trailing dislocations (Fig. 12.59).

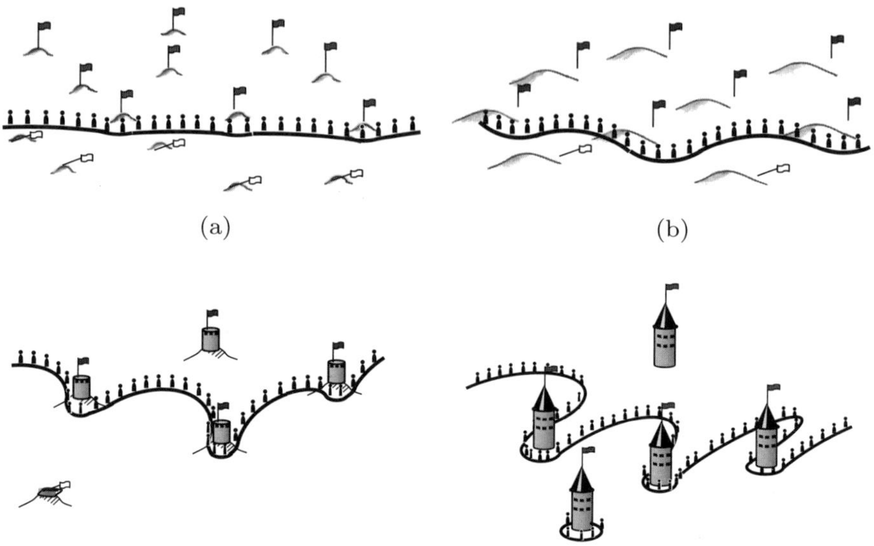

(a) (b) (c) (d)

Fig. 12.60 (a) is solid-solution hardening. Individual solute atoms are dispersed in the matrix, so that while the matrix is strained (compared to unevenness of the ground) a dislocation can overcome the solute without changing its $\boldsymbol{b}$, (b) is early stage of precipitation hardening (GP zone). A group of solute atoms coagulate to form GP zones, which have crystal sructure similar to the matrix. Thus, the strain increases (unevenness becomes stronger), but a dislocation is still able to overcome without changing its $\boldsymbol{b}$ significantly, (c) Peak of the precipitation hardening. Precipitates are now incoherent to the matrix and have different crystal structure from the matrix. Therefore, a dislocation cannot overcome the precipitates without changing its $\boldsymbol{b}$. Precipitates are fortified, albeit small, and it is difficult to overcome them with a thermally activated process, (d) Overaging: Precipitates are now castles, and dislocations give up overcoming and surmout by Orowan's mechansim.

Hardness peak Hardness shows a peak somewhere between stages of the initial precipitation and overaging softening. At this stage, precipitates are too large for a dislocation to proceed by cutting and too densely distributed for Orowan's mechanism to operate.

Mechanisms for solid solution hardening and precipitation hardening are schematically illustrated in Fig. 12.60. Here a dislocation is moving from bottom to top. (a) corresponds to solid solution hardening, (b) to the initial stage of precipitation hardening, (c) to the peak hardening and (d) to overage softening.

Dispersion hardening Precipitates grow during aging. This means that, when a precipitation-hardening alloy is used at high temperature, precipitates may grow during service to such an extent that the alloy softens. In order to cope with this difficulty, chemically stable particles such as oxides are dispersed mechanically in the matrix. Typical alloy of this kind is an oxide dispersion alloy. Oxides do not deform plastically, so that Orowan's mechanism is the only way for a dislocation to overcome the dispersed oxides

12.7.3 *Radiation hardening*

When energetic particles bombard metals, internal defects are formed, which hinder dislocation motion. This is referred to as radiation hardening. Energetic particles include neutrons, electrons and ions.

In electron radiation, when the energy of electrons exceeds a threshold value, an atom on the lattice is knocked out. This is referred to as a knock-on-collision (knock-on-collision: KOC) (see Fig. 12.61(a)). The knocked-out atom wanders in the crystal and stays as an interstitial atom (I). On the other hand, the lattice site out of which the atom is knocked out as an interstitial is now vacant: This vacant point is referred to as vacancy (V). Thus, a pair of I and V is formed. This is referred to as a Frenkel pair.

In contrast, in neutron radiation the kinetic energy of a neutron is so large that many V and I are produced simultaneously and clusters of

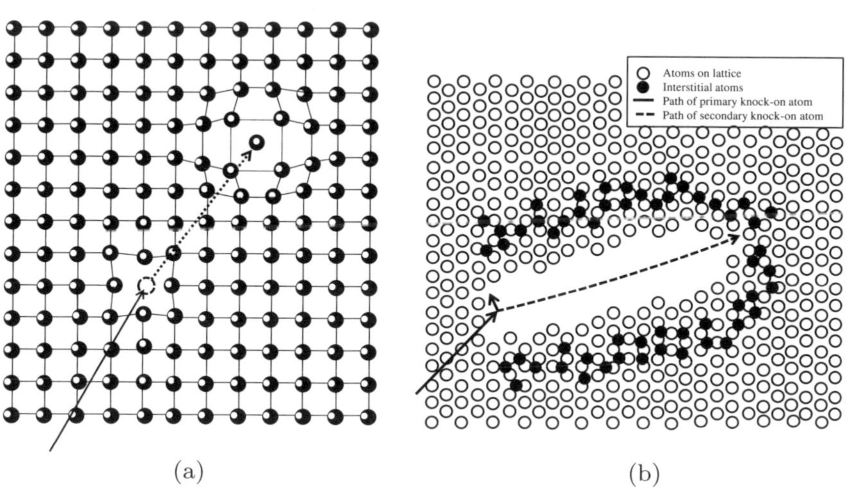

Fig. 12.61 (a) Frenkel pair, (b)Spike (Reproduced with permission from [45]).

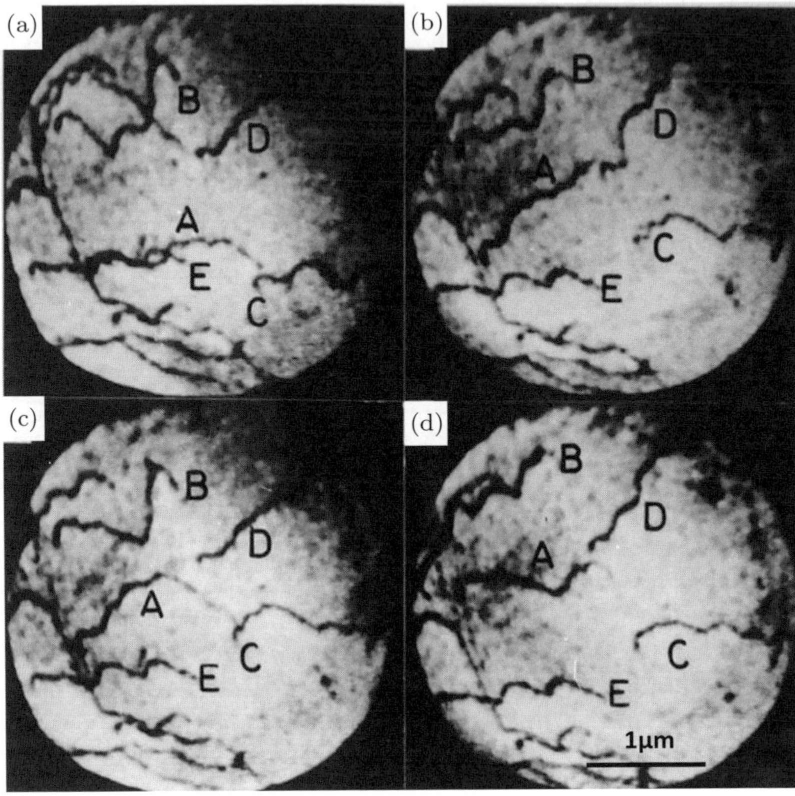

Fig. 12.62 Motion of dislocations in electron-irradiated Ni (Reproduced with permission from [46]).

point defects (spike) are formed (Fig. 12.61(b)). When spikes are formed successively, cascades of spikes are formed.

Figure 12.62 shows dislocation motions in pure Ni as observed by an *in situ* deformation experiment in a HVEM. The specimen was irradiated by electrons in a HVEM and many defects of I type are introduced. Note dislocations A, B, C and D move while interacting with fine (not seen) barriers.

Radiation hardening is essentially different from solid-solution hardening, precipitation hardening and dispersion hardening in that the obstacles formed by radiation can be annihilated by reactions with moving dislocations. A variety of mechanisms have been proposed to explain annihilation of radiation damages. When a dislocation sweeps out the radiation damages, resulting in a drastic decrease in the density of the radiation damages,

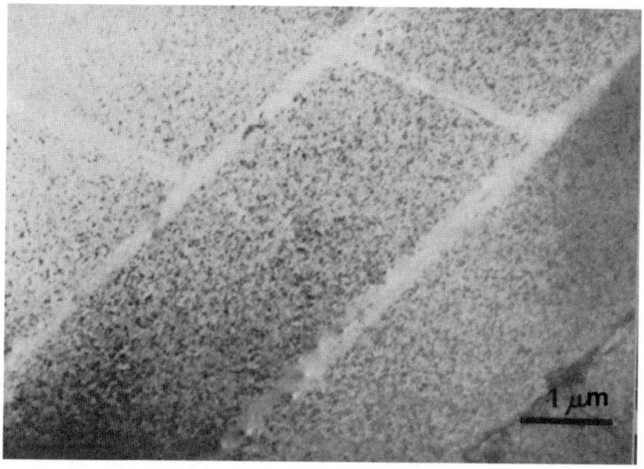

Fig. 12.63 Dislocation channels in neutron radiated Cu (Reproduced with permission from [47]).

the trailing dislocation can glide much more easily, thereby reducing further the density of radiation damage. Local softening takes place and the glide is concentrated in this region. This is referred to as dislocation channelling (Fig. 12.63).

References

1. Schmid, E. and Boas, W. *Plasticity of Crystals*, Chapman & Hall, London; 1950.
2. Barrett, C. S. *Structure of Metals, Crystallographic Methods, Principles, and Data*, 2nd ed. McGraw-Hill (1952).
3. Saka, H., Zhu, Y. M. Kawase, M., Nohara, A. and Imura, T. *Phil. Mag.* A,**51** (1985), 365.
4. Johnston, W. G., Gilman, J. J. *J. Appl. Phys.* **30** (1959), 129.
5. Johnston, W. G. *J. Appl. Phys.* **33** (1962), 2716.
6. Suzuki, T. Strength of Metals (in Japanese), in *Physical Properties of Metals*, ed. by Physical Society of Japan (Shokabo, Tokyo 1968), 491.
7. Nakagawa, K., Maeda, K. and Takeuchi, S. *J. Phys. Soc. Jpn.* **49** (1980), 1909.
8. Saka, H., Noda, K., Matsumoto, K. and Imura, T. *Scripta Met.* **10** (1976), 29.
9. Kuramoto, E. and Suzuki, H. *Trans. JIM.* **9** (1968), 697.
10. Hong, M. H. and Saka, H. *Phil. Mag.* A, **74** (1996), 509.
11. Hong, M. H., Doctorial Thesis, Nagoya University 1997.
12. Hong, M. H. and Saka, H. Proc. 4th Conf. on Zn and Zn alloy coated steel sheet, Galvatech '98 (1998), 248.
13. Hong, M. H. and Saka, H. *Acta Mater.* **45** (1997), 4225.
14. Okada, K., Nohara, A. and Saka, H. *Phil. Mag. Lett.* **79** (1999), 699.
15. Hirose, S., Itoh, T., Makita, S., Arai, S., Sasaki, K. and Saka, H. *Intermetallics* **11** (2003), 633.

16. Saka, E., Hayakawa, T., Nakamura, T. *et al. Phil. Mag.* **A68** (1993), 871.
17. Stein, D. F and Low, Jr., J. R. *Acta Met.* **14** (1966), 1183.
18. Saka, H., Noda, K., and Imura, T. *Crystal Lattice Defects* **4** (1973), 45.
19. Imura, T., Noda, K., Matsui, H., Saka, H. and Kimura, H. *Dislocations in Solids*, (University of Tokyo Press, 1985) 287.
20. Taylor, G. and Saka, H. *Kristall u Technik*, **14** (1979), 1249.
21. Vitek, V., Perrin, R. C. and Bowen, D. K. *Phil. Mag.* **21** (1970), 1049.
22. Hirsch, P. B. *Trans. JIM* **9** (1968), 30.
23. Suzuki, H. *Introduction to Dislocation Theory* (in Japanese) Agne, 1967.
24. Saka, H. and Taylor, G. *Phil. Mag.* **A43** (1981), 1377.
25. Taylor, G. Thermally-activated deformation of BCC metals and alloys, *Progress in Materials Sci.* **36**, 1992, 29.
26. Statham, C. D., Vesely, D. and Christian, J. W. *Acta Met.* **18** (1970), 1243.
27. Taoka, T., Takeuchi, S. and Furubayashi, E. *J. Phys. Soc. Jpn.* **19** (1964), 701.
28. Sestak, B. and Zarubova, N. *Phys. Stat. Soli.* **10** (1965), 239.
29. Takeuchi, S., Furubayashi, E. and Taoka, T. *Acta Met.* **15** (1967), 1179.
30. Spitzig, W. A. and Keh, A. S. *Metall. Trans.* **1** (1970), 2751.
31. Argon, A. S. and Maloof, S. R. *Acta Met.* **14** (1966), 1449.
32. Saka, H. in *Modern Metallurgy*, Vol.3, Atomic theory of strength (in Japanese), (Japan Inst. Metals, 1985).
33. Garratt-Reed, A. J. and Taylor, G. *Phil. Mag.* **A39** (1979), 597.
34. Saka, H., Noda, K., Imura, T., Matsui, H. and Kimura, T. *Phil. Mag*, **34** (1976), 33.
35. Saka, H. and Tsujimoto, K. Unpublished work.
36. Yoshida, K. and Gotoh, Y. *Bulletin of Japan Inst. Metals* (in Japanese) **7** (1968), 666.
37. Chalmers, B. *Physical Metallurgy* (Wiley, 1962) Redrawn based on Figs. 8.36 and 8.37).
38. Mitchell, T. E., Foxall, R. A. and Hirsch, P. B. *Phil. Mag.* **8**, 1963, 1895.
39. Friedel, J. *Dislocations*, Pergamon Press, Oxford, 1964.
40. Saka, H., Noda, K. and Matsui, H. Unpublished work.
41. Suzuki, H., *Strength of Metals* (in Japanese) Agne, 1972, 59 (Fig. 1.56).
42. Humphreys, J. F. in *Understanding Materials*, ed. by C. J. Humphreys, Festschrift for Sir Peter Hirsch, Maney Publishing Co, for The Inst. of Materials, 2002, London).
43. Bonar, L. G. and Kelly, A. cited by Kelly, A., Nicholson, R. B., Precipitation Hardening, in *Progress in Materials Science*, **10**, 1961, 313.
44. Fisher, J. C., Hart, E. W. and Pry, R. H. *Acta Met.* **1** (1953), 336.
45. Takamura, J. Point Defects (in Japanese), *in Physical Properties of Metals*, ed. by The Physical Society of Japan (Shokabo, Tokyo1968), 455. See also Brinkman, J. A., *J. Appl. Phys.* **25** (1954), 961.
46. Saka, H. Noda, K. Matsumoto, K. and Imura, T. *Scripta Met.* **9** (1975), 499.
47. Sharp, J. V. in *The World through the Electron Microscope*, Metallurgy III, compiled by S. Koda, (Japan Electron Optics Laboratory Co.Ltd, 1965) 23. See also *Phil. Mag.* **16** (1967), 77.

Chapter 13

Dislocations in Thin Foils

13.1 Misfit Dislocations in Epitaxial Layers

13.1.1 *Critical thickness*

Let us suppose that a thin crystal is deposited onto a substrate crystal. When the deposited (thin) crystal is coherent to the substrate crystal, this growth is referred to as epitaxial growth and the deposited crystal is referred to as an epitaxial layer (in the contracted form an epilayer). The deposited crystal can either be the same material as the substrate (referred to as homoepitaxy) or a different material but with the same (or similar) crystal structure with different lattice parameters (referred to as heteroepitaxy).

Let the lattice parameter of the substrate crystal be $a_{\rm sub}$, and consider an epitaxial growth of a thin crystal with lattice parameter $a_{\rm ep}$, following Stach *et al.* [1].

Generally, $a_{\rm sub} \neq a_{\rm ep}$: Let $a_{\rm sub} < a_{\rm ep}$. The thin crystal, which is growing epitaxially on the substrate crystal, shrinks horizontally (in directions parallel to the interface between the substrate and the film) at the very beginning of the growth to maintain coherency. At the same time, it is elongated vertically. As a result, the horizontal lattice constant changes to $a_{//}$ and the vertical lattice constant to $a_{\perp}$ (Figs. 13.1(a) and 13.1(b)). In order to maintain coherency not only the epilayer shrinks but also the substrate is elongated, however, since the substrate is much thicker than the epilayer, let us assume that only the epilayer shrinks. That is, let us assume $a_{\rm ep} = a_{\rm sub}$.

While the epilayer is thin enough, the coherency is maintained by elastic deformation of the epilayer, however, on increasing the thickness of the epilayer, the coherency cannot be maintained by elastic deformation and a

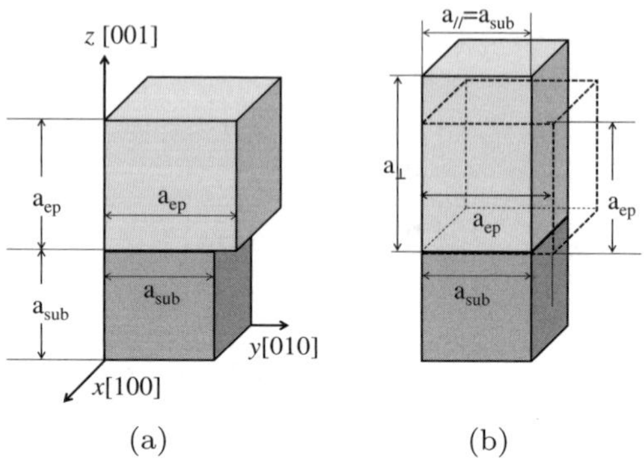

Fig. 13.1 (a), (b)

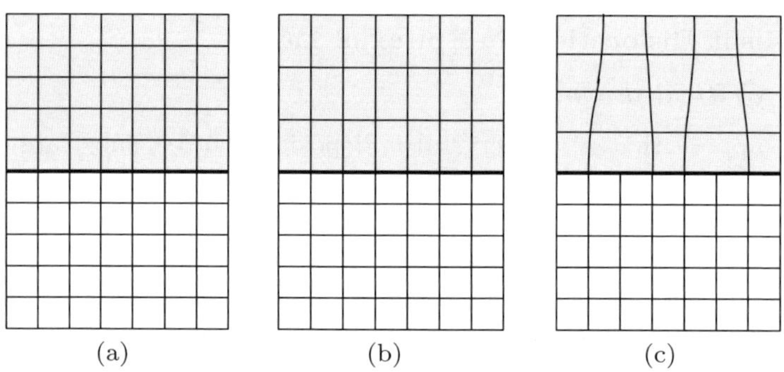

Fig. 13.2 (a)–(c)

dislocation is introduced at the interface (Fig. 13.2(a)–(c)). This dislocation is referred to as a misfit dislocation[1] and the thickness at which the dislocation is introduced is referred to as the critical thickness. The misfit dislocations deteriorate the quality or performance of the epilayer. So the critical thickness is of great importance in the process of epitaxial growth.

13.1.2 *Estimate of the critical thickness*

There are two methods to estimate the critical thickness.

(1) Introducing a misfit dislocation releases an elastic strain inside an epilayer, resulting in a decrease of the elastic energy. However, this

[1]Misfit dislocations are parallel to the interface. On the other hand, those (nearly) perpendicular to the interface are referred to as threading dislocations (TDs).

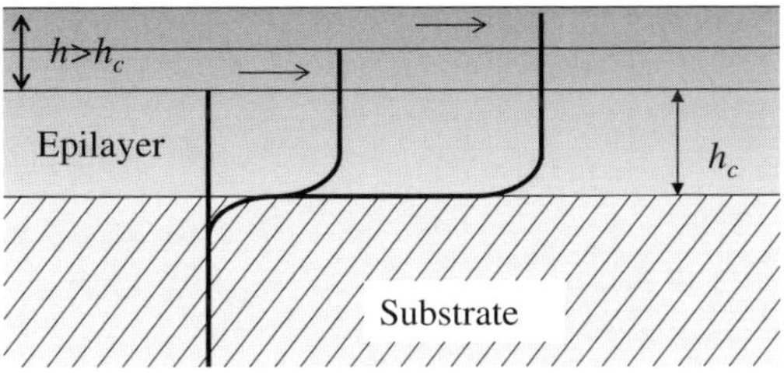

Fig. 13.3

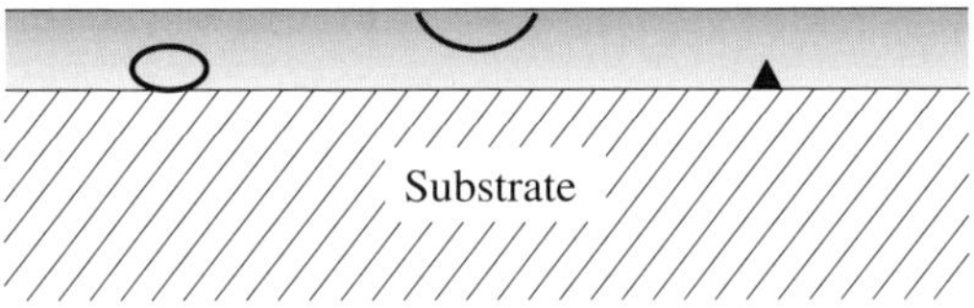

Fig. 13.4

introduces self-energy of the misfit dislocation. It is when these energies become equal that the critical thickness is reached.

(2) There are two possible origins of misfit dislocations.
In the first one, a dislocation pre-existing in the substrate is inherited to the epilayer and spreads laterally in parallel to the interface (Fig. 13.3). The driving force for this lateral motion is elastic energy stored in the epilayer, i.e., an elastic stress. In the other one, the substrate crystal is perfect and a dislocation is nucleated inside the epilayer (Fig. 13.4). Even in this case, the lateral motion of the newly nucleated dislocation, that is, introduction of a misfit dislocation, is driven by an elastic strain (or equivalently stress) stored in the epilayer. This is nothing but a transition of the epilayer from elastic region to plastic region. In other words, at the critical thickness the epilayer yields and deforms plastically.

Anyway, it is of necessity to know the elastic properties of an epilayer. Let us estimate the elastic stress and strain of an epilayer in the absense of dislocations. This is a typical exercise of plane stress (see 3.4.2), and stresses containing z-component can be neglected. Letting $\sigma_{33} = 0, \sigma_{13} = 0$ and $\sigma_{23} = 0$ in Eq. (3.29) and also $\sigma_{11} = \sigma_{22} = \sigma_o$, we obtain

$$\varepsilon_{11} = \frac{1}{E}\left[\sigma_{11} - \nu\sigma_{22}\right]$$

$$= \frac{1}{E}\sigma_{11}(1-\nu)$$
$$= \frac{1}{E}\sigma_0(1-\nu)$$
$$= \frac{(1-\nu)}{2\mu(1+\nu)}\sigma_0,$$
$$\therefore \sigma_0 = \frac{E}{1-\nu}\varepsilon_{11} = 2\mu\frac{1+\nu}{1-\nu}\varepsilon_{11} = 2\mu\frac{1+\nu}{1-\nu}\varepsilon_0. \qquad (13.1)^2$$

Here,

$$\varepsilon_0 = \frac{a_{\mathrm{ep}} - a_{\mathrm{sub}}}{a_{\mathrm{sub}}},$$

where, a_{ep} and a_{sub} are the lattice constants of the epilayer and the substrate, respectively. In the case of a cubic crystal, the lattice constants are usually referred to [100] direction, however, in the diamond or zincblende structure, the coordinate system where x- and y-axes are parallel to $\langle 110\rangle$ is more practical (z-axis is parallel to [001]). Thus, let x′-axis be parallel to [110] and y′-axis be parallel to $[\bar{1}10]$. Transforming from the Cartesian coordinates (where $x = [100], y = [010], z = [001]$), we obtain (Fig. 3.6(again)),

$$\varepsilon'_{11} = \Sigma_{l,m}T_{1l}T_{1m}\varepsilon_{lm}$$
$$= T_{11}T_{11}\varepsilon_{11} + T_{12}T_{12}\varepsilon_{22}$$
$$= \frac{1}{2}\varepsilon_{11} + \frac{1}{2}\varepsilon_{22}$$
$$= \varepsilon_{11}.$$

Note that strain remains unchanged on the transformation of the coordinate.

In what follows, omitting ′, let [110] be x-axis, $[\bar{1}10]$ be y-axis and [001] be z-axis. By putting $\varepsilon_{11} = \varepsilon_{22} = \varepsilon_0$, and applying Hooke's law we obtain

$$\varepsilon_{33} = -\frac{\nu}{E}(\sigma_{11} + \sigma_{22})$$

2

$$E = \frac{\mu(3\lambda + 2\mu)}{\mu + \lambda} = \frac{9\mu B}{3\lambda + \mu} = 2\mu(1+\nu). \qquad (3.31)$$

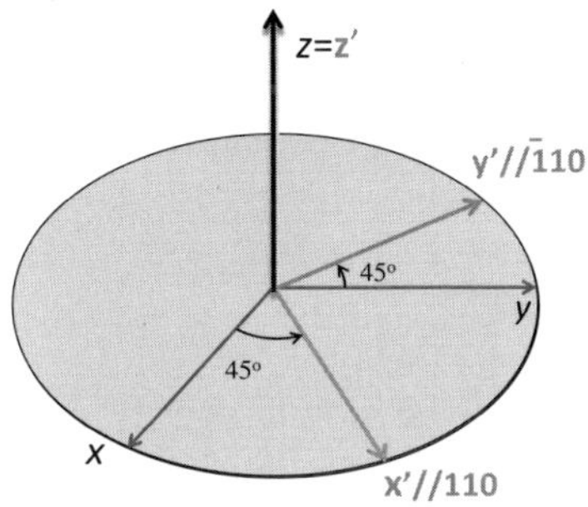

$$T_{ij} \quad \begin{bmatrix} \frac{\sqrt{2}}{2} & \frac{\sqrt{2}}{2} & 0 \\ -\left(\frac{\sqrt{2}}{2}\right) & \frac{\sqrt{2}}{2} & 0 \\ 0 & 0 & 1 \end{bmatrix}$$

Fig. 3.6 (again)

$$= -\frac{\nu}{E}\left(\frac{E\varepsilon_{11}}{1-\nu} + \frac{E\varepsilon_{22}}{1-\nu}\right)$$

$$= -\frac{2\nu}{1-\nu}\varepsilon_0. \tag{13.2}$$

Therefore,

$$a_{//}/a_{\mathrm{ep}} = 1 + \varepsilon_0 \text{ (where } \varepsilon_0 < 0,\ i.e., \text{ compression)}$$

$$a_{\perp}/a_{\mathrm{ep}} = 1 + \varepsilon_{33} = 1 - \frac{2\nu}{1-\nu}\varepsilon_0 \text{ (where } \varepsilon_{33} > 0, \text{ tension)}$$

$$\therefore \frac{a_{\perp}}{a_{//}} = \left(1 - \frac{2\nu}{1-\nu}\varepsilon_0\right) \Big/ (1+\varepsilon_0)$$

$$\cong (1-\varepsilon_0)\left(1 - \frac{2\nu\varepsilon_0}{1-\nu}\right)$$

$$= 1 - \frac{2\nu\varepsilon_0}{1-\nu} - \varepsilon_0 + \frac{2\nu}{1-\nu}\varepsilon_0^2$$

$$= 1 - \frac{1+\nu}{1-\nu}\varepsilon_0. \tag{13.3}$$

Taking Poisson ratio ν to be 1/3, i.e., $\nu = \frac{1}{3}$, the vertical strain ε_{33} is twice as large as the in-plane strain ε_0.

13.1.3 *Energy consideration*

The elastic energy stored in an epilayer (with unit area × thickness h) before a misfit dislocation is created, is given by

$$E_{\mathrm{el}} = \frac{\mu\varepsilon_0^2}{2}h. \tag{13.4}$$

Next, consider an increment in the elastic energy caused by creating one misfit dislocation. In the diamond and zincblende structures, only screw or 60° dislocations are to be considered. Among them, a screw dislocation cannot accommodate a strain in compression or tension, therefore is not considered here. Also, for the sake of simplicity, we consider only an edge dislocation instead of a 60° dislocation.

The elastic strain energy of an edge dislocation per unit length is given by

$$\begin{aligned}\frac{W}{L} &= \frac{\mu b^2}{4\pi(1-\nu)}\ln\frac{R}{r_o}\\ &= \frac{\mu b^2}{4\pi(1-\nu)}\ln\frac{h}{r_o},\end{aligned} \tag{13.5}$$

where we put $R = h$ and $L = 1$.

Density of misfit dislocations can be calculated as follows.

Suppose that n unit cells in the substrate is equal to $(n-1)$ unit cells in the epilayer in longitude (Fig. 13.5(a)).

$$\begin{aligned}na_{\mathrm{ep}} &= (n+1)a_{\mathrm{sub}},\\ n &= \frac{a_{\mathrm{sub}}}{a_{\mathrm{ep}} - a_{\mathrm{sub}}} = \frac{1}{\varepsilon_0}.\end{aligned}$$

Therefore, the spacing between the misfit dislocations is given by (Fig. 13.5(b)),

$$\begin{aligned}na_{\mathrm{ep}} &= (n+1)a_{\mathrm{sub}} \cong nb\\ &= \frac{b}{\varepsilon_0}.\end{aligned}$$

Thus, the density of misfit dislocations is given by $N = \varepsilon_o/b$.

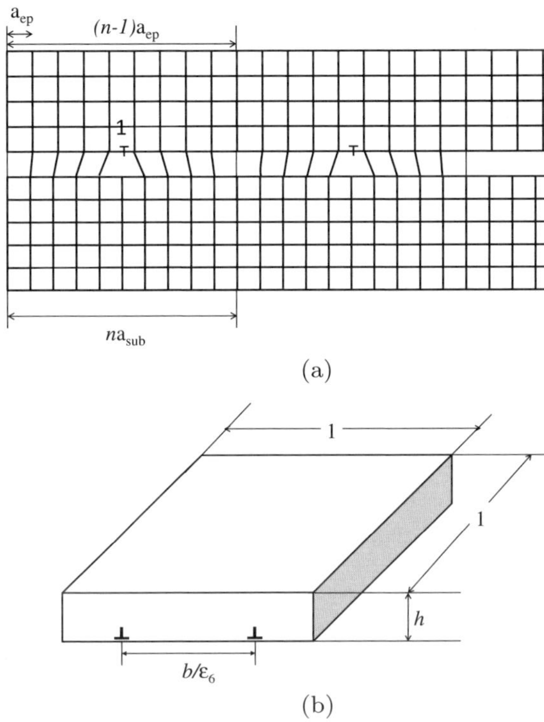

Fig. 13.5 (a), (b)

In other words, $N = \varepsilon_o/b$ dislocations are created, so that the total strain energy is given by

$$E_{\rm dis} = \frac{W}{L} \times N = \frac{\varepsilon_0 \mu b}{4\pi(1-\nu)} \ln \frac{h}{r_o}. \tag{13.6}$$

Putting $E_{\rm dis} = E_{\rm el}$, Eqs. (13.4) and (13.6) give

$$\frac{\mu\varepsilon_0^2 h_c}{2} = \frac{\varepsilon_0 \mu b}{4\pi(1-\nu)} \ln \frac{h_c}{r_o},$$

$$\frac{h_c}{b} = \frac{1}{2\pi\varepsilon_0(1-\nu)} \ln \frac{h_c}{r_0},$$

$$\approx \frac{1}{4\varepsilon_o}\left(\ln \frac{h_c}{b} - \ln \alpha\right). \tag{13.7}$$

13.1.4 *Onset of slip*

In Eq. (13.1)

$$\sigma_0 = \sigma_{11} = \sigma_{22} = 2\mu\frac{(1+\nu)}{(1-\nu)}\varepsilon_0, \qquad (13.1') \text{ (again)}$$

give the normal stresses along x' and/or y' ($\langle 110\rangle$) directions in Fig. (13.6), i.e., tensile or compression stress. Consider the process by which a dislocation glides under the action of these stresses. For the sake of simplicity, a uniaxial stress (σ_0) along the x'-direction will be considered (Fig. 13.7). Let the slip plane be $(\bar{a})$ and the direction be BD.

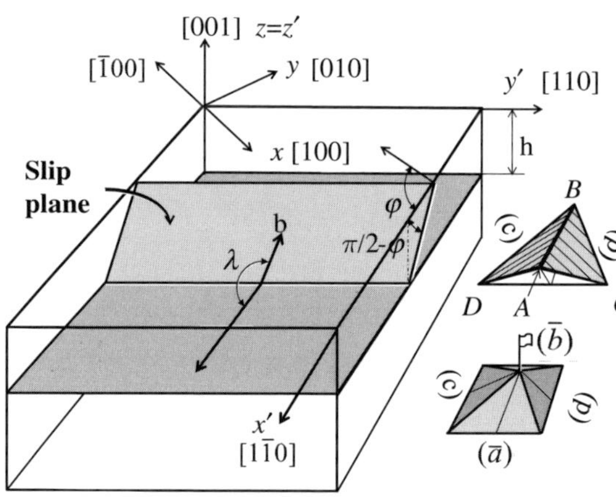

Fig. 13.6

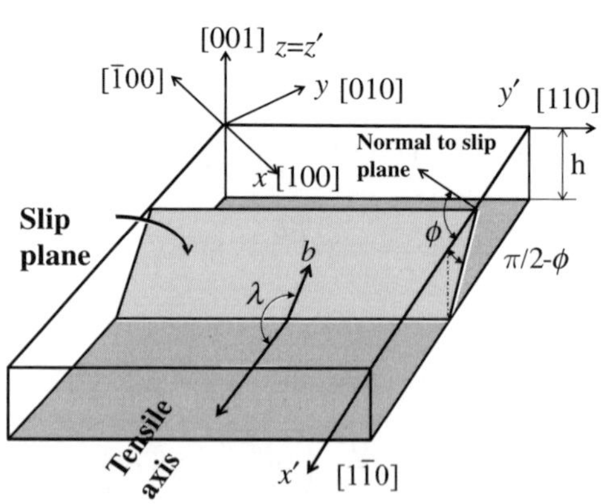

Fig. 13.7

Exercise 13.1 Show that we must consider BD and BC slips on (d) plane and AD, AC slips on (b) planes.

For the glide motion of a dislocation, shear stress is important. Thus, the uniaxial stress σ_0 must be resolved to shear stress τ_a in the slip system with slip plane (a) and slip direction BD. For this transformation the Schmid factor described in Chapter 12 is used, and we obtain

$$\tau_a = 2\mu\frac{(1+\nu)}{(1-\nu)}\varepsilon_0 \cos\lambda\cos\phi, \tag{13.7$'$}$$

where λ and ϕ are angles shown in Fig. 13.7.

Force exerting on the dislocation F_a is given by Eqs. (4.4) and (4.16) as follows.

$$\begin{aligned} \frac{F_a}{L} &= \tau_a b \\ &= 2\mu b\frac{(1+\nu)}{(1-\nu)}\varepsilon_0\cos\lambda\cos\phi, \\ \text{where,}\quad L &= \frac{h}{\cos\phi}, \\ \therefore\quad F_a &= 2\mu b\frac{(1+\nu)}{(1-\nu)}\varepsilon_0 h\cos\lambda. \end{aligned} \tag{13.8}$$

(see Fig. 13.8).

When a dislocation moves under the action of a force F_a, it bows out. This results in an increase in its length, giving rise to a back stress. The back stress can be estimated in the following two ways.

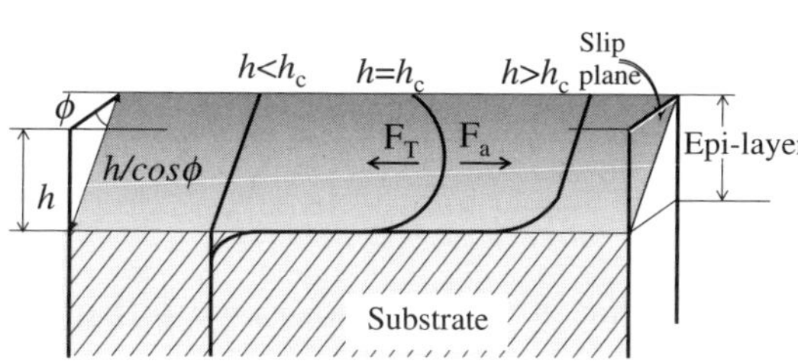

Fig. 13.8

① Elastic energy of a dislocation per unit length is given by

$$\frac{W}{L} = \frac{\mu b^2}{4\pi}\left(\cos^2\beta + \frac{\sin^2\beta}{1-\nu}\right)\ln\frac{R}{r_o}, \qquad (4.19)\text{ (again)}$$

where β is an angle between the dislocation line and $\boldsymbol{b}$ (for edge $\beta = 90°$ and for screw $\beta = 0°$) and r_o is the cut-off radius. Equating R to the thickness of the epilayer $h(R = h)$,

$$\begin{aligned}\frac{W}{L} &= \frac{\mu b^2}{4\pi}\left(\cos^2\beta + \frac{\sin^2\beta}{1-\nu}\right)\ln\frac{h}{r_o}\\ &= \frac{\mu b^2}{4\pi}\frac{(1-\nu\cos^2\beta)}{1-\nu}\times\ln\frac{h}{r_o}.\end{aligned}$$

The back stress to hinder the bowing-out of the dislocation (F_T) is given by

$$\begin{aligned}F_T &= \frac{d}{dL}\left(\frac{W}{L}\times L\right)\\ &= \frac{\mu b^2}{4\pi}\frac{(1-\nu\cos^2\beta)}{1-\nu}\times\ln\frac{h}{r_o}.\end{aligned} \qquad (13.9)$$

When the effective force acting on the dislocation F_a is lower than F_T, i.e., $F_a < F_T$, the dislocation does not move. When $F_a > F_T$, the dislocation moves, reducing ε and at the same time destroying the coherency between the substrate and the epilayer. Therefore, the critical thickness h_c is given by

$$\begin{aligned}F_T &= \frac{\mu b^2}{4\pi}\frac{(1-\nu\cos^2\beta)}{1-\nu}\times\ln\frac{h_c}{r_o}\\ &= F_a = 2\mu b\frac{(1+\nu)}{(1-\nu)}h_c\varepsilon_o\cos\lambda.\end{aligned}$$

Thus,

$$\begin{aligned}h_c &= \frac{b}{8\pi\varepsilon_o}\frac{(1-\nu\cos^2\beta)}{1+\nu}\times\ln\frac{h_c}{r_o}\\ &\approx \frac{b}{35\varepsilon_o}\left(\ln\frac{h_c}{b} - \ln\alpha\right).\end{aligned} \qquad (13.10)^3$$

Here, it is assumed that $\beta = 60°$. Also, $r_o = \alpha b$ (Eq. (4.5′)).

[3]Here, it is assumed that $\beta = 60°$.

② In Sec. 6.2.1 the back stress to hinder the bowing-out of a dislocation was approximated as follows. Let the line tension of the dislocation be T, then Eq. (6.2) gives

$$T \approx \frac{\mu b^2}{2}, \qquad \text{(6.2) (again) (see Fig. 6.8(a))}$$

Substituting into Eq. (6.3),

$$\begin{aligned}\tau_{\text{back}} &= \frac{T}{Rb} = \frac{\mu b^2}{2Rb} = \frac{\mu b}{2R} \\ &= \frac{\mu b \cos\phi}{h} \end{aligned} \qquad \text{(13.11) (see Eq. (6.2))}$$

Here, $2R = h/\cos\phi$.

Equating τ_{back} to τ_a,

$$\begin{aligned}\tau_{\text{back}} &= \frac{\mu b \cos\phi}{h} = \tau_a = 2\mu b \frac{(1+\nu)}{(1-\nu)} \varepsilon_o \cos\lambda \cos\phi \\ \frac{h_c}{b} &= \frac{(1-\nu)}{2(1+\nu)\varepsilon_o \cos\lambda} \approx \frac{1}{2\sqrt{2}\varepsilon_o}, \end{aligned} \qquad (13.12)$$

where $\cos\lambda \approx \frac{1}{\sqrt{2}}$.

13.1.5 *Summary*

To summarize,

① Energy consideration

$$\frac{h_c}{b} \approx \frac{1}{4\varepsilon_o}\left(\ln\frac{h_c}{b} - \ln\alpha\right). \qquad \text{(13.7) (again)}$$

② Balance between forces 1

$$\frac{h_c}{b} \approx \frac{1}{35\varepsilon_o}\left(\ln\frac{h_c}{b} - \ln\alpha\right), \qquad \text{(13.10) (again)}$$

③ Balance between forces 2

$$\frac{h_c}{b} \approx \frac{1}{2\sqrt{2}\varepsilon_o}. \qquad \text{(13.12) (again)}$$

In ① and ② the results depend on the concrete value of α. In theoretical papers treating the critical thickness it was assumed that $\ln\alpha = -1$. This corresponds to taking $\alpha = 1/e = 0.38$ in the cut-off parameter $r_o(= \alpha b)$, where e is natural logarithm, i.e., $e = 2.71$. Experimental estimate has been

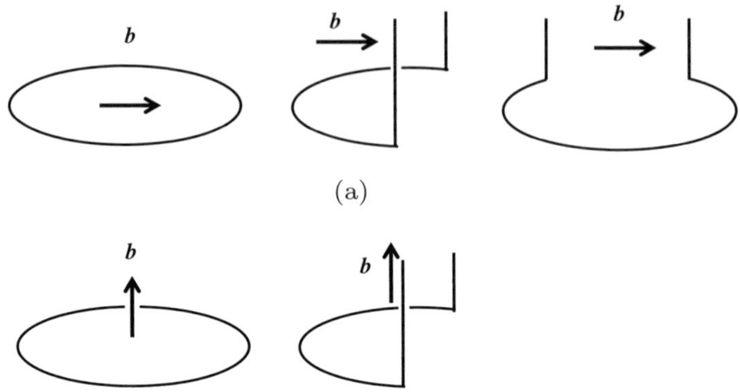

Fig. 13.9 Threading dislocation (TD). (a) TED, (b) TSD.

carried out on a Cu–Al alloy, which showed that $r_o \doteqdot 3b$ (that is $\alpha = 3$). This is in sharp contrast to the aforementioned assumption.

Furthermore, in ② and ③ it is assumed that the motion of dislocations triggers yielding of the epilayer, however, the yield stress of the epilayer critically depends on whether the substrate crystal is perfect or not. If a dislocation-free perfect crystal (for instance Si) is used as a substrate, an epilayer in the initial stage, that is, below the critical thickness, misfit dislocations are not present. Thus, yielding of the epilayer is controlled by nucleation of dislocations in a perfect crystal. This can be compared to yielding of a whisker. On the other hand, when a substrate crystal contains dislocations, which are inherited to the epilayer, yielding of the epilayer corresponds to yielding of the conventional crystals which contains dislocations. As is shown in Fig. 12.46 in Chapter 12, there is a big difference between these two phenomena of yielding. In conclusion, the critical thickness is to be considered as a qualitative measure.

13.2 Threading Dislocations in Epilayers

In Fig. 13.3(a) vertical dislocation segments pierce an epilayer. Those dislocations piercing an epilayer are referred to as threading dislocations (TDs). TDs are classified into two groups: one is a threading edge dislocation (TED) and the other is a threading screw dislocation (TSD).

TED is formed as a result of climb or cross-slip of a glide dislocation lying at the interface between the substrate and the epilayer (Fig. 13.9(a)).

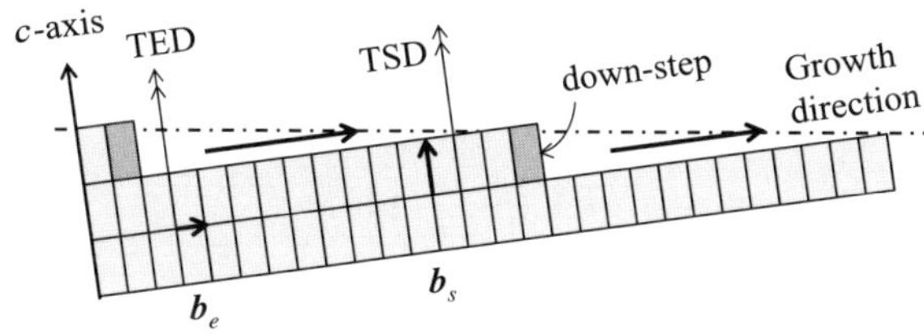

Fig. 13.10 Step-controlled epitaxial growth.

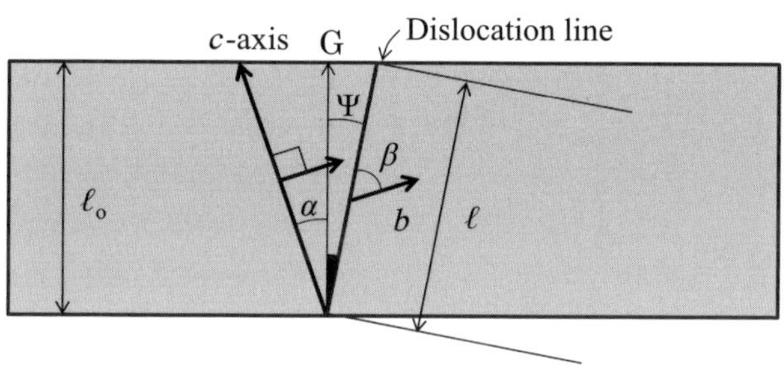

Fig. 13.11

TSD is formed as a result of the pencil glide of an edge dislocation loop (Fig. 13.9(b)).

TD is likely influenced by the surface and changes its orientation. Especially those in an epilayer grown by a step-controlled epitaxial growth technique inevitably deviate from the exact edge or screw orientations. Figure 13.10 illustrates the principle of the step-controlled epitaxial growth [2]. A substrate crystal has a surface orientation slightly deviated from a low-index plane (for instance (0001) plane in hexagonal close-packed (HCP) crystals). On the surface of the substrate, steps are introduced intentionally so that the growth direction can be controlled as shown by the arrows. The epilayer grown by this technique is schematically illustrated in Fig. 13.11.

The growth orientation (G) is tilted by α (usually $\alpha = 4°$ or $8°$) from c-axis. α is referred to as a cut-off angle. $\boldsymbol{b}$ is the Burgers vector of a glide dislocation in the basal plane and therefore perpendicular to c-axis. Now, suppose that a dislocation line is tilted from the growth direction by an angle Ψ. Let the angle between $\boldsymbol{b}$ and the dislocation line be β, that is, $\alpha + \Psi + \beta = 90°$.

The strain energy of a dislocation line *per unit length* is given by Eq. (4.19).

$$\frac{W}{L} = \frac{\mu b^2}{4\pi}\left(\cos^2\beta + \frac{\sin^2\beta}{1-\nu}\right)\ln\frac{R}{r_0}. \qquad (4.19)\text{ (again)}$$

L is the length of the dislocation line. The length of a dislocation line in an epilayer with thickness of ℓ is given by

$$\ell = \frac{\ell_0}{\sin\psi} = \frac{\ell_0}{\cos(\beta+\alpha)}$$

Therefore, setting the proportional coefficient $\frac{\mu b^2}{4\pi}\ln\frac{R}{r_0}\times(\ell_0)$ to unity, the total energy is expressed as follows:

$$\begin{aligned}
E &= \left(\cos^2\beta + \frac{\sin^2\beta}{1-\nu}\right)\times\ell \\
&= \left(\cos^2\beta + \frac{\sin^2\beta}{1-\nu}\right)\times\frac{1}{\cos(\beta+\alpha)} \\
&\approx \left(\cos^2\beta + 1.5\sin^2\beta\right)\times\frac{1}{\cos(\beta+\alpha)} \\
&= \frac{1+0.5\sin^2\beta}{\cos(\beta+\alpha)}.
\end{aligned}$$

E is calculated as a function of θ_ξ for $\alpha = 4°$ and $8°$, and shown in Fig. 13.12. Here, $\theta_\xi = \alpha + \Psi$, that is, the angle between TED and c-axis. It is obvious that a pure edge dislocation is energetically unfavourable.

Consider a pure edge dislocation parallel to c-axis in Fig. 13.11. At this point, the strain energy per unit length (W/L) is a maximum. When this dislocation is tilted slightly either to the right or to the left, W/L decreases. If tilted to the left, W/L decreases but the total length of the dislocation ℓ increases. W/L depends on β only insensitively around $\beta = 90°$. In other words, a gain (negative) in energy by tilting to the left is not so large as a loss (positive) due to an increase in the total length. On the other hand, if tilted to the right, both W/L and ℓ decrease. The result is that the dislocation is tilted to the right.

Exercise 13.2 Perform a similar calculation for a threading screw dislocation and show that the dislocation line is not coincident to c-axis nor the growth direction G nor ***b***. (Hint) $\beta = \alpha + \Psi$.

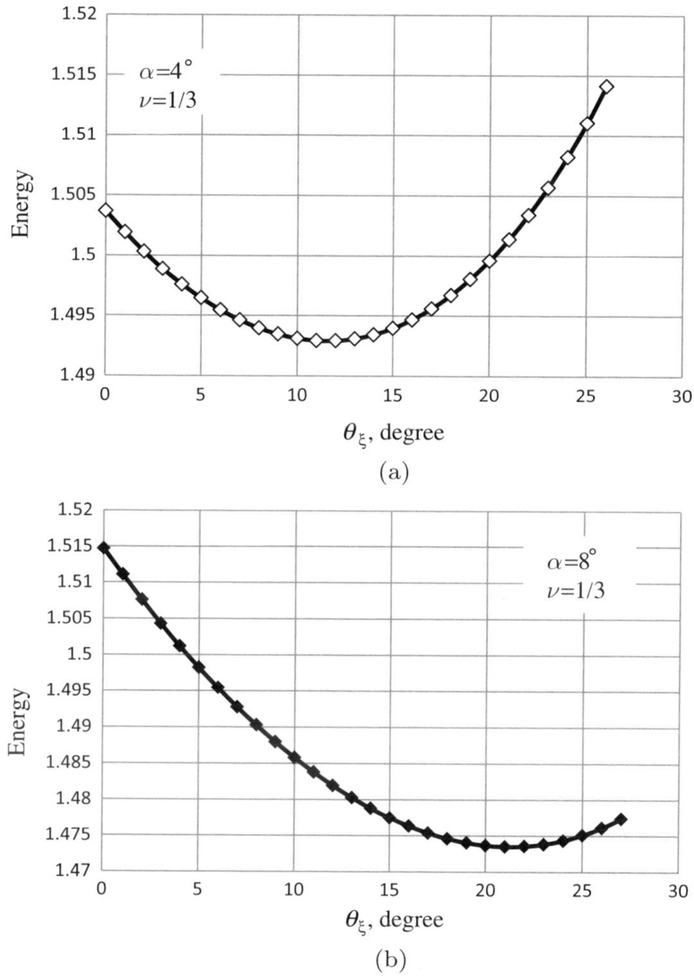

Fig. 13.12 (Reproduced with permission from [3]).

13.3 Dissociated Dislocations in Thin Crystals

13.3.1 *Bending of dislocations in the vicinity of crystal surface*

Let us consider a single crystal shown in Fig. 13.13. The slip plane is inclined by approximately 45° with respect to the tensile (vertical) direction. For the sake of simplicity, let a dislocation be bent sharply as E→O→S. OE is an edge component and OS is a screw one. Screw component OS can cross slip and reach OS″.

Here, let us assume that only the segment within the depth of ℓ_0 underneath the surface is affected by the image force from the crystal surface (Fig. 13.14). The screw component makes an angle of λ to the surface (top face in this case). The elastic energy of a dislocation per unit length is given by Eq. (4.19) as follows.

$$\frac{W}{L} = \frac{\mu b^2}{4\pi}\left(\cos^2\beta + \frac{\sin^2\beta}{1-\nu}\right)\ln\frac{R}{r_0}. \qquad (4.19\ \text{(again)})$$

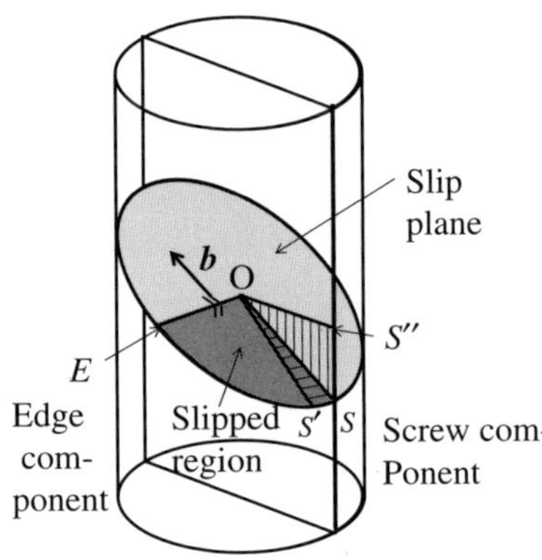

Fig. 13.13 Lohne force.

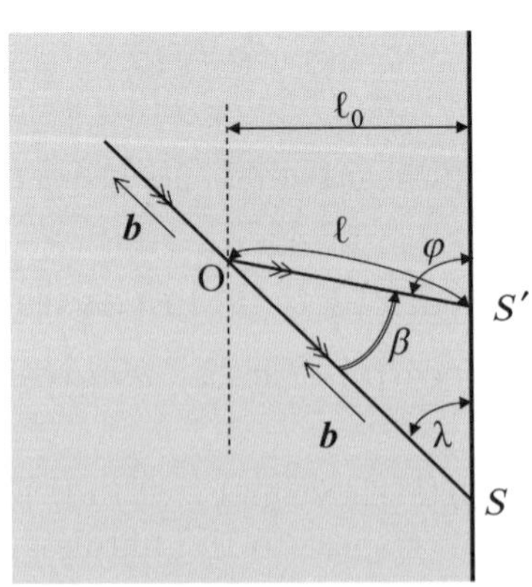

Fig. 13.14

The length of the bent segment (OS′) ℓ is given by

$$\ell = \frac{\ell_0}{\sin\varphi} = \frac{\ell_0}{\sin(\beta+\lambda)}.$$

Neglecting the proportional coefficient in Eq. (4.19),

$$\begin{aligned} E &= \left(\cos^2\beta + \frac{\sin^2\beta}{1-\nu}\right) \times \ell \\ &= \left(\cos^2\beta + \frac{\sin^2\beta}{1-\gamma}\right) \times \frac{1}{\sin(\beta+\lambda)}. \end{aligned} \tag{13.13}$$

Here, we assume that $\ell_0 = 1$.

λ is approximately 45° according to Schmid's law. Let $\lambda = 45°$ and $\nu = 1/3$. Equation (13.14) is given by

$$\begin{aligned} E &= \left(\cos^2\beta + \frac{3\sin^2\beta}{2}\right) \times \ell \\ &= \left(\cos^2\beta + \frac{3\sin^2\beta}{2}\right) \times \frac{1}{\sin\beta\cos\lambda + \sin\lambda\cos\beta} \\ &= \cos^2\beta + \frac{3\sin^2\beta}{2\left(\sin\beta \times \frac{1}{\sqrt{2}} + \cos\beta \times \frac{1}{\sqrt{2}}\right)} \\ &= \frac{\left(\cos^2\beta + \left(\frac{3}{2}\right)\sin^2\beta\right)\sqrt{2}}{\sin\beta + \cos\beta}. \end{aligned} \tag{13.14}$$

When E is plotted against β in Fig. 13.15, a minimum is present at $\beta = 25°$. In other words, a screw component along OS is bent by approximately 25° toward OS′ in the vicinity of surface, that is, it cross-slips away from the primary slip plane (Fig. 13.14). The driving force for this is referred to as Lohne force [4].

Exercise 13.3 Calculate E as a function of β (or ϕ) for $\lambda = 40°$, 50° and show that a minimum is present at around $\beta \fallingdotseq 25°$.

When the cross-slip plane on which the screw segment cross-slips contains the tensile direction as shown in Fig. 13.16(b), this segment OS cannot move, and the screw part remains on the P_1 plane (left side of OS in Fig. 13.13). If the cross-slip plane is inclined as shown by P_2 in Fig. 13.16(c) and has a Schmid's factor large enough, a slip on P_2 is preferred to P_1 beneath the surface. In a foil specimen, whose thickness is comparable to l_0, P_2 is activated over the thickness, resulting in a failure of Schmid's law.

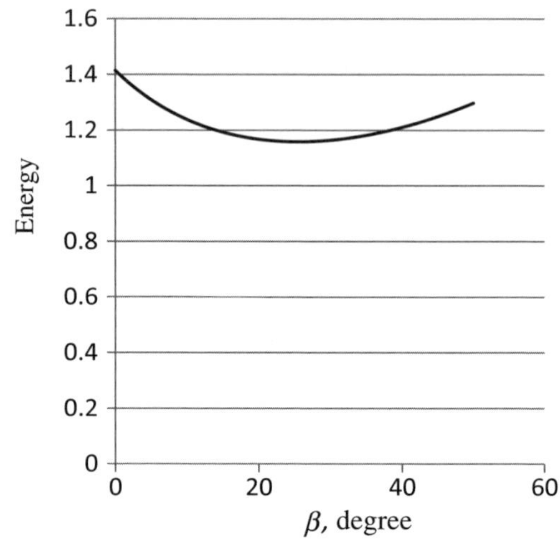

Fig. 13.15

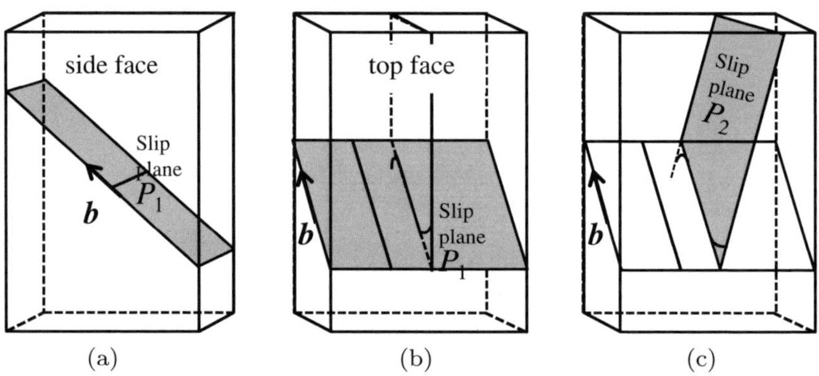

Fig. 13.16 Lohne force [5].

13.3.2 *Effect of surface on the dissociation distance of a dissociated dislocation*

For a dissociated dislocation in face-centered cubic (FCC) crystals, the following must be taken into account.

① Dissociation distance of a dissociated dislocation is determined by the balance between a repulsive force between the two Shockley partials and an attractive force due to the stacking fault (Fig. 13.17(a)). However, strictly speaking, this holds only in an infinite crystal. In a thin foil, the

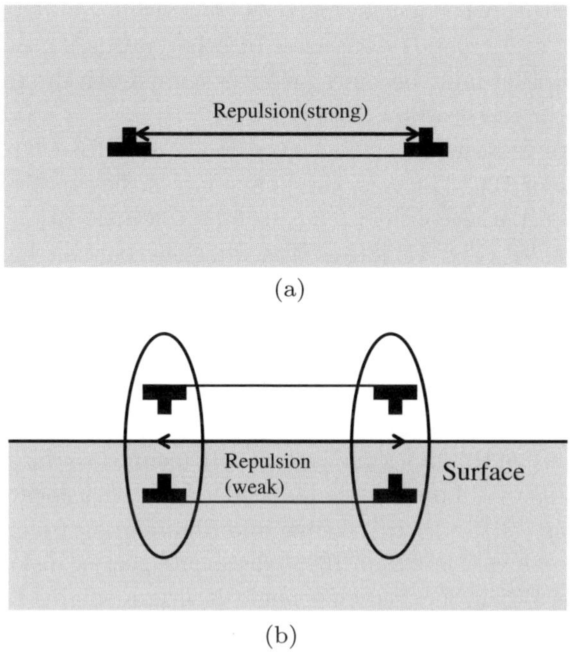

Fig. 13.17 (a), (b)

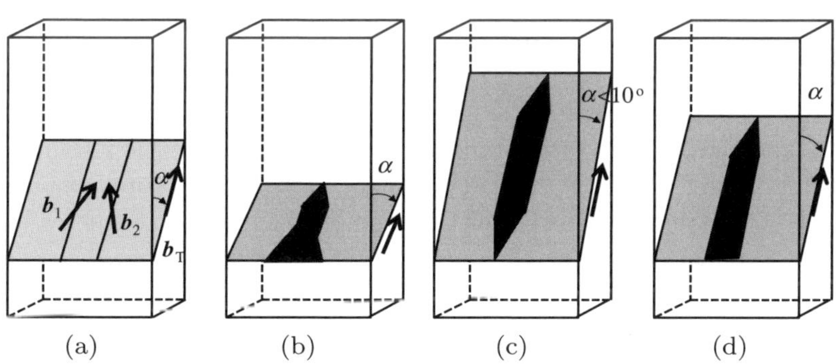

Fig. 13.18 Effect of Lohne force on Shockley partials of a dissociated dislocation.

dislocation receives mirror forces from its imaginary dislocation. This is illustrated schematically in Fig. 13.17(b); Shockley partial in the real crystal and its imaginary counterpart form a dipole. The stress fields of the two dislocations (with $+\boldsymbol{b}$ and $-\boldsymbol{b}$) comprising the dipole cancel each other, so that a long-range stress decreases rapidly with the distance: The repulsive force between the two Shockley partials comprising the dissociated dislocation (in the real crystal) decreases. In other words, an attractive force due to the stacking fault becomes stronger comparatively, thus leading to narrowing of the dissociation distance.

② Burgers vecttors $\boldsymbol{b}$ of Shockley partials of a dissociated dislocation make an angle of 120°. That is, the Lohne forces they receive are different from each other. Let us consider a pure total screw dislocation with the total Burgers vector $\boldsymbol{b}_T$ (Fig. 13.18(a)). The Shockley partials have $\boldsymbol{b}_1$ and $\boldsymbol{b}_2$, respectively (Fig. 13.18(a)). Lohne force acts on a near-screw dislocation so as to rotate the dislocation toward the pure screw direction. That is, partials tend to align themselves parallel to $\boldsymbol{b}_1$ and $\boldsymbol{b}_2$. As a result, on one surface the dissociated dislocation shrinks and on the other it expands (Fig. 13.18(b)).

However, when the slip plane is shallowly inclined to the surface, mirror force ① prevails over Lohne force ②. Then, a screw dislocation shrinks at both ends (Fig. 13.18(c)). Hazzledine and Shaibani [6] predict that, when the glancing angle α is less than 10°, a dissociated screw dislocation shrinks at both ends. When the glancing angle is intermediate, it shrinks only at one surface (Fig. 13.18(d)). Figure 13.19(a) shows a dark-field weak-beam (DFWB) micrograph of dissociated dislocations corresponding to Fig. 13.18(a) [7].

In Fig. 13.19(b) all the dislocations ①–⑧ shrink at the upper surface. In the lower surface the dissociation distance does not change. This corresponds to Fig. 13.19(d).[4] The cross-section of the specimen is wedge-shaped and the thickness increases from dislocation ① to ⑧ (Fig. 13.19(c)). It is noted that the dissociation distance even in the middle of the foil increases with increasing the local thickness. That is, the dissociation distance is influenced by the surface. It is of vital importance to use as thick a specimen as possible to measure the stacking fault energy from the dissociation distance (Sec. 8.2).

Supplement to Chapter 13 Mechanism of Transformation of BPD to TED in 4H–SiC (Fig. 11.27)

When a BPD is dissociated on the basal plane, as is often observed in 4H–SiC, the dissociated BPD must constrict in order to transform itself to TED. Understanding the mechanism of constriction necessitates understanding

[4]Dislocation ⑦ shrinks at the surface and cross-slips off the original {111} plane.

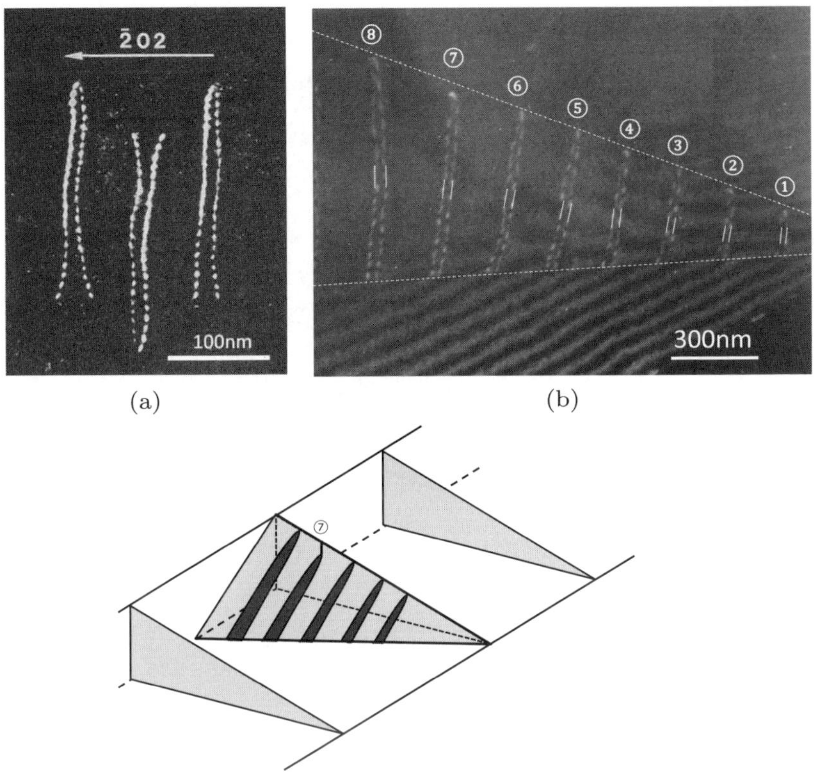

Fig. 13.19 (a) (Reproduced with permission from [7]), (b) (Reproduced from [8]) (c) Schematic illustration.

the behaviour of a dissociated dislocation threading at a glancing angle in a thin foil. We follow Hazzledine–Shaibani [6].

Let us consider a perfect screw dislocation on $\{111\}$ slip plane (or (0001) basal plane in HCP) threading at a glancing angle α in a foil with the thickness t (Fig. 13.20(a)). Let the (total) Burgers vector of the perfect screw dislocation be $\boldsymbol{b}_T$, then the Burgers vectors of Shockley partials $\boldsymbol{b}_p$ are inclined by $\pm 30^\circ$ with respect to the direction ξ of the total dislocation line. Figure 13.20(a) shows one of the two Shockley partials. This Shockley partial receives a Lohne force from the surface but its motion is restricted on the $\{111\}$ slip plane. This corresponds to the bending of O–S to O–S$''$

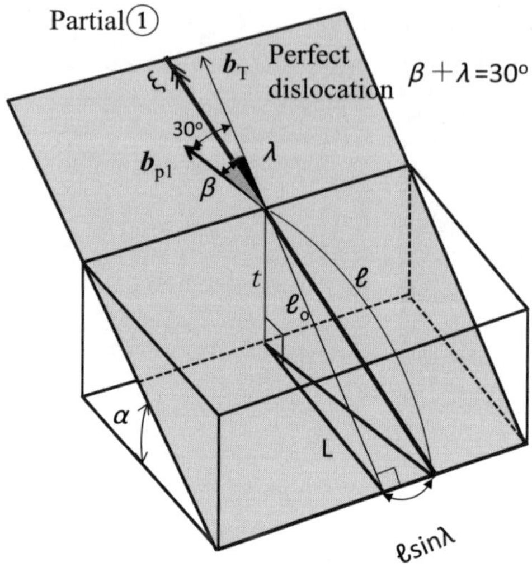

Fig. 13.20 (a) Dissociated screw dislocation in a thin foil.

in Fig. 13.13. Energy of this 30° partial is given by

$$\begin{aligned} E &= \left(\cos^2\beta + \frac{\sin^2\beta}{1-\nu}\right) \times \ell \\ &= \left(\cos^2\beta + \frac{\sin^2\beta}{1-\nu}\right) \times \frac{\ell_o}{\cos\alpha} \\ &= \frac{\cos^2\beta + \frac{3}{2}\sin^2\beta}{\cos\gamma} \times \ell_0 \\ &= \frac{2+\sin^2(30-\gamma)}{2\cos\gamma} \times \ell_0. \end{aligned}$$

β is an angle between the direction of the partial dislocation line and its Burgers vector $\boldsymbol{b}_p$. Also we assume that $\nu = 1/3$. E is plotted as a function of γ in Fig. 13.20(b). It shows a minimum at $\gamma = \sim 14°$. That is, the Shockley partial is tilted toward $\boldsymbol{b}_p$ by $\sim 14°$.

The other partial of $\boldsymbol{b}_p = -30°$ is tilted by $-14°$. That is, as shown in Fig. 13.18(b), the dissociation distance becomes narrower on one surface of the foil and wider on the other surface. There are the effects of Rohne force.

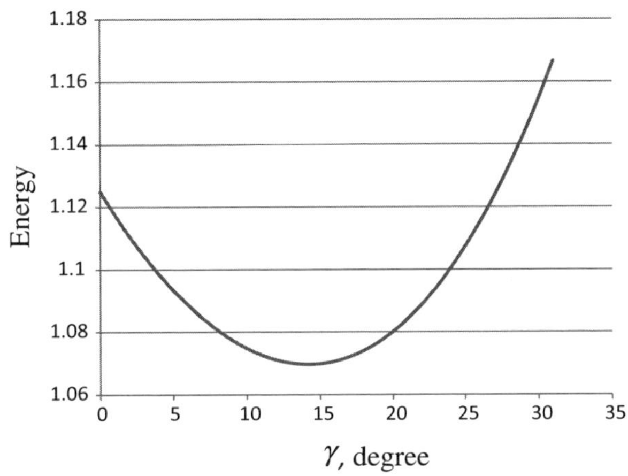

Fig. 13.20 (b) Energy of a dislocation in a thin foil.

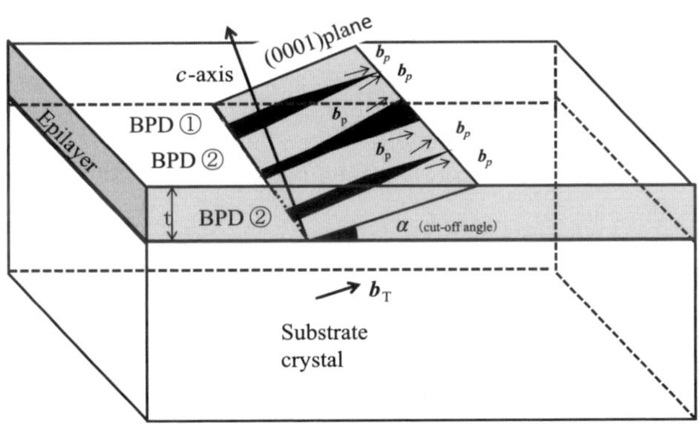

Fig. 13.21 Behaviour of BPD near the free surface.

Now let us apply this result to the growth process of an epilayer by step-control epitaxial technology. Redrawing Figs. 13.10 and 13.18, we obtain Fig. 13.21.

That is, a BPD in the dissociation mode denoted by BPD ① shrinks at the surface of the growing epilayer, while a BPD in dissociation mode denoted by BPD ② expands at the surface.[5]

[5]Refer to Fig. 11.24 for dissociation modes in 4H–SiC and GaN.

However, an image force from the imaginary dislocation is superimposed on Lohne force (Fig. 13.17). The image force always makes the dissociation distance narrower. As a result, BPD ① shrinks perfectly at the surface of an epilayer. According to Hazzledine–Shaibani [6], when α (i.e., cut–off angle) $< 10°$, even BPD ② shrinks at the surface because the image force prevails over the Lohne force.

In summary, in a step-control epitaxial technology with a cut-off angle (α) 4° or 8° BPD always shrinks at the surface of a growing epilayer. This is not to be attributed to an effect of kinetics of the crystal growth but to energetic at equilibrium. Once BPD shrinks, it transforms to TED since TED is much shorter than BPD, i.e., its energy $((W/L) \times L)$ is lower than that of BPD.

References

1. Stach, E., Hull, R., Tromp, R. M., Ross, F. M., Reuter, M. C. and Bean, J. C. *Phil. Mag. A* **80** (2009), 2159.
2. Matsunaga, H. *Physica B* **185** (1996), 65.
3. Saka, H., Watanabe, H., Kitou, H., Hirose, F. and Onda, S. *Jpn. J. Appl. Phys.* **53** (2014), 111302.
4. Lohne, O. *Phys. Stat. Soli.* (a)**18** (1973), 403.
5. Hazzledine, P. M. Private communication.
6. Hazzledine, P. M., Shaibani, S. J. *Strength of Metals*, ICSMA (Aachen), 1978, p. 45.
7. Hazzledine, P. M., Karnthaler, H. P. Winter, E. *Phil. Mag.* **32** (1975), 81.
8. Saka, H. *Res. Mechanica*, **11** (1980), 211.

Appendices

A.1 Reducing Dislocation Density

Dislocations contribute to strengthening in metallic materials while they deteriorate performance of semiconductors. For reducing the dislocation density in crystals thermal cyclic annealing, developed by Dr. Kitajima of Kyushu University, is very effective[1]. In this method, crystals are annealed in a temperature range changing periodically. This method was first applied to FCC and HCP metals where the P–N force is low, but later successfully applied to BCC metals and Si where the P–N force is high [2, 3].

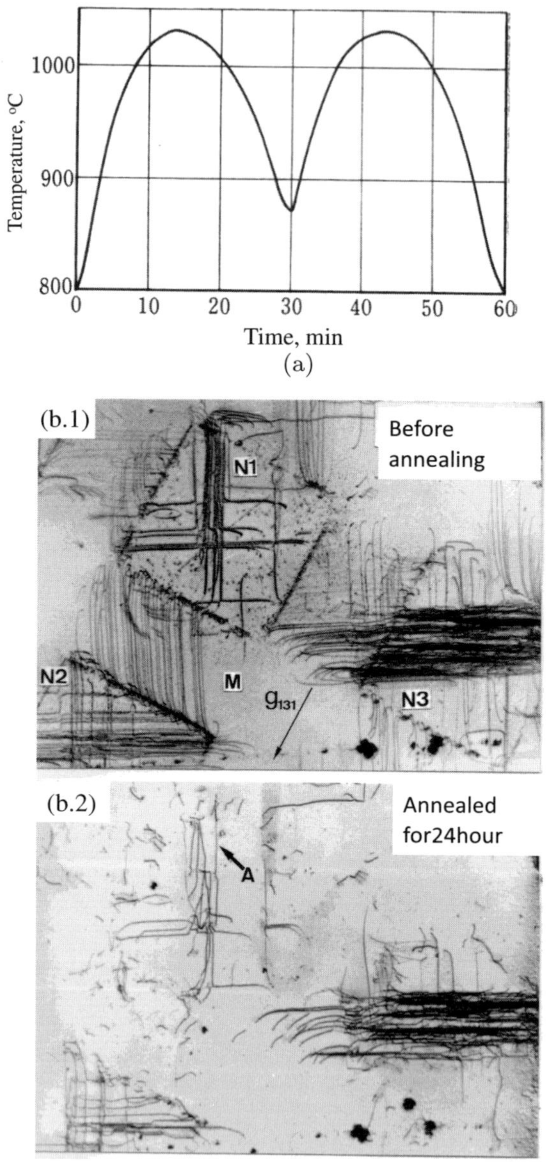

(b)

Fig. A.1 (a) Cycle annealing method for Cu (Reproduced with permission from [1]), (b) Effect of cycle annealing on reduction of dislocations in Si. In (b.1) many dislocations are present, however, after cyclic annealing for 24 hours (b.2) the dislocation density decreases significantly (Reproduced with permission from [2]).

A.2 Crystal Model of HCP

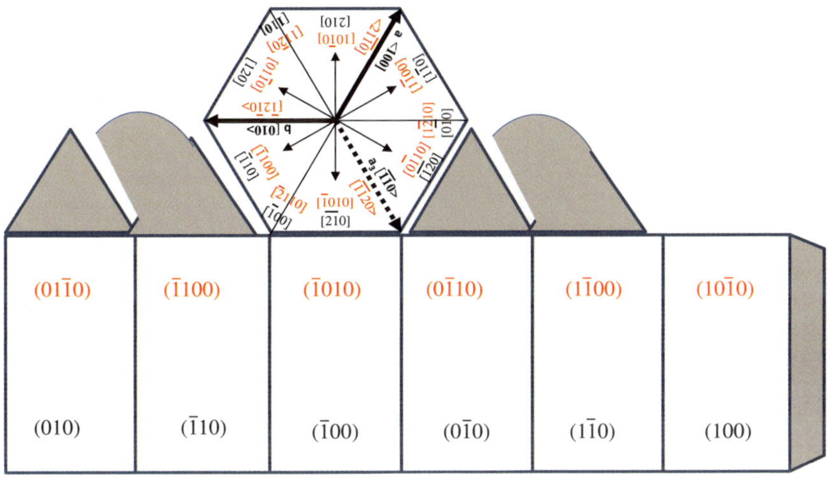

Fig. A.2

A.3 Model of a Screw Dislocation

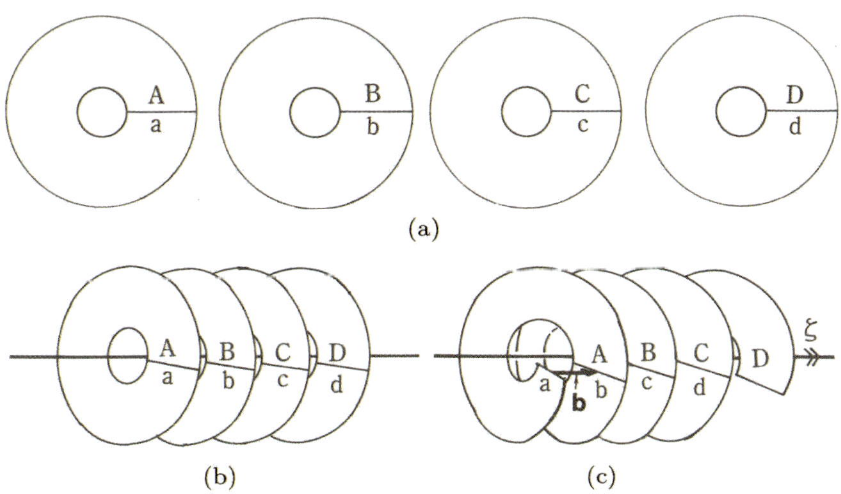

Fig. A.3 Model of a screw dislocation (Reproduced with permission from [4]).

A.4 Stereoprojection

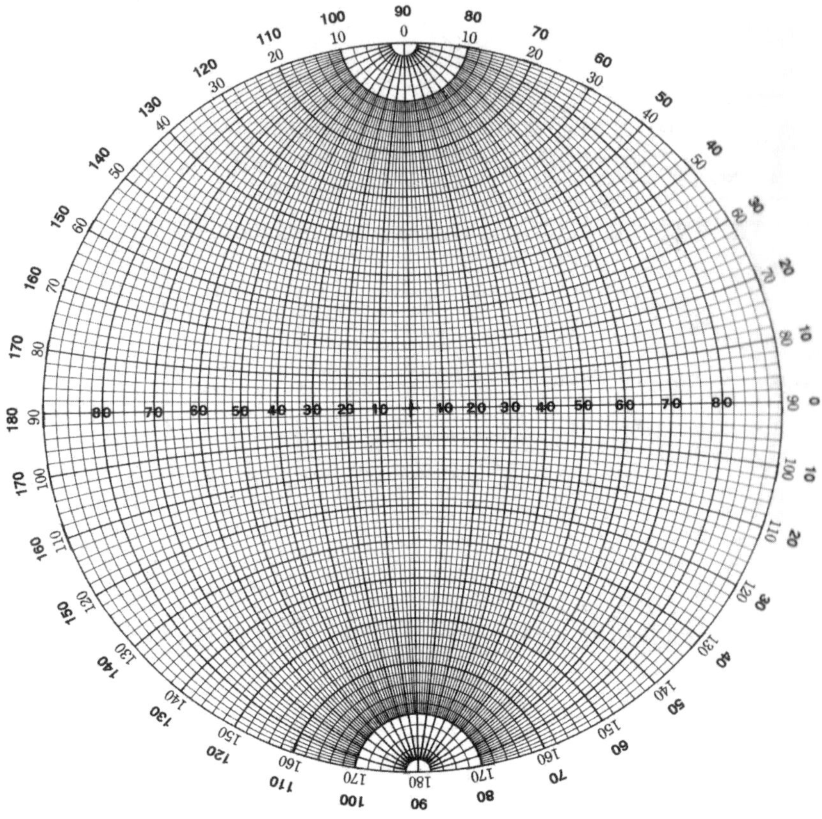

Fig. A.4 (a) Wulff net.

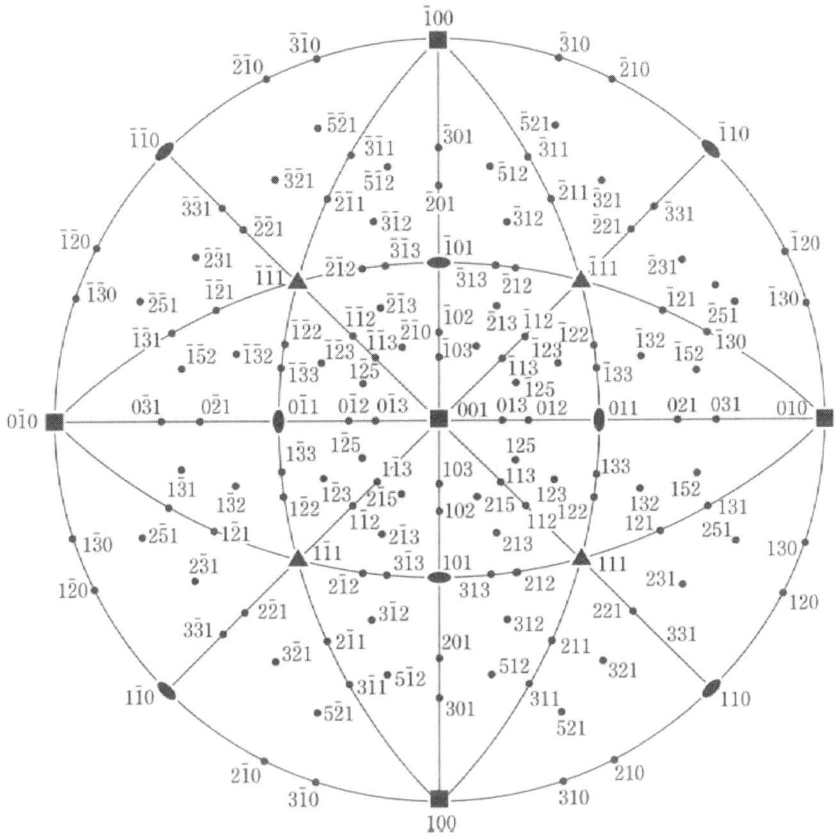

Fig. A.4 (b) 001 standard projection (cubic).

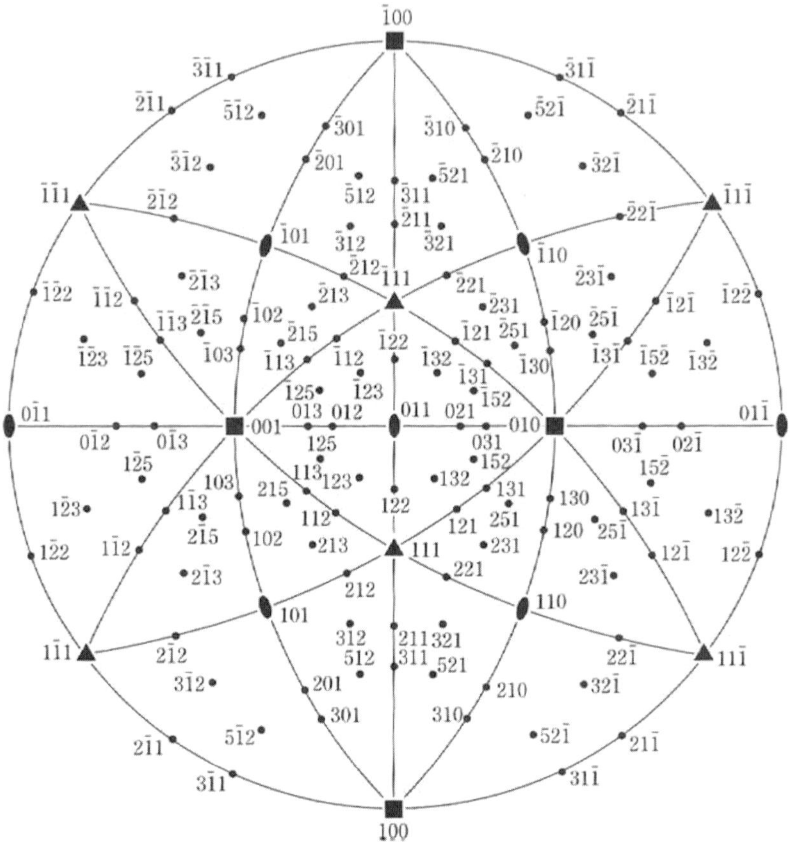

Fig. A.4 (c) 011 standard projection(cubic).

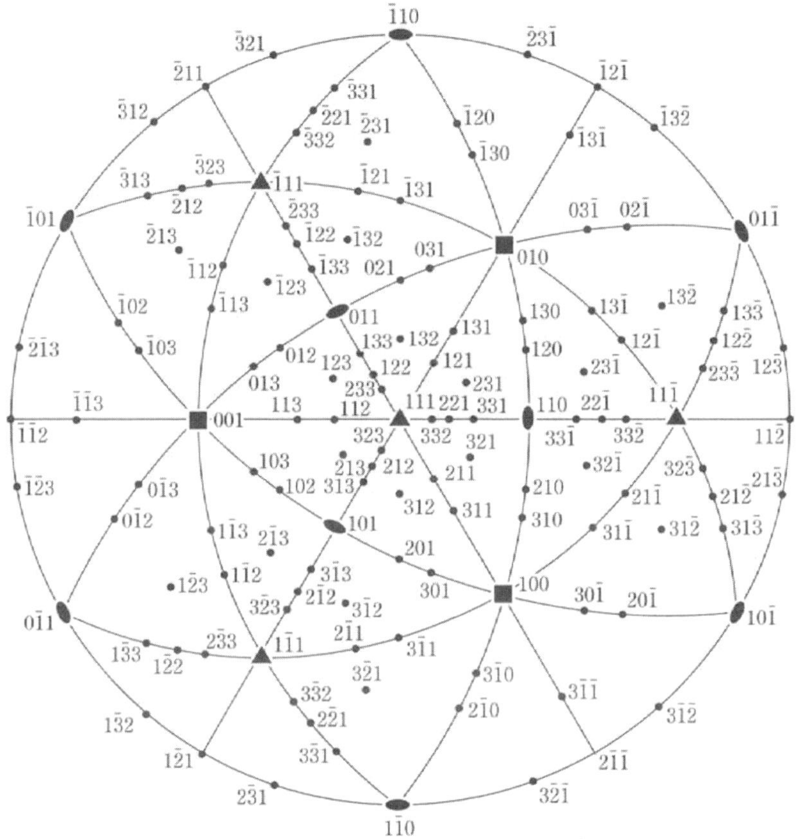

Fig. A.4 (d) 111 standard projection(cubic).

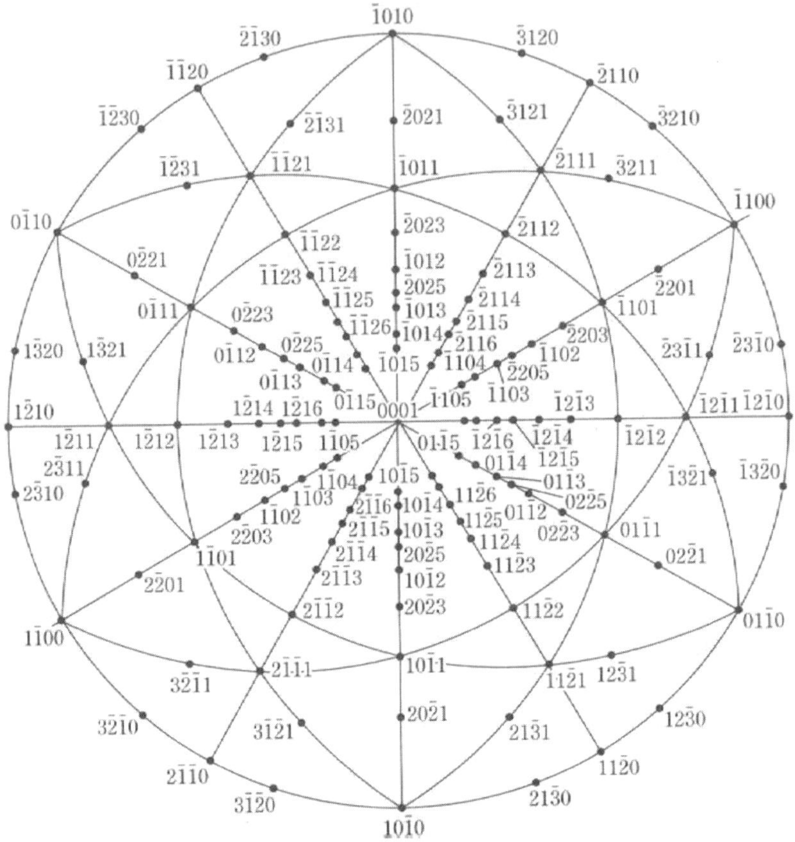

Fig. A.4 (e) 0001 standard projection (hexagonal)(c/a = 1.633) (Reproduced with permission from [5]).

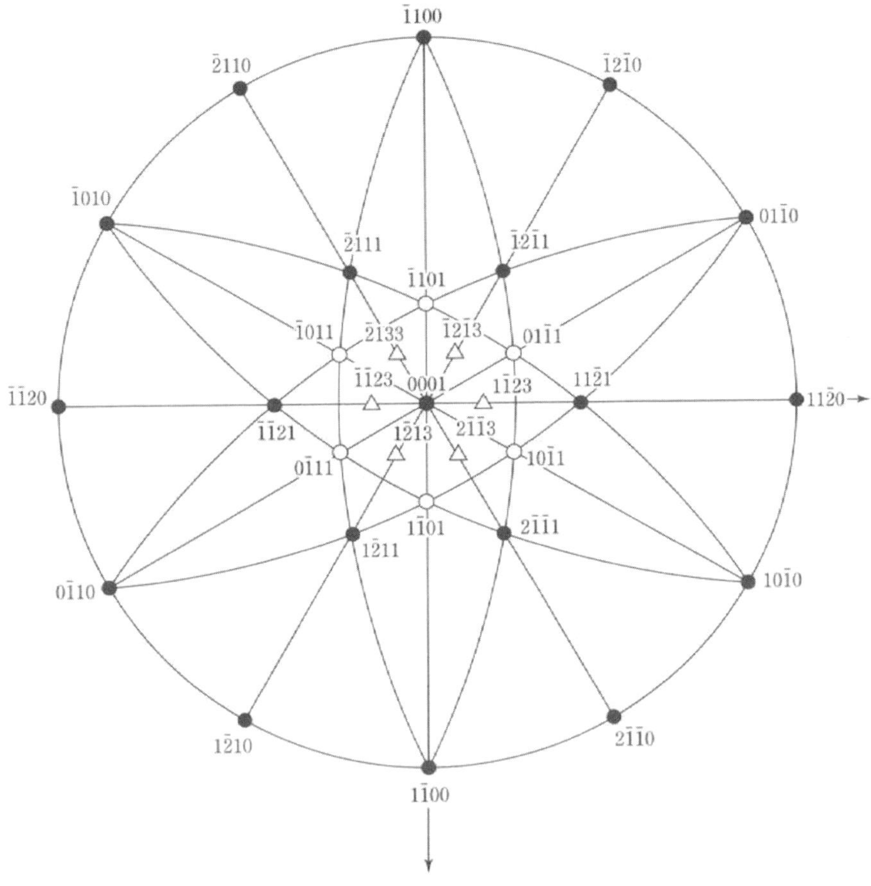

Fig. A.4 (f) 0001 standard projection (4H-SiC).

A.5 Determining the Slip System

A.5.1 *Slip direction*

See Fig. 12.6.

A.5.2 *Slip plane*

Slip plane is determined by observing slip lines on crystal surfaces. It is necessary to observe at least two surfaces. Let us suppose that in Fig. A.5(a) slip lines on (010) and (001) planes are observed, their directions being AA and BB as shown in Figs. A.5(b), (c). The normal (P) of the slip plane is perpendicular to both AA and BB. However, here, (b) is a stereoprojection

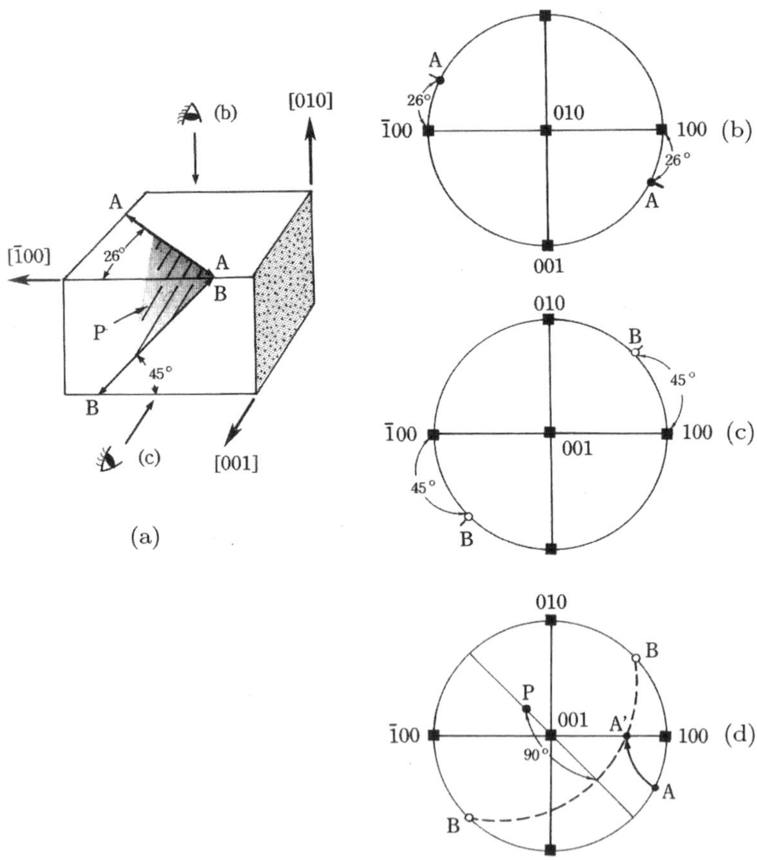

Fig. A.5 Analysis of a slip plane from slip traces on two surfaces (Reproduced with permission from [4]).

onto [010], while (c) is onto [001]. We must have both AA and BB on one stereoprojection. To do so, if (b), for instance, is rotated along [100] by 90°, pole A is shifted to pole A' on 001 projection, as shown in (d). Then, in (d) the direction P which is perpendicular both to B and A' is the normal of the slip plane. Here, for the sake of clarity, (010) and (001) surfaces were chosen as typical examples, however, needless to say, any two surfaces will work. However, the error is minimized when the two surfaces make an angle 90°.

To improve the preciseness many (more than two) surfaces are to be analyzed. For this purpose a cylindrical specimen is suitable. Figure A.6 shows a method to observe surfaces of a cylindrical specimen examined under an optical microscope. The cylinder is approximated by a polygon composed of a, b, c, d and e. Here a and e correspond to (001) and (010)

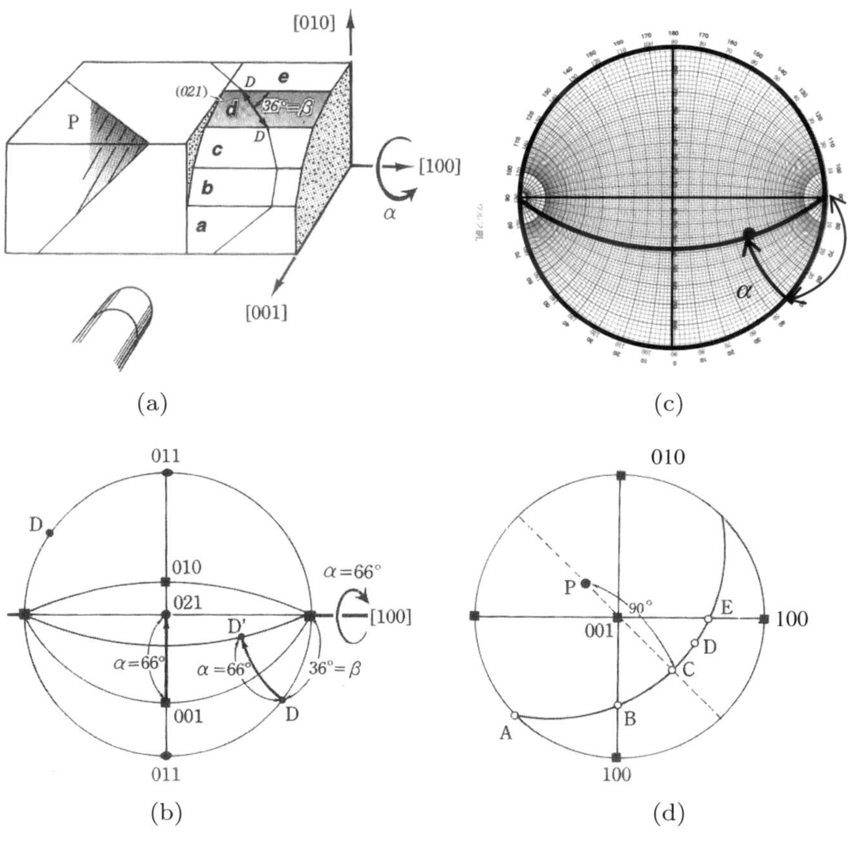

Fig. A.6 Analysis of a slip plane (2) (Reproduced with permission from [4]).

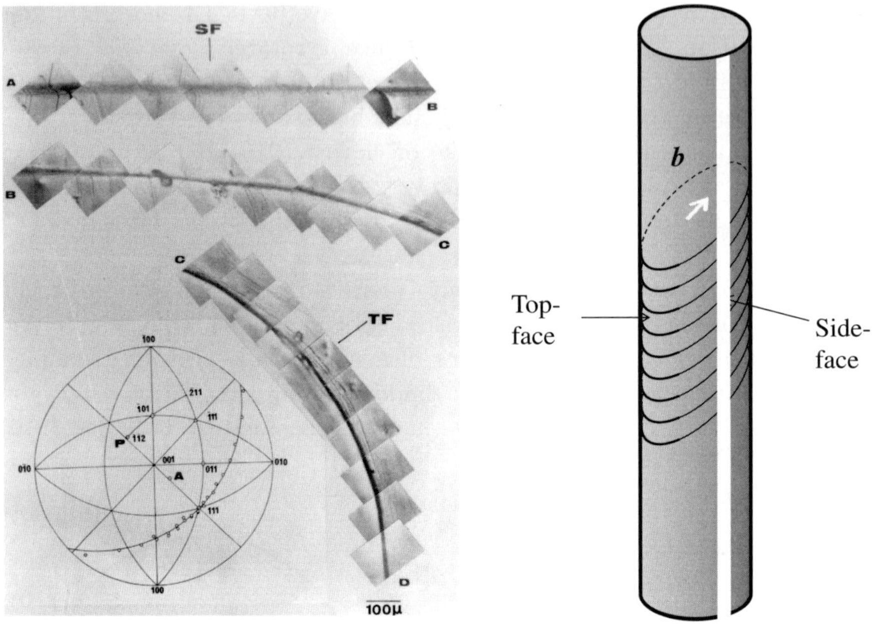

(a) (b)

Fig. A.7 (a) Analysis of a slip plane (3) (Reproduced with permission from [6]), (b) Schematic illustration of side-face and top-face.

in Fig. A.6, respectively. The specimen is rotated counterclockwise around [100] axis. Suppose that d plane is now being examined. For the sake of clarity, let the Miller index of d be (021). The plane d, i.e., (021) makes an angle of $\alpha = 66°$ with (010). Slip line on plane d is inclined by $\beta = 36°$ with respect to the [100] direction. This direction is indicated by $D{-}D$ on a stereoprojection along [021] (Fig. A.6(b)). Here, this stereoprojection is rotated clockwise along [100] direction by 66° back to the [001] stereoprojection (c). If similar operations are carried out for a, b, c and e, then P is determined precisely as shown by P in (d).

Figure A.7(a) shows an example. Here, at SF the slip lines appear only weakly, while at TF they appear strongly. At SF (i.e., side face) $\boldsymbol{b}$ is parallel to the surface, so that steps associated with slip do not appear, while at TF (i.e., top-face) the steps become a maximum. In other words, information on the slip direction is also obtained by this method.

A.6 Thompson Tetrahedra

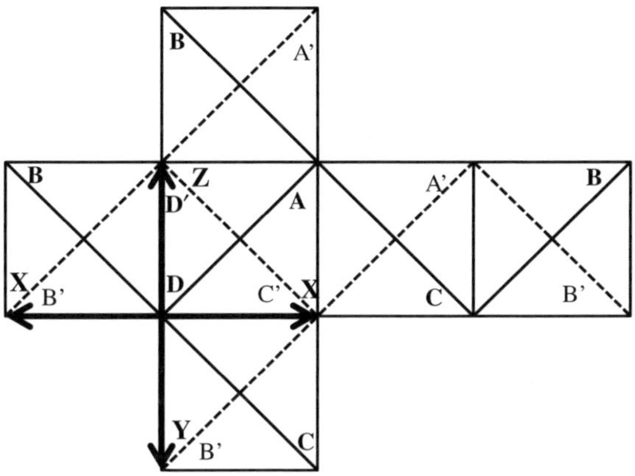

Fig. A.8

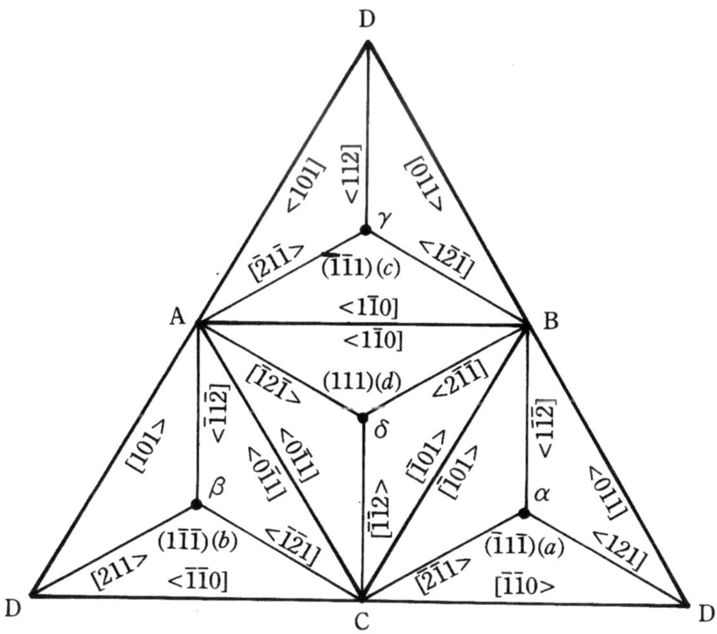

Fig. A.9

A.7 Transformation of Stress (Eq. 3.18)

$$T_{ij} = \begin{pmatrix} T_{11} & T_{12} & T_{13} \\ T_{21} & T_{22} & T_{23} \\ T_{31} & T_{32} & T_{33} \end{pmatrix},$$

$$\sigma'_{ij} = \sum_{lm} T_{il} T_{jm} \sigma_{lm}$$

$$\begin{aligned} \sigma'_{\square\square} &= T_{\square 1} \cdot T_{\square 1}\sigma_{11} + T_{\square 1} \cdot T_{\square 2}\sigma_{12} + T_{\square 1} \cdot T_{\square 3}\sigma_{13} \\ &\quad + T_{\square 2} \cdot T_{\square 1}\sigma_{21} + T_{\square 2} \cdot T_{\square 2}\sigma_{22} + T_{\square 2} \cdot T_{\square 3}\sigma_{23} \\ &\quad + T_{\square 3} \cdot T_{\square 1}\sigma_{31} + T_{\square 3} \cdot T_{\square 2}\sigma_{32} + T_{\square 3} \cdot T_{\square 3}\sigma_{33} \end{aligned}$$

$$\begin{aligned} &= (\quad)(\quad)\sigma_{11} + (\quad)(\quad)\sigma_{12} + (\quad)(\quad)\sigma_{13} \\ &\quad + (\quad)(\quad)\sigma_{21} + (\quad)(\quad)\sigma_{22} + (\quad)(\quad)\sigma_{23} \\ &\quad + (\quad)(\quad)\sigma_{31} + (\quad)(\quad)\sigma_{32} + (\quad)(\quad)\sigma_{33}. \end{aligned}$$

A.8 Inverse transformation of stress (Eq. 3.21)

$$\sigma_{ij} = \sum_{l,m} T_{li} T_{mj} \sigma'_{lm}$$

$$\begin{aligned} \sigma_{\square\square} &= T_{1\square} \cdot T_{1\square}\sigma'_{11} + T_{1\square} \cdot T_{2\square}\sigma'_{12} + T_{1\square} \cdot T_{3\square}\sigma'_{13} \\ &\quad + T_{2\square} \cdot T_{1\square}\sigma'_{21} + T_{2\square} \cdot T_{2\square}\sigma'_{22} + T_{2\square} \cdot T_{3\square}\sigma'_{23} \\ &\quad + T_{3\square} \cdot T_{1\square}\sigma'_{31} + T_{3\square} \cdot T_{2\square}\sigma'_{32} + T_{3\square} \cdot T_{3\square}\sigma'_{33} \end{aligned}$$

$$\begin{aligned} &= (\quad)(\quad)\sigma'_{11} + (\quad)(\quad)\sigma'_{12} + (\quad)(\quad)\sigma'_{13} \\ &\quad + (\quad)(\quad)\sigma'_{21} + (\quad)(\quad)\sigma'_{22} + (\quad)(\quad)\sigma'_{23} \\ &\quad + (\quad)(\quad)\sigma'_{31} + (\quad)(\quad)\sigma'_{32} + (\quad)(\quad)\sigma'_{33}. \end{aligned}$$

References

1. Kitajima, S., Tonda, H. and Kaieda, J. *Japan. Inst. Met. Mater.* **32** (1968), 164 (in Japanese).
2. Saka, H., Sakai, A., Imura, T., Nakatani, S. and Kikuta, S. *Phys. Stat. Sol.* A **101** (1987), 1.
3. Sakai, A., Saka, H. and Imura, T. *Phys. Stat. Sol.* A **97** (1986), 57.
4. Saka, H. *Electron Microscopy of Crystals* (in Japanese), (Uchidarokakuho, 1997).
5. Andrews, K. W., Dyson, D. J. and Keown, S. R. *Interpretation of Electron Diffraction Patterns* (Plenumn, 1967).
6. Saka, H. and Taylor, G. *Phil. Mag.* A **43** (1981), 1377.

Index

$\psi - \chi$ curve, 287, 288, 296
definition of, 289

A- or plus (+) surface, 230
activation area, 277
activation parameter, 281
calculation of, 279
activation volume, 277, 280
of functional material, 281
anisotropic strain, 95
anisotropy, 88
factor, 88
ratio, 55
anomalous slip, 296, 298
anti-phase boundary (APB), 190, 191
energy, 192, 200
tube, 199, 200
anti-twinning sense, 293–295
Arrhenius equation, 278
athermal barrier, 270, 271
long-range barrier, 271
athermal stress, 272
austenitic stainless steel, 164, 165

B- or minus (−) surface, 230
B2 barrier, 152
B2 structure, 189, 190, 194, 200
dislocation in, 194
B4 barrier, 152
back stress, 108–110, 325
due to the line tension, 109
Bain's relationship, 96, 97
Bardeen–Herring source, 103, 111–113
basal plane dislocation (BPD), 243
BCC, 2
bending of dislocation
in the vicinity of crystal surface, 331
BPD (basal plane dislocation), 243
near the free surface, 339
brittle fracture, 10
bulk modulus, 56
Burgers circuit, 16, 17
Burgers vector, 16, 18, 22, 124
definition of, 18
of mixed dislocation, 22
of dissociated dislocation, 132
of Frank partial dislocation, 130

cell wall, 305
chemical interaction, 172
climb, 35, 37
of dissociated dislocation, 154, 157, 159
of edge dislocation, 35
of screw dislocation, 37
climb dissociation, 206–208
climb motion, 36, 74
of dissociated dislocation, 154
climb-dissociated dislocation
motion of, 209
climb-down, 36

climb-up, 36
coherency, 317
compressibility, 56
conjugate slip system, 260, 261
conservative motion, 35, 36
coplanar slip system, 261
core cut-off radius, 151
core energy, 103
core structure
 of screw dislocation, 296
Cottrell atmosphere, 94, 95, 307
Cottrell effect, 91, 95, 300
critical resolved shear stress, 253, 254
 law of, 253
critical slip system, 260, 261
critical thickness, 318, 319
cross slip, 24, 274
 of screw dislocation, 24
cross slip system, 261
CRSS (critical resolved shear stress), 254, 293
crystal globe, 6
crystallographic direction, 1, 2, 7
crystallographic plane, 3
Curie point, 189, 190
Curie temperature, 189
curvilinear coordinate
 definition of σ_{ij} in, 61
cut-off angle, 329
cut-off parameter, 66, 150, 327
cut-off radius, 326
cutting
 of solute atoms, fine precipitates and forest dislocation, 274
CuZn, 25, 26, 88, 192, 193, 200, 205, 211, 212, 257, 259
cycle annealing method, 342
cylindrical coordinate, 58

dangling bond, 230
DBTT (ductile-brittle transition temperature) in Si, 200, 222–24
deformation twin in BCC, 294
diamond structure, 2, 215, 320
 coordinate of atoms in, 216
 dislocation in, 215
 P-N potential in, 220
 60° dislocation in, 141, 220
dipole, 82, 335
dislocation, 15, 16, 28–30, 88
 in anisotropic crystal, 88
 crossing of, 28–30
 definition of, 15, 16
 dipole, 82, 335
 multiplication of, 103
 pileup, 115–118
 polarity of, 236
 self-energy of, 319
 in SiC, 241
 in zincblende structure, 232
dislocation cell wall, 271, 273
dislocation channelling, 315
dislocation core, 64
 energy, 65, 103
 cut-off radius, 151
dislocation density, 341
dislocation dipole, 27
dislocation loop, 20, 88
dislocation reaction, 25, 28
 between two dislocations, 25
 between two dissociated dislocations, 149
 $\langle 111\rangle = \langle 001\rangle + \langle 110\rangle$, 210
dislocation source, 106
disordered alloy, 189
dispersion hardening, 313
displacement, 41, 48
 transformation of, 48
 around screw dislocation, 64
dissociated dislocation, 123, 126, 130, 139
 Burgers vector of, 128
 Cherns–Hirsch–Saka model for climb motion of, 161
definition of, 123
 jog on, 138
 reaction between, 148
 in thin crystal, 331
dissociation distance, 136, 138, 139, 171
 effect of surface on, 334

distinguishing between roles played by edge and screw dislocations, 212
DO_3, 195
  superdislocation in, 196
double kink, 106, 107
double–kink mechanism, 105
ductile fracture, 10
ductile–brittle transition temperature in Si, 220
dynamic recovery, 306

easy glide region, 302
easy sense, 293, 295
edge dislocation, 20, 69, 93
  definition of, 20
  displacement and stress around, 69
  force acting on, 73
  glide motion of, 73
  imaginary dislocation of, 75
  strain energy, 71
  stress field, 72
effective stress, 270, 278
ehp (extra half plane), 25, 35, 37, 38, 230
elastic constant, 14, 43, 45, 51
elastic deformation, 9, 11
elastic energy, 103
  in epilayer, 322
  of dislocation, 332
elastic interaction, 91
  between dislocation and solute atom, 91
electron concentration, 163
epilayer, 317, 319
  grown by step-controlled epitaxial growth technique, 329
  yielding of, 328
epitaxial growth, 317, 318
epitaxial layer, 317
Eshelby twist, 69
etch pit, 265, 266
extended dislocation, 123
extended SF (stacking fault) node, 150, 151
external force, 275, 276
extra half plane (ehp), 23, 25, 35, 37, 230

F–R (Frank-Read) source, 103, 106–109, 111, 116, 303
FCC, 2
force, 78
  acting on edge dislocation, 63, 73
  acting on the mixed dislocation, 78
  acting on screw dislocation, 66
forest dislocation, 199
fracture stress, 221
Frank criterion, 127, 250
Frank loop, 146
Frank partial, 131, 132, 144
Frank symbol, 131, 240
  new (extended), 179–181
Frank–Read source, 103, 106–109, 111, 116, 303
  single-ended, 111
  modification of, 111
free energy versus composition curve, 162
Frenkel pair, 313
Frenkel's model, 12, 13, 15
friction theory, 306, 308, 309
FS/RH(Perfect), 16, 18, 19
  definition of, 19

GaAs, 228–239
  stacking-fault energy in, 167
GaN, 240
GD (glide-set dislocation), 219, 222
GD (glide-set dislocation) and SP (shuffle-set) dislocations
  mixture of, 226
GD (glide-set dislocation) in Si
  dissociated, 224
  widely dissociated, 224
Gibbs dyadic notation, 50, 79
glide, 11
glide dislocation, 32
glide dissociation, 206–208

glide-set dislocation (GD), 153, 218, 232
grain boundary, 68
grown-in APB, 194, 198
  reactions with, 197
Guinier–Preston zone, 309, 312

hard sense, 293, 295
HCP, 2
Heidenreich-Shockley partial dislocation, 126
helical dislocation (helix), 37–40
  transformation from screw dislocation, 37
heteroepitaxy, 317
hexagonal Index, 3–6
  prismatic plane, 6
  pyramidal plane, 6
  transformation between the three and four indices, 5
Hirsch model of screw dislocation in BCC, 285
Hirth Lock (HL), 152, 153
Hirth–Lothe axiom, 238
Hirth–Lothe's Axiom I, 135, 143
Hirth–Lothe's Axiom II, 144
homoepitaxy, 317
Hooke's law, 9, 14, 43, 44, 51, 320
  in isotropic solid, 51
  matrix notation of, 44
Hume-Rothery's law, 162

ideal strength, 12, 14
image force, 67, 75
  on screw dislocation, 67
imaginary dislocation, 67, 75, 335
  of edge dislocation, 75
  of screw dislocation, 67
imperfect dislocation, 126
Instron-type testing machine, 264
interaction energy, 101
  between screw dislocation and C atom, 101
interaction of dislocation with solute atom, 308
intermetallic compound, 189
internal stress, 270, 278
interstitial, 32, 33, 39
inverse temperature dependence of strength, 200
  in β-brass, 205
  in Ni_3Al, 201
  mechanism of, 203
inverse transformation
  of coordinate, 47
  of stress, 49, 354
  of strain, 49
inverse Wulff construction, 88

jog, 28, 37, 38, 140
  constricted, 145
  extended, 140
  motion, 145
  non-conservative motion of, 274
  on dissociated screw dislocation, 143
Johnston–Gilman theory, 264, 267, 270
  modification to, 267

Kear-Wilsdorf's mechanism, 204
kink, 140
Kirchhoff's law, 24–27, 39, 116, 141
knock-on-collision, 313

$L1_2$, 195, 197, 200
  $\langle 110\rangle/\{001\}$ slip in, 200
  $\langle 110\rangle/\{111\}$ slip in, 202
Lame constant, 55
lattice constant, 1
line tension, 108, 110, 327
linear hardening region, 302
Lohne force, 332–334, 336
Lomer-Cottrell (LC) barrier, 152, 153
long-range barrier, 270
long-range stress, 270
lower yield point, 299

martensite, 97
microplasticity, 273
Miller index, 3, 4
mirror force, 335, 336

misfit dislocation, 317, 318
mixed dislocation, 20, 22, 75
  definition of, 20
  displacement, stress and strain of, 22
  force acting on, 78
mobility of dislocation, 266, 267
  in a variety of crystals, 268
  of edge dislocation in BCC, 283
  of screw dislocation in BCC, 284
multiple dipole, 303, 304

Ni_3Al, 200
normal force, 254

order-disorder transformation, 189, 210
ordered alloy, 189
  dislocation in, 194
Orowan's by-pass mechanism, 309, 310, 313
Orowan's loop, 311
overaging, 309, 312

P–N (Peierls-Nabarro) force, 103–105, 109, 218, 285, 296
  in FCC, 104
  in HCP, 105
  in BCC, 282, 296
P–N (Peierls-Nabarro) potential, 103, 105, 274, 282
  overcoming of, 105
partial dislocation, 126
  Frank, 143
  Shockley, 126, 132, 144, 146, 238, 243
Peach–Koehler formula, 78, 137
  application of, 78–86
Peierls (-Nabarro) force, 103–105, 282
  in BCC, 282, 296
  in FCC, 104
  in HCP, 105
  of the glide-set dislocation, 218
  of the shuffle-set, 218
Peierls–Nabarro potential, 103, 105, 274, 282
pencil glide, 35
perfect dislocation, 126, 138
permanent strain, 10
phase diagram, 95, 96
  of the Fe–C binary system, 95
  of the Fe–C system, 96
  of Fe-X binary, 166
pileup of dislocation, 115–118
pinning theory, 306, 307
plane strain, 57
plane stress, 57
plastic deformation, 10, 11
  elemental process of, 11
point defect, 33
Poisson ratio, 55, 321
polarity, 228
  of {111} plane, 229
  of dislocation core, 232
  of Shockley partial in zincblende, 239
  of Shockley partial in GaN, 242
polygonization, 117
polytype, 243
precipitation hardening, 309, 312
primary slip system, 261
principal axis, 91, 98
  of strain, 91
prismatic dislocation loop, 32, 33–35, 38, 39
  Burgers vector of, 34
  formation by precipitation of point defect, 32
prismatic plane, 3
pyramidal plane, 3

radiation hardening, 313, 314
radiation-enhanced diffusion, 174
radiation-induced segregation, 174
resolved shear force, 254

Schmid factor, 254, 260, 293, 325
Schmid law, 254, 265, 333
  failure of, 291, 296, 333
screw dislocation, 20, 31, 66, 95
  core structure of, in BCC, 293
  crossing of, 31

crystallographic symmetry of, in BCC, 284
displacement, strain and stress, 63
force on, 66
image force on, 67
jog on, 141
model of, 343
self-energy of, 110
strain energy of, 64
transformation into helical dislocation, 37
sessile dislocations, 153
SF (stacking fault), 127, 130, 136
extrinsic, 132
Frank, 130, 131
intrinsic, 131
extended node, 151
SFE (stacking-fault energy), 138, 162
composition dependence of, 161
temperature dependence of, 165
thermodynamic significance of, 161
SFT (stacking fault tetrahedron), 146–148
shear modulus, 55
shear strain, 256
shear stress and shear strain
derivation of, 255–257
Shockley partial dislocation, 126, 132, 144, 146, 238, 243
effect of Lohne force on, 335
short-range barrier, 270, 273, 280
short-range stress, 270
shuffle-set (SP) dislocation, 218, 230
in Si introduced at 77 K, 225
shuffle-to-glide transformation, 227
SiC, 241–252
β-SiC, 243
2H–SiC, 243
4H–SiC, 245–252
6H–SiC, 244
polytypes of, 241
stacking fault energy in, 243
side face, 352
slide, 11
slip, 11
slip direction, 350
slip line
in BCC, 289
on crystal surface, 350
slip plane, 23, 350
slip plane in compression
rotation of, 258
slip system, 350
in FCC, 259, 260
small angle tilt boundary, 117
solid solution, 91, 94, 95, 97
complete, 306
interstitial, 91, 95, 306
substitutional, 91, 94, 306
supersaturated, 97
solid solution hardening, 306, 307, 309, 312
mechanism for, 312
solute atom, 93
strain field of, 93
SP (shuffle-set dislocation), 219, 223, 224
introduced just below DBTT in Si, 226
spherical coordinate, 60, 61
spike, 313
stacking fault (SF), 127, 130, 136
extrinsic, 130, 136
Frank, 130, 131
intrinsic, 130, 135
stacking fault energy (SFE), 130
in 4H-SiC, 243
temperature dependence of, 167, 168
temperature dependence of, in alloy, 166
temperature dependence of, in austenitic stainless steel, 169
temperature dependence of, in pure material, 165
stacking fault tetrahedron (SFT), 146–148
stacking of (111) plane in FCC, 125

stacking order (sequence), 125, 126, 130, 341
of HCP, 130, 131
of FCC, 125, 131
stage I of work hardening, 302, 303
stage II of work hardening, 304
stage III of work hardening, 306
stair-rod dislocation, 138, 141, 146, 150, 152, 153
on extended jog, 140, 142
standard projection, 6, 8
4H-SiC, 349
001, 345
0001, 348, 349
011, 346
111, 347
step-controlled epitaxial growth, 329, 340
stereographic projection, 6, 344
strain, 9, 42
inverse transformation of, 49
strain and stress, 48
transformation of, 48
strain energy, 64, 71, 78
of dislocation line per unit length, 330
of edge dislocation, 71
of screw dislocation, 64
strain rate, 263, 265, 269
strengthening by ordering, 197
stress, 9, 43
definition, 43
inverse transformation of, 49
stress–strain curve, 9, 10, 220, 267, 300
at different stages of precipitation, 311
in Si, 221
of mild steel, 300, 301
of Mo single crystal, 292
of Nb, 302
of whisker, 300
substrate, 317
superdislocation, 189, 192
in $L1_2$, 197
reaction with APB, 193
superlattice, 189
superpartial dislocation, 189, 191
separation of, 192
surface layer, 68
Suzuki atmosphere, 172, 307
Suzuki effect, 170, 172, 174
experimental confirmation of, 172
in austenitic stainless steel, 173
mechanism of, 170
Suzuki model of screw dislocation in BCC, 285–288
Suzuki segregation, 171, 172

TD, 329
tensile axis
rotation of, 258, 259
tetragonal strain, 97, 102
thermal activation, 278
thermal energy, 274–277
thermal stress, 270, 272
thermally activated process, 277
of dislocation motion, 270
thin foil, 317–340
Thompson tetrahedron, 123–127, 132, 133, 151, 353
opened up at corner D, 124
viewed from the inside, 134, 135
viewed from the outside, 134, 135
Thompson vector, 126, 127, 132
vector algebra for, 132
threading dislocation (TD), 246, 328
inclination of, 331
threading edge dislocation (TED), 246, 328
threading mixed dislocation 1(TMD), 246, 250
threading mixed dislocation 2(TMD'), 246
threading near screw dislocation (TnSD), 246, 250
threading pure screw dislocation (TpSD), 246, 250

threading screw dislocation (TSD), 246, 328, 329
    dissociation of, 246
three-stage hardening, 301–306
tilt boundary, 119
top face, 352
transformation of BPD to TED in 4H–SiC, 336
transformation of coordinate, 46
transformation of stress and strain, 46, 354
twinning, 293
twinning sense, 294, 295
twist boundary, 120

unit cell, 1
    of Γ, 281
upper yield point, 299

vacancy, 32, 39
    equilibrium concentration of, 146

weak-beam, 150
whisker, 299
work hardening, 301–305
Wulff net, 344
wurtzite, 239, 243
    Shockley partial dislocation in, 239

X-ray topograph, 112, 221

yielding, 10
    of BCC Metals, 282
    of mild steel, 300
    phenomenology of, 260
    of single crystal, 253
Young modulus, 55

zincblende, 2, 215, 320
    dislocation in, 228
    nomenclature of dislocation in, 232
zone axis, 24